The World of Science

The World of Science

by
Alfredo Porati
Roberto Favilla
Massimo Gori
Sergio Tagliavini
Cesare Lamera
Luigi Vacchi
Maria Teresa Merella

**Translated from the Italian by
Simon Pleasance**

**Book Club Associates
London**

Art Director
Enzo Orlandi

Editorial Director
Giorgio Marcolungo

Editor
Italo Bosetto

English-language edition
edited by
Raymond E. Maddison

Picture Editors
Ettore Mochetti
Raffaello Segattini

Picture Researchers
Paola Brunetta
Luciana Conforti
Gioietta Gioia
Franco Minardi
Nicoletta P. Tanucci
Katherine Siberblatt

This edition published 1978
by Book Club Associates
By arrangement with Macdonald and Jane's Publishers Limited
Originally published in Italian under the title
COLORAMA *Il Mondo della Scienza*

Filmset by Keyspools Ltd, Golborne, Lancs.
Printed and bound in Italy
by Arnoldo Mondadori Editore, Verona

CONTENTS

MATTER AND ENERGY

PHYSICS

Physics, unlike, for example, medicine or botany, is a division of science the limits of which are rather difficult to define. The word comes from the Greek *physis*, which means 'nature', but in a sense all science is the study of nature, or natural things, and thus the original meaning of the word is of little help. Science has become, over the last two centuries, much more complex than it was, and many particular scientific disciplines have been established, so that today there are geologists, astronomers, chemists, biologists, microbiologists, and so on; they all study some part of nature. Of course, 200 years ago, most scientists were described merely as 'natural philosophers' – to a large extent because they studied the laws of Nature, and their interests generally ranged over what today would be eight or nine particular scientific disciplines.

Nevertheless, most of the great natural philosophers began their studies by looking closely at very simple everyday occurrences in the world about them, and asking themselves why things happened as they did: why objects fell to the ground, for example, why a cannon ball fired from a gun went a certain distance before falling to earth. It is the examination of apparently elementary, simple things like this that led to the establishment of basic scientific laws which are still at the root of modern science, and it is the study of these elementary laws that is still the concern of physics.

In fact many of these simple things are much more complicated scientifically than they appeared to be at first sight, and their examination, as will later become evident, has led physicists to develop some very complicated theories. The subject of investigation, however, remains those common everyday phenomena which everyone sees or experiences.

The theories that physicists have developed to explain these events have been of great value in many cases, in explaining problems in other scientific disciplines. In this way modern physics has contributed greatly to the work of, for example, chemists. Thus the quantum theory, explained below, was first developed by physicists to explain the patterns they observed when studying the way electromagnetic waves were emitted by different atoms or molecules, but this theory has now been employed by chemists to lead to a much greater understanding of the composition of chemical materials.

Because of the nature of its investigations into seemingly simple phenomena, physics has developed traditional methods which are applied to whatever problem is under consideration. Essentially, physicists *measure* things – so that they acquire a body of statistical data from which conclusions are drawn. Thus, if a physicist is presented with the phenomenon of gas in a balloon and it appears that the gas expands as the temperature increases, then as a physicist he will subject the gas in the balloon to a series of carefully controlled experiments, in which the temperature is different every time, and he will measure the expansion in every case. He will thus finally accumulate data which will relate temperature to degree of expansion, and from this he will be able to deduce a relationship between the two. Of course this is an over-simplified example, but it can well serve to illustrate the fundamental pattern of a physicist's investigations. This emphasis on measurement means that physics is closely related to, and indeed increasingly dependent on, mathematics – and mathematics is the only other scientific discipline of which physics has need, although, as suggested above, the converse is not true: many other scientific disciplines depend a great deal on the principles established by physics.

Physics and mathematics have drawn close to each other, even though physics is essentially a very practical subject, and mathematics traditionally is a very abstract one. This contradiction, however, is more apparent than real: pure mathematics is certainly an abstract science, but just as the principle of physics can be applied to other scientific discipline, so the principle of mathematics can be applied to the data-collecting techniques in physics. Hence 'applied mathematics' is a name often used for some aspects of physics. Certainly mathematics, from experience, is the best tool to use in observing, describing and analyzing the physical world and the natural phenomena in it. Thus one of the two pillars on which the whole edifice of modern physics is built is mathematics: the other is experimental research. The working pattern of the physicist is as follows: the experimental observation of some (generally simple) phenomena leads him, after analysis of the data he has collected, to formulate a theory which explains the phenomena observed. Very often this theory will be developed from, or built on, other already accepted principles, although occasionally, as in the case of the work of Planck or Einstein, the physicist will introduce some quite novel theory which upsets much that is generally accepted. In either case, the new theory formulated by the physicist will imply other consequences – that is, if it is true, then such and such should happen in other particular circumstances. These consequences are accordingly tested by experiment and, if the results bear out the prediction of the theory, then the theory is accepted. It will, however, be accepted only for so long as further experiments and observations confirm it as apparently true. Almost certainly, sooner or later, some experiments will show that the theory is not correct – or not correct in all circumstances – and then a revision will be undertaken and a new theory formulated.

This is how progress is made in physics, and it emphasizes the limitations of experiments. Experiments never can confirm the absolute truth of a theory: they can only confirm that, so far as current knowledge goes, it still holds good – there is always the possibility present in the research physicist's mind that round the next corner, as it were, he will find new experimental data that will force him radically to change the principles he currently accepts. This is worth emphasizing because, although in everyday speech it is customary to talk of such and such a scientific explanation or theory as 'true', it is more accurate to speak of the theory as 'appearing to explain all known relevant data and as yet disproved by no experiment'.

Sometimes a well-established theory which has to be revised to accommodate newly observed phenomena is found nevertheless still to be valid within certain limits. Thus the theories developed by Newton in the field of mechanics were shown by Einstein to be incorrect when applied to some phenomena observed in the Universe. But Newton's theories still hold good to explain many phenomena in the everyday world. In this case, research showed not that the older theory was wrong but that it had a more limited application than had previously been believed.

Thus is progress made in physics and the great researchers of today, even when they must change or revise the theories of their predecessors, remain fully aware of the cultural heritage of those on whose work they build, and retain a due sense of humility in the knowledge that next month, or next year perhaps, new experiments may totally reverse their own work.

MECHANICS

To many people the term *mechanics* may suggest an immediate mental picture of spanners, crowbars, cranes and similar tools or machines. This is not, of course, what the word means in science, although tools and machines are in fact linked to mechanics in the scientific sense because they are practical everyday examples of the application of scientific mechanical principles. Scientifically, mechanics is that part of physics that deals with the study of the action of forces on bodies (bodies, of course, used in the sense of objects and not restricted to the human body). Why and how a body moves when hit by another, why and how a long lever will move a heavy body more easily than a short lever: these are very simple mechanical problems. These kinds of problems have occupied the curiosity of men for a very long time. The great Austrian physicist Ernst Mach fittingly described mechanics as 'that part of physics which is the most ancient and also the most simple, and is as a result a necessary basis for the understanding of many other branches'.

Today, mechanics is divided into three main areas: kinematics which studies the relationship between distance, time, speed and acceleration; dynamics which studies the way forces give rise to motion; and statics which studies the effect of forces on a motionless body.

Yet although elementary application of mechanical principles was used by both the Greeks and the Romans, there was very little study made, except by Archimedes, of any of the principles themselves. In the succeeding centuries, although practical craftsmen worked wonders with only the most elementary mechanical aids, and obviously knew how to use them, there was still little advance in the working out of any theory to explain the reasons why these simple machines worked. Not until the time of Leonardo da Vinci did the world find a successor to Archimedes.

Leonardo was the first true forerunner of modern mechanics. Born in 1452, he was a most remarkable figure in many fields of human activity: a mathematician, physicist, civil and military engineer, naturalist and outstanding painter, as well as being a deeply fascinating man, both exceptionally handsome and endowed with remarkable physical strength. His knowledge was not derived from studying ancient authorities (if we exclude Archimedes for whom he had a deep admiration) but was developed from studying practical technology (technics) and to the same practical matters he subsequently applied the new theories he had developed. Only relatively few of his manuscripts survive (although more are constantly being discovered and published) but it is virtually certain that many of the concepts later clearly defined by Galileo and Newton had already been perceived by Leonardo, albeit sometimes imprecisely. For example, he clearly outlines a kind of principle of relativity and says that someone who found himself on the Moon would think of himself as stationary, and see all the other bodies in motion, thus concluding that they were, on the Moon, at the centre of the Universe. Today we may raise our eyebrows and smile at the 'geocentric' theory his words imply (i.e. the theory which placed the Earth at the centre of the solar system), but we must bear in mind the extent of the authority of the ancient writers, and Aristotle in particular, so that rather than laugh at Leonardo for not challenging the then almost universally accepted geocentric theory, one

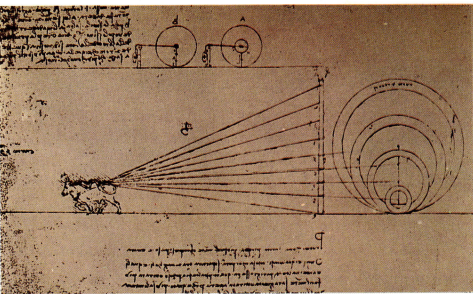

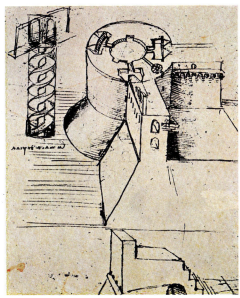

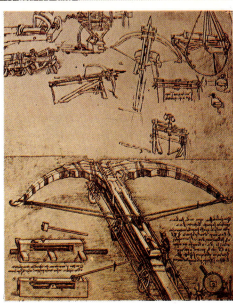

Drawings by Leonardo. Left: self-portrait in red crayon. Top: drawings for a study of traction force. Below centre: the application of Archimedean screws. Right: crossbow with inclined wheels.

must admire his ability to see its full implications for scientific theory. Leonardo's greatness lies essentially in the daring logic of his thought and his greatness as a physicist lies in his being the first person to recognize that '... mechanics is the paradise of the mathematical sciences because in it one reaches the core of mathematics'.

Leonardo made distinguished contributions, above all, to statics (the study of the equilibrium of bodies) and to applied mechanics. His knowledge of the movement of bodies (dynamics) was less advanced. The true founder of the science of dynamics was Galileo Galilei, born in Pisa in 1564, who became a professor at Padua and Florence. Galileo's major contribution was the introduction of the experimental method to test the truth of commonly accepted ideas. This caused him to reject the principle of authority – the idea that because a particularly distinguished teacher said something was true, it must be so. This earned Galileo the condemnation of the Church. Faced with the belief of the ancients that if two bodies were dropped from the same height, the first to reach the ground is the heavier, Galileo did what no one before him had done – *he carried out the experiment*, and actually dropped two bodies. The result was his realization that the assertion made by the ancients was false.

13

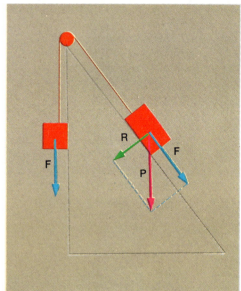

Galileo did not develop a complete theory of mechanics – that was the later work of Newton – but his *method* was new and has basically remained unaltered throughout the entire subsequent history of physics. Experiment must test the validity of theory. It is to Galileo also that we owe the first clear formulation of the *principle of inertia*, which had already been to some extent outlined by Leonardo. This principle lays down that 'if a body on which no forces are acting is stationary to begin with, it remains stationary, and if it is moving, it remains moving at a constant identical speed along a straight line.'

This assertion seems obvious at first sight but in fact is anything but obvious and is perhaps the most important principle in mechanics and in the whole of physics. Little progress was made in the development of modern physics until the full implications of this principle were appreciated. It was a principle that, a moment's reflection will show, is far from obvious because it seems to conflict with everyday experience: perhaps this is why its formulation was so long in the making. In Aristotle's view the opposite principle was the valid one, that is: 'a moving body stops moving after a while if the force acting on it ceases so to act.' This does seem obvious: every cyclist, for example, knows that if he is pedalling along a perfectly flat road, and stops pedalling, then he comes to a

Left: Galileo's study at Pisa. Top right: an experiment being carried out in the presence of Giovanni de' Medici. Below: the inclined plane and the resolution of forces. Using the parallelogram of forces, the vertically directed force of weight is resolved into two forces: one perpendicular to the plane, and the other parallel.

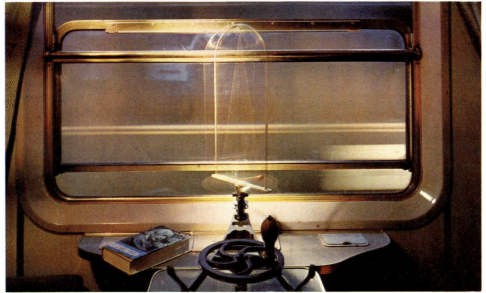

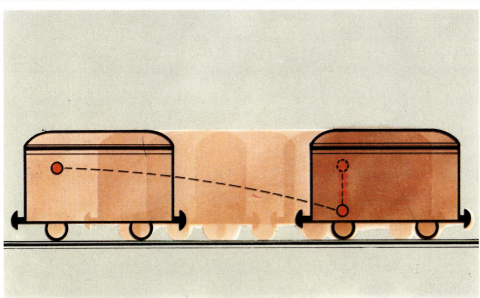

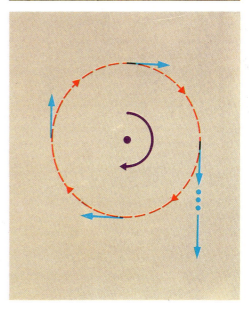

The principle of Galilean relativity says that physical laws are the same for two observers moving with a relative straight and uniform motion; the movement of a pendulum in a train (top left) travelling at a constant speed along a straight track will obey the same laws as it will on the ground. Below: the composition of movements; the vertical trajectory of a mass inside a railway carriage is seen by an outside observer as a parabola. Right: an illustration of the principle of inertia: when a body is not subject to forces, it moves at a constant speed and in a straight line; thus a discus released by a discus-thrower (top) moves along a straight line on the tangent (below).

halt before very long; if he wants to carry on along the road at a constant speed, then he has no option but to pedal. In other words, he has to produce a force which will push him continuously forward. To discover where the fallacy lies now suppose that the cyclist wants to cover the longest possible distance once he has stopped pedalling, still on a perfectly flat road. First, he might oil the bicycle, or replace the ball bearings with better ones, and then he would indeed travel a little further. Then he might try and repeat the experiment on a smoother surfaced road ... and he would gain a few more metres; then he might bend lower over the handlebars, thus reducing the air resistance ... and gain yet more metres.... Ultimately, if the cyclist had an ideal bicycle in which every moving part functioned perfectly, and cycled along a perfectly flat road, in a totally airless environment, he would travel forward for ever. That is the principle of inertia. Why then, in real life, does the bicycle sooner or later come to a standstill?

It comes to a halt because various forces are at work on it – the friction of the gears, the friction of the wheels on the road, and the friction of the wind against the cyclist's body. These forces do not help to maintain the speed of the bicycle, they in fact cause a variation in its speed: just as the force of pedalling increases

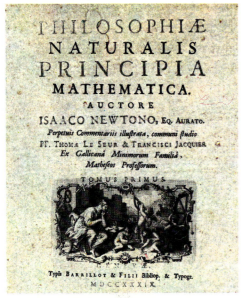

the speed, so these forces reduce its speed. There is therefore a direct link not between force and speed (as Aristotle believed) but between force and *variation* in speed. The forces working against the bicycle tend to reduce its speed, and the more powerful the forces, the more the speed is reduced – or varied from what it would otherwise be.

The principle of inertia allows several other phenomena to be explained. Why, for example, if we are in a car and we apply the brakes, do we feel ourselves being thrust forward? The answer is quite simply that because we tend to maintain our speed, and carry on travelling forward if the car brakes (i.e. slows down), we therefore tend to 'overtake' it and as a result feel ourselves being thrust forward. Similarly, we feel ourselves being thrust backward if the car accelerates; or being thrust sideways if the car rounds a corner, for then we tend to proceed along a straight line at the speed at which we were originally travelling, even though the car has turned sideways.

If Galileo laid down the vital premises for the foundation of mechanics as a science, it was with Isaac Newton that mechanics became organized on a more or less definitive basis. In fact it was not until 1900 – until the work of Einstein – that anything substantially new emerged after Newton. But Einstein's

Newton (right) was the true founder of mechanics. Top left: his major work the Principia Mathematica. *Below: instruments used by him in his research at Trinity College, Cambridge.*

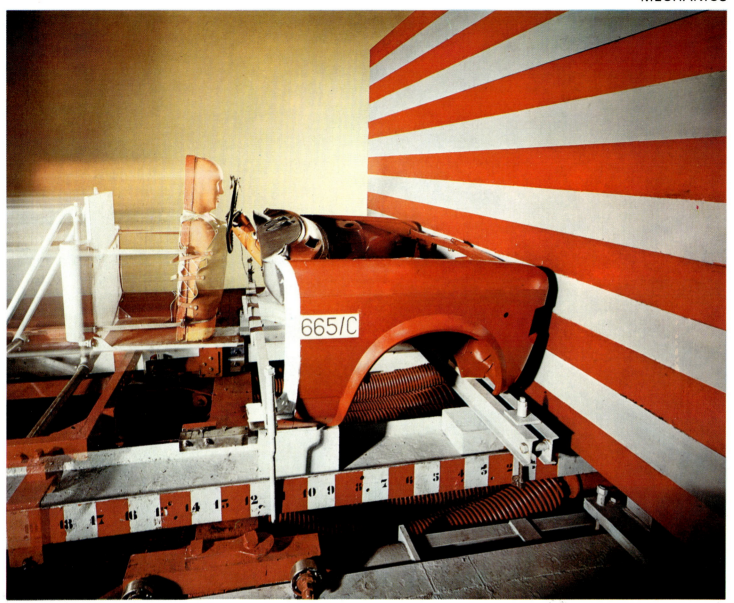

The violent impact of a dummy against a fixed object not only demonstrates the law of inertia but is also an extremely useful test for the safety of modern vehicles.

mechanics did not do away with Newton's mechanics, but adopted them as true in special cases. Newtonian mechanics are still studied today and advances are made in understanding them, but they are a system that cannot be used when dealing with problems that involve very high speeds (comparable to the speed of light) or huge distances (like inter-galactic distances) or minute distances (as between individual atoms for example). Newtonian mechanics are still 'man-sized'.

Newton was born in England, near Nottingham, in 1642 (the year in which Galileo died). During his lifetime his genius was recognized and he received public acclaim and honour. The inscription on his tomb reads: *humani generis decus* (the ornament of mankind).

Newton's name is linked with what came to be known as the three laws of motion: the first, to which reference has been made is the principle of inertia, that a body will continue in its state of motion unless impelled by a force to act otherwise; the second law is the principle that the rate of change of motion (the acceleration or deceleration) is proportional to the force applied, and occurs in the direction of the force applied; the third law is that to every action of a force there is an equal and opposite reaction. Where he did not establish the essential

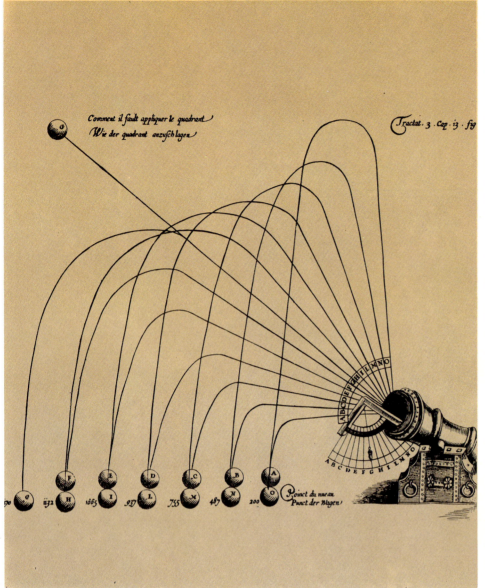

Comment il fault appliquer le quadrant
(Wie der quadrant anzuschlagen)

Tractat. 3 . Cap . 13 . fig

Poinct du nuean
Punct der Bügen

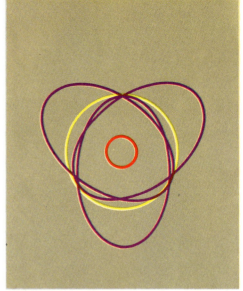

principle involved, Newton systematized and regularized mathematically the work of his predecessors.

Almost every advance in mechanics after Newton and before Einstein was an elaboration of these three laws.

The second law, the basic law of dynamics, deserves more explanation. It means that if a force is applied to a body, the body undergoes a variation in speed (i.e. an acceleration) which corresponds to the force applied. If the force is double, the acceleration will be double, if triple, triple, and so on. In mechanics there is understood to be a variation in speed – i.e. acceleration – not only when the speed increases (for example from 50 to 100 kph [31–62 mph]) but also when, at a constant speed of 50 kph (31 mph), there is a change of direction (this occurs, for example, when a corner is taken at a constant speed).

Thus a physicist is not only concerned with the magnitude of a force, but also the direction in which it acts on a body; likewise the direction in which a body is travelling is as important as its speed. Thus the physicist will speak not only of the force or speed but of its direction as well: this compound, speed/direction, or force/direction, is called a vector, and is represented diagrammatically by an arrow.

Left: parabolae of projectiles in a seventeenth-century print. Top right: the orbits of the electrons in an atom. Below: the parabolic orbit of a jet of water. All these trajectories can be explained by the three Newtonian laws of dynamics.

This stroboscopic photo of a pole-vault clearly shows the parabola made by the athlete.

The relationship between a force (F) acting on a body of mass (m) and giving rise to an acceleration (a) is expressed by the formula $F = m \times a$. The 'mass' of the body is difficult to define: it suggests the 'amount of matter' in the body. The exact meaning is a complex question beyond the scope of this book, but if scientifically innaccurate it is still largely true – and certainly helpful – to think of mass as weight.

The formula $F = m \times a$ is, quite rightly, the most famous formula in the whole of physics.

From it can be deduced the movements of bodies once their masses and the forces acting on them are known although the calculations are often very difficult, or, the forces can be determined once the movements and the masses are known. This formula also embraces, as a special case, the principle of inertia: if there is no force acting on a body, then according to this formula the body does not undergo any variation in speed; but the only movement in which there is no variation in speed is movement at a constant speed in a straight line.

By means of this law one can explain why a projectile makes a parabolic arc, and why its maximum range is obtained with a 45° sight; but one can also understand the laws governing the movement of the planets round the Sun and,

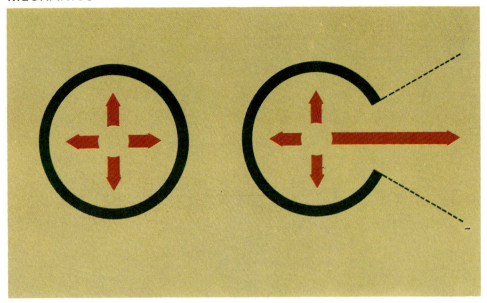

in part at least, the movements of electrons around the nuclei of atoms.

Newton's third law, that of action and reaction, can be expressed as follows: 'If Body A exerts a certain force on Body B, Body B exerts another force on Body A, directed in the opposite direction but of the same intensity.' Thus if we push against a wall with a certain force, the wall in turn reacts by pushing towards or against us with the *same* force, but in the opposite direction. If we now combine this third law with the second, we are in a position to explain all the phenomena of mechanics. Let us return to the example of the wall: if I push the wall (on the basis of the second law) the wall should move with an acceleration equal to the force applied by me, divided by the mass of the wall itself, however, the wall does not move. Why? Because the wall is attached to the Earth, and I am in practice not only pushing the wall but the whole Earth, the mass of which is so great that the resulting acceleration is negligible. So the wall does not move: the force (my pushing) is not enough. But why, then, do I not move, given that the wall exercises the same force on me? Because I too am (more or less) fixed to the Earth, by my feet. If I had a pair of well-oiled roller-skates on my feet, then I would soon be aware that I was undergoing an accelerative thrust pushing me away from the wall! One could mention many

The principle of action and reaction, or the third principle of dynamics, lays down that if one body exercises a force on another, the latter exercises on the former an equal force but in the opposite direction. Top left: the outlet of gas to the right causes the receptacle to be thrust towards the left. Below: an 'impossible machine'; the boat cannot move forward because the force exercised on the sail is cancelled out by the opposite force exercised on the boat by the boatman. Right: a jet-propelled missile being launched at Cape Canaveral.

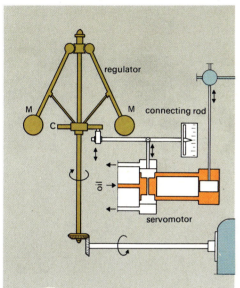

Centrifugal force is an apparent force directed outwards which is felt by a person making a curved trajectory. Top left: a rotating space station in which the centrifugal force causes an artificial gravity (from the film: "Space Odyssey"). Below: James Watt's automatic speed regulator. Top right: when taking a bend, motor-cyclists lean inwards to counter the effect of centrifugal force. Below: an experiment with centrifugal acceleration used in the training of astronauts.

examples – some of them paradoxical – to which the third law applies. Many man-made machines function on this principle – not only jet-planes but also simple row boats. It applies equally to our own actions. When we walk we are pushing against the Earth backwards and the Earth is pushing us forward, thus enabling us to proceed.

The third law also leads to the 'principles of conservation' of which there are many practical applications. One law which emerged is that if there are two isolated bodies, which only interact between each other (because there are no external forces acting on them), then, if stationary to begin with, after the interaction has occurred, the bodies will have a speed inversely proportional to their masses. Thus, if a gun weighing one kilogramme (2·2 lbs) fires a bullet weighing one gramme, at a speed (or velocity) of 1000 kph (620 mph), the gun itself will move backwards at a speed of 1/1000 × 1000, that is one kph; this is a phenomenon well known to everyone who has fired a gun and felt the recoil. Similarly, centrifugal force (the pull felt as a weight is swung round, for example, at the end of a string) is also explained by the law of action and reaction, for it is nothing more than the force of reaction to the real, centripetal force pulling the weight towards the centre of the circle, which itself is what

causes the movement to be a curved one.

With Newton's three laws it was possible to explain the movements of bodies once we knew their masses and the forces acting on them. It is fairly evident that when these laws were clearly understood and well assimilated by the scientific community (which was in the meantime expanding at a remarkable pace and on an international scale) there was a great upsurge of study and research, aimed at applying these laws to practical problems. Thus many advances were made in technology which, in turn, fostered the industrial revolution.

In addition new branches of physics arose which applied the three laws to a wide variety of bodies. The application of Newtonian principles to the behaviour of liquids led both to a fuller understanding of well-known phenomena such as the way objects float in liquids, and also to the formulation of new laws, such as those of the Swiss scientist Bernoulli about the pressure of liquids in movement. Other researchers were thinking in terms of applying the laws of mechanics to living organisms.

In nature, the forces with which one has to cope are numerous and varied, but the application of these three, almost miraculous, laws led to the greater

According to the laws of mechanics, if a force is applied to a system, and the force sets up oscillations, or wave movements, which are the same as the wave movements in the system itself, the wave movements increase until the system collapses, as in the case of this bridge, where vibrations from traffic matched the vibrations of the bridge itself. This phenomenon is called oscillations.

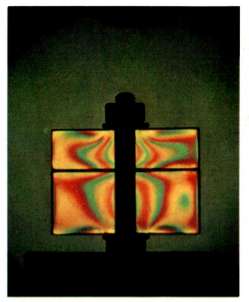

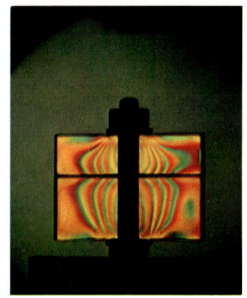

Left: internal stresses in plexiglass parallelepipeds under pressure, photographed in polarized light. Top right: an apple struck by a projectile 'explodes' because of the sudden increase of internal pressure. Below: a marine siphon in the Tonga islands. The complicated explanations of all these phenomena are derived solely from the laws of dynamics.

understanding of many problems in the life sciences.

The complexity of the problems studied stimulated more and more advanced mathematical research, so the 'physicist-mathematician' came into his own. Some of these problems were very complicated indeed, for example, the theory of elasticity. Here the forces involved are proportionate to the change in position of the body in question. By introducing proportionate forces into the equation for the second law, movements which were periodic resulted in movements that were repetitive and characterized by the interval between them (the time taken by the body to return to the point of departure) and their 'amplitude' (the maximum shift from the position of equilibrium). These are intricate complex matters and a considerable area of the most advanced mathematics has developed from research of this kind.

Another type of force which was studied in depth was surface tension, produced in fact by attraction among the molecules forming a liquid. In this field, and still on the basis of the three laws, fairly complex phenomena were explained, such as capillarity, the phenomenon whereby a liquid contained in a very narrow capillary tube tends to move upwards. It was thus realized how plants, which are full of capillary vessels, manage to draw water up from the

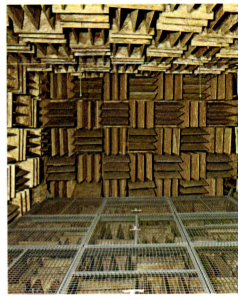
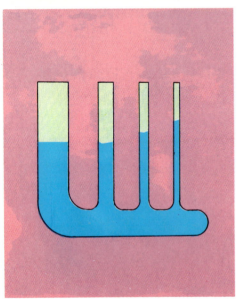
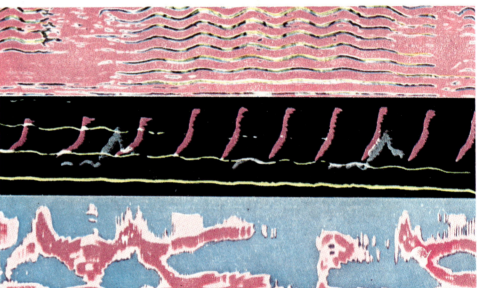

soil to their very topmost leaves, in some cases many feet above ground level.

Acoustics, one of the traditional branches of physics, also became associated with mechanics: sound is nothing more than waves of alternately compressed and rarefied air which spread at a certain velocity. When they reach the human ear, these waves cause the ear drum to vibrate and thus sound is heard. Physicists tested the validity of Newton's laws by transmitting sound waves through substances other than air, and the results of these experiments bore out exactly the predictions made according to the laws of motion.

In all these fields of research the concept of 'work' was very useful, as was the related concept of 'energy', which made it possible to establish, without going beyond the three laws, further principles of conservation on the basis of which it was then possible to solve often highly complex problems. In physics every term has a very precise meaning, and work does not mean what it does in everyday language: 'work' to a physicist is the product of force by movement in the direction of the force. Thus if a stone is lifted or is pushed a certain distance, the physicist will say that the force required to move or push the stone that particular distance is the work done; moving the stone twice as far will require twice as much work. The resource available to do that work is called by the

The study of the movement of masses in constant motion, such as liquids and gases, is just another chapter in mechanics. Top, left to right: a water-gnat on the surface of the water, supported in this position by the effect of surface tension; waves on the water; a echoless chamber for the absorption of sound. Below left: an illustration of the phenomenon of capillarity and (below right) sonograms, i.e. diagrams which visualize complex sounds, like birdsongs.

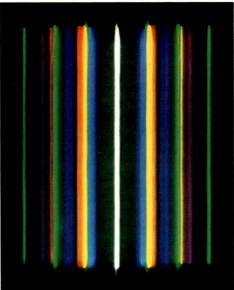

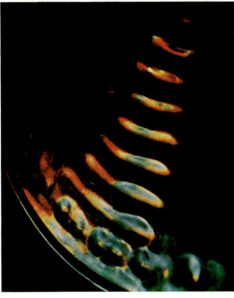

Ultra-sounds – sounds with frequencies higher than those audible to the human ear (about 16,000 vibrations per second) – are used effectively to increase the combustion of fuels (left and right). Top centre: an ultrasonic spectrum. Below: ultrasonic wavefronts on the surface of a liquid.

physicist energy.

The three laws developed by Newton served science well for centuries and still today form the basis of the scientific explanation of the way the world we see around us behaves. Newton's laws, however, do not solve all the problems for scientists studying the cosmos or the behaviour of those minute particles that make up the atom.

Towards the end of the nineteenth century many scientists studying natural phenomena became aware that the strange and inexplicable effects were noted when, for example, observations were affected by the speed of light. Light rays travel so quickly that in the ordinary world they can be regarded as moving instantaneously – there is no appreciable time lag for example in the beam from a lighthouse reaching a ship at sea. However, the speed of light does make an appreciable difference when, for example, a star many millions of miles away from the Earth is studied.

The additional laws needed by physicists to assist in explaining the phenomena observed in the behaviour of the very small and the very large were developed by one of the greatest of all scientists, Albert Einstein. Born of Jewish parentage at Ulm in Bavaria in 1879, he studied at Munich and Zürich

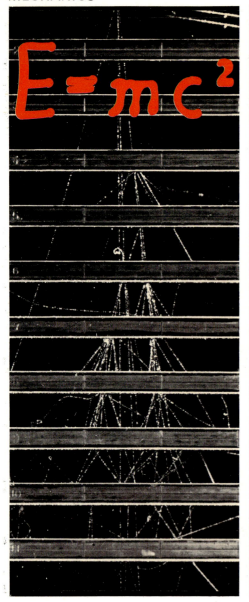

and in 1905 published his 'special theory of relativity': in 1916 he announced a further development of that special theory, the 'general theory of relativity'.

The arguments and the mathematics on which these theories were based are very complicated and perhaps not very easy for the layman to understand. Einstein began his argument from the premiss that the motion of any object must be measured relative to some other object. In the whole Universe nothing is absolutely at rest so that every movement is relative to some other movement: the movement of a spaceship for example is a movement relative to the movement of the Earth. The emphasis on *relative* motion is why the theory is called the theory of relativity. It may seem when speed is measured that the Earth is stationary and the spaceship moving, but, so far as the laws of nature go, it would make no difference if the spaceship were thought of as being at rest and the Earth moving at the speed of the spaceship.

Einstein further laid down that the speed of light is always found to have the same value no matter what is the movement of the source of light or of the observer.

Many very important consequences flow from the ideas of Einstein. Perhaps the most startling is that as its speed increases, so does the mass of a body. Thus

Albert Einstein (right) was the great innovator in mechanics at the turn of the century. His 'special theory of relativity' foresaw, among other things, the equivalence between mass and energy, using the famous formula $E = mc^2$. Einstein's theoretical prediction was later resoundingly confirmed by experiment. Left: a high-energy gamma ray with no mass creates two particles which do have a mass, i.e. an electron and a positron, which in turn emit gamma rays, and so on.

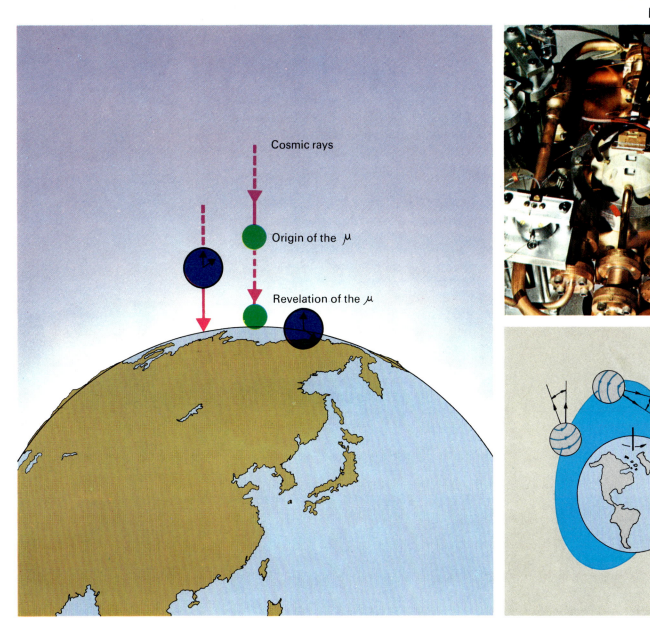

Cosmic rays

Origin of the μ

Revelation of the μ

Another disconcerting, but subsequently confirmed prediction of the theory of relativity is the variability of the duration of time intervals. Left: a mu meson as it were 'immortalized'; normally these unstable particles are so short-lived that they disintegrate before passing through the atmosphere. Einstein also postulated a wider theory, the general theory of relativity, but experimental confirmation of this is still not conclusive; numerous experiments are currently in progress (to verify or refute it) which use gyroscopes. Top right: section of a gyroscope consisting of a small sphere of quartz. Below: a plan of the experiment.

the mass of a body varies according to its motion. Of course, at the speeds normally experienced and observed, the increase in the mass of a body is so small as to be unnoticed but as objects approach the speed of light, the increase in mass occurs very rapidly. Einstein thus established his now famous connexion between mass and energy, in the formula familiar to all physicists, $E = mc^2$ (where E is the energy, m the mass and c the speed of light).

Not only did Einstein throw new light on the behaviour of nature but by successfully reducing his theories to mathematical terms, re-asserted the traditional links between mechanics and that austere discipline. His theories were startlingly novel, but other scientists were able to show that many of Einstein's theoretical predictions, based on his theory, were correct experimentally and his ideas, although not the whole answer to the mystery of the Universe have undoubtedly contributed significantly to the development of modern physics.

Not all the practical results of Einstein's work have been beneficial to mankind. His thesis that mass can be changed into energy and energy into mass is in fact the theoretical principle on which the fearsome destructive power of the atom bomb is based.

HEAT AND ENERGY

One natural phenomenon which, despite its day-to-day connotations, has considerably taxed the minds of scientists over recent centuries is heat, because there are some curious things about it.

In ancient Greece, on the basis of a theory outlined by Aristotle, people thought that there were four basic substances: earth, air, water and fire, and accordingly even up to the beginning of the nineteenth century, it was still believed that heat was a specific substance, a sort of weightless fluid which was distributed throughout a body in minute cavities, and could be measured, according to the amount there was present. The way one body grew warmer when in contact with a body hotter than itself was attributed to the movement of part of the 'heat-fluid' from the hotter to the colder body; the process of heating by friction was explained by suggesting that friction forced heat-fluid inside the body being rubbed to come to its surface.

But the scientific method of investigating natural phenomena, which had already borne important fruit in the field of dynamics, in the end disproved these various theories about the nature of heat, which had no experimental basis, and eventually physicists explained the phenomena associated with heat, according to the laws of physics.

As early as the beginning of the seventeenth century, an initial step in the right direction had been made by the Belgian Johann Baptist Van Helmont (1577–1644). Although still steeped in mediaeval ideas, he was an accurate observer and experimenter, and the first person to understand that there are gases which are quite separate from air (it would seem that it is to him that we owe the very term 'gas'). The significance of this discovery for an explanation of heat lies in the fact that it destroyed the Aristotelian theory of the four elements (air, fire, water, earth). Moreover, indirectly it cast doubt on the legitimacy of considering heat as a fluid (that concept had derived directly from the Aristotelian theory of four elements). A little later the Englishman Robert Boyle (1626–1691) appeared on the scene. This truly modern scientist categorically rejected the concept of four basic elements and formulated the concept of the chemical element – a concept in which both heat and fire could not easily be included.

But the old-fashioned method of thinking based on the proposition of theories which were more or less fanciful and not derived from experimental work led to a new and infertile hypothesis to explain the nature of heat, which earned a huge following among seventeenth- and eighteenth-century chemists. To explain the phenomenon of combustion, they devised a poorly defined heat-principle, the 'phlogistic' theory. The hypothesis here was that when combustion occurred an element 'phlogiston' became separated from the substance which was burning, and by becoming separate produced fire, heat and light.

The great French chemist Antoine Laurent Lavoisier (1743–1794) was responsible for permanently demolishing the phlogiston theory. He first sensed and later verified by experiments that combustion was the combination of the oxygen in the air with the inflammable or combustible substance being burnt. Despite this crucial discovery, the nature of heat remained shrouded in mystery and even Lavoisier himself continued to consider it a substance.

When a car brakes, i.e. changes from one speed to another, it loses kinetic energy, which is transformed into heat by friction, as is clearly visible in this photo of a racing car braking. The tread of modern tyres for racing cars is designed to produce friction and disperse heat.

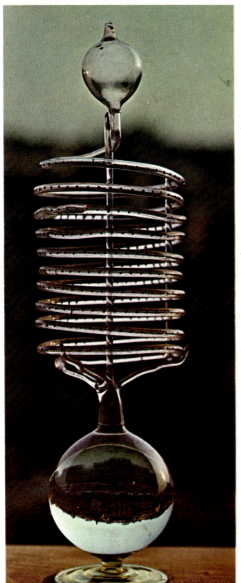

The key to the interpretation of heat was not to be supplied by chemistry, although, as we have seen, chemistry had provided the bases for the rejection of the various ancient theories. The key emerged from studies which were more directly concerned with phenomena of a mechanical nature. In the latter years of the eighteenth century the Anglo-American physicist Benjamin Thompson, Count Rumford (1753–1814) was engaged in studies concerned with the heating of metal bodies by friction. On the basis of experimental data, he came to the conclusion that it was inadmissible to maintain that the heat produced by friction was generated at the expense of a heat-reserve present in the body to which friction was applied. Broadly speaking, he reasoned as follows: because it can be observed that by rubbing pieces made of appropriately insulated material together new amounts of heat are produced indefinitely, it must follow that a body of limited size is an inexhaustible source of heat. This appears to be absurd, whereas, conversely, it appears quite reasonable to think that heat is an aspect of that which is *transmitted* by friction (or rubbing), which is nothing other than *movement*.

At almost precisely the same time another English scientist, Sir Humphrey Davy (1778–1829), carried out experiments in which two pieces of ice melted

Left and top right: two ancient thermometers from the Accademia del Cimento in Florence. Below right: a recording telethermometer and a vertical thermometer with holder.

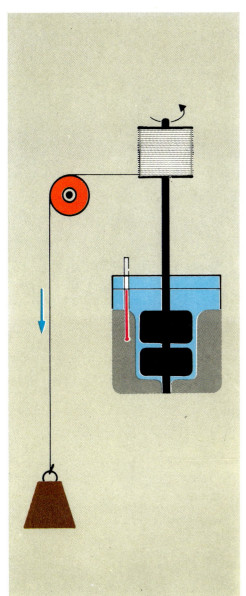

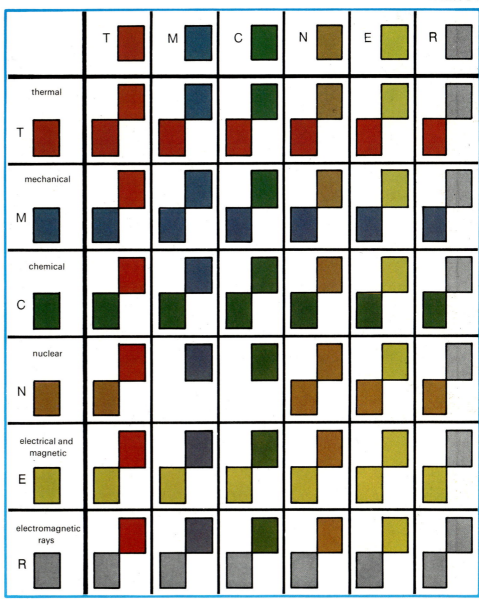

Left: a diagram showing Joule's mill, an experiment for demonstrating the equivalence between heat and mechanical work. Right: a diagram of the major possible transformations of energy; energy develops by passing from one to another of the forms in which it manifests itself.

when rubbed together in a vacuum, although, obviously enough, there was no possibility of deriving the heat from the ice. This heat had, therefore, to come from the friction itself.

The experiments carried out by Rumford and Davy thus showed quite clearly that heat is a consequence of movement. However, the problem had still not been studied in such a way as to provide that experimental data from which, mathematically, a law could be worked out.

The basic problem therefore remained: to determine the mechanical equivalent of heat, that is the amount of work required to produce a particular quantity of heat. A very interesting attempt in this direction was made by the German physicist Julius Robert von Mayer (1814–1878) who assessed the heat produced in the compression of gases. However, the crucial experiments which in effect accurately and precisely established the equivalence between mechanical work and heat were made by the English physicist James Prescott Joule (1818–1889). He carried out these experiments using extremely accurate and precise methods of measurement. The most famous apparatus used by him consisted of a small wheel with blades, placed in a cylindrical container filled with water. The wheel was turned by dropping a weight on it; inside the

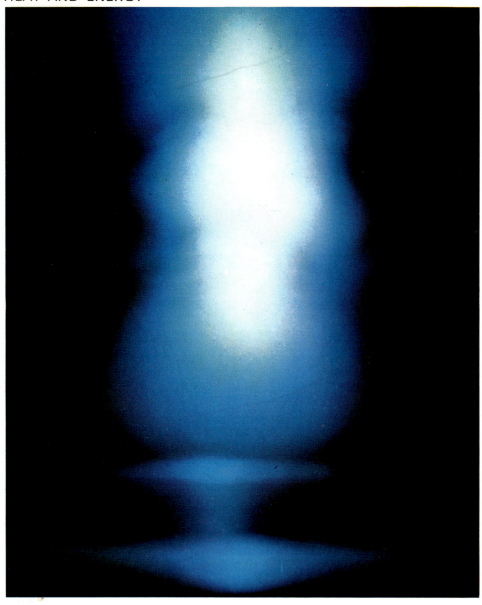

receptacle further fixed blades stopped the water following the rotation of the wheel, so causing friction between the water and the blades.

The experiment showed that, gradually, as the wheel rotated, the temperature of the water increased. From the increase in temperature it was possible to calculate, theoretically, the amount of heat supplied to the water by the 'work' of the wheel. In turn this 'work' could be calculated when the exact weight of the weight dropped on the wheel was known. By comparing the heat produced with the work needed to produce it, Joule was able to establish the amount of work required to produce a unit of heat.

In this way, by establishing the equivalence between mechanical work and heat, it was shown once and for all that heat was not some rather mysterious substance but merely a particular form of energy. This discovery embodied the principle of an extremely important physical law, the First Principle of Thermodynamics, and showed that energy as a whole cannot be increased or decreased: it can only be changed from one form to another, so that the energy lost by the wheel is the energy gained by the water as heat. A branch of physics was thus born, dealing specifically with phenomena in which transformations of work into heat occur, and *vice versa*.

Left: plasma produced by heating a very low-pressure gas to more than 100,000° C (180,000° F). Top right: a Zeta machine at Harwell, England, which has produced temperatures of up to 5,000,000° C (9,000,000° F) for a fraction of a second. Below: sunspots (caused by hydrogen explosions).

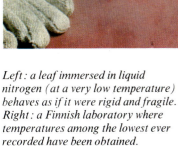

Left: a leaf immersed in liquid nitrogen (at a very low temperature) behaves as if it were rigid and fragile. Right: a Finnish laboratory where temperatures among the lowest ever recorded have been obtained.

It may be helpful at this point to draw attention to the difference between two terms, *heat* and *temperature*, which are often used rather loosely by laymen. They are of course quite different. To the physicist the heat involved in any particular circumstances is an actual quantity of energy – and represents the amount of work that has to be done to make the body on which the work is done, hotter. Temperature is no more than a measure of hotness, or coldness. Temperature is a measure used to indicate the effect of heat on the body. Thus if a kettle of water is made to boil, the gas or electricity used to make the water hot represents the heat – the energy used – and temperature is the measure used to indicate the degree of hotness of the water as it becomes warmer and warmer, through the work done by the heat energy applied to it.

This difference is easier to remember if the units of measurement used for heat and temperature are borne in mind. The traditional measure of heat is the *calorie*, which is defined as the amount of heat needed to raise the temperature of one gm of water through one degree on the centigrade temperature scale. There is now another unit used increasingly in science, the *joule*, but although this has a more complicated definition, the principle is the same – it is the exact amount of energy used, for example, by one watt in one second: that definition

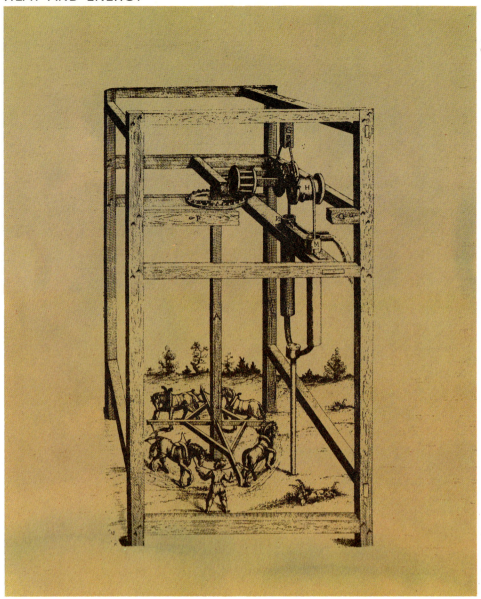

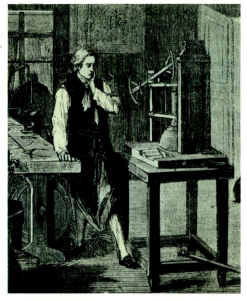

may not be very clear until the reader has studied this book more fully. But the essential fact to remember is that both calorie and joule are measures of particular quantities of energy. One calorie will raise one gm of water through 1° Centigrade; two calories will raise two gm of water through 1° Centigrade.

Temperature, however, is not a quantity in the same sense: temperature is merely a convenient scale to measure changes in hotness or coldness, and although used both by scientists and ordinary people in everyday life, and understood by all, they are really quite arbitrary scales. The Centigrade scale, for example, has 0 as the freezing point of water, and 100 as the boiling point of water, and each 1/100 part of hotness or coldness between freezing and boiling is 1° Centigrade. On the Fahrenheit scale 32 is the freezing point, equivalent to 0 on the Centigrade scale and 212 the boiling point, equivalent to 100 Centigrade. These scales are well known and therefore more or less widely used but have no justified theoretical scientific basis as, for example, have the exact measurements such as the calorie or joule.

Another important temperature scale used by scientists is the 'absolute' scale. Here zero ('absolute zero') corresponds to about −273° C (−459° F), whereas the freezing point of water is about 273° (absolute). The highest

Left: a horse-powered pump; when animal energy was replaced by the thermal energy produced by steam, the term 'horsepower' came into being. Top right: James Watt who developed the steam-engine. Below: a vehicle designed by N. J. Cugnot in 1765 which was powered by a steam-engine.

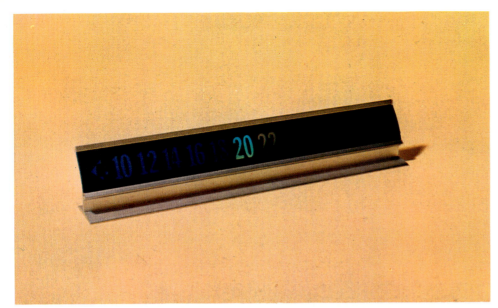

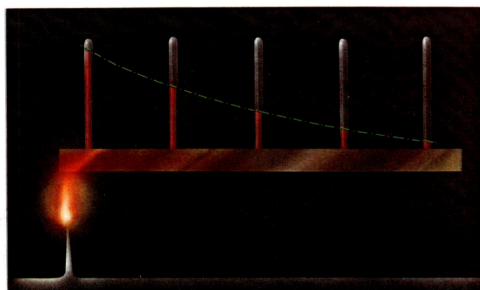

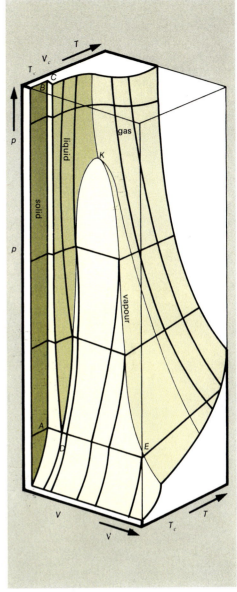

Top left: a liquid crystal thermometer. Below: a heat conduction experiment; five thermometers placed in a receptacle which is heated on the left-hand side show a progressive rise in temperature as the heat spreads along the base. Right: a spatial diagram (pressure-volume-temperature) for a pure substance.

temperatures obtained by man are those reached in the centre of a hydrogen bomb, at around 400,000,000° C (720,000,000° F); in the laboratory a temperature of 50,000,000° C (90,000,000° F) has been produced for a very short time. The lowest temperature ever obtained experimentally, on the other hand, is in the region of 1/1,000,000th of 1° (absolute), that is almost, but not quite, absolute zero.

Heat can be transferred from one body to another in a variety of ways: by contact, radiation, conduction or convection. Heat can pass by contact from one body to another when the bodies are touching one another; it can pass by radiation, from one body which emits heat-rays to another which receives them (thus the Earth receives heat from the Sun). And heat can transmit itself within one and the same body by means of conduction as when, if one end of a rod is heated in a fire, the other end will also ultimately become hot.

On the other hand, the application of the same amount of heat – the same number of calories – to different substances does not raise their temperature by the same amount. The size, the chemical composition and the physical state of the substance will all affect the degree of change it undergoes. Similarly, when a certain temperature is reached, many substances change their state; water

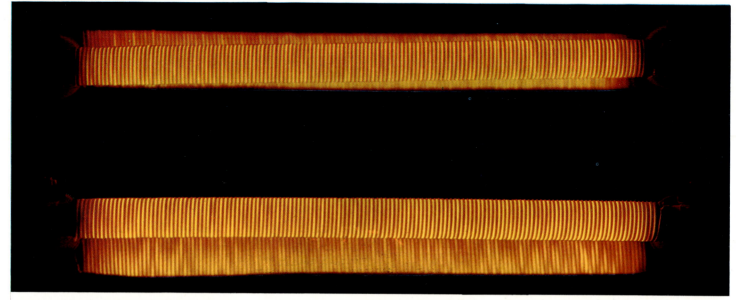

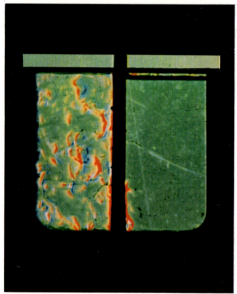 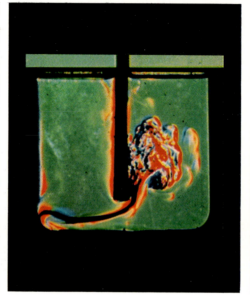 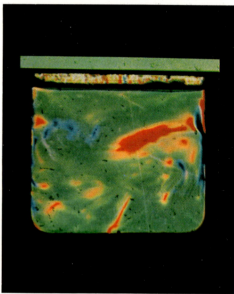

becomes ice, ice becomes water; water becomes steam or steam condenses into water. This change of state itself uses up energy, and while that change of state is taking place, energy will be used by the substance without any increase or decrease in temperature.

The fact that some bodies are colder than others – that is, are at different temperatures, and that in the natural world the effect of the Sun is to change temperatures – became of importance to physicists because a great deal of the practical importance of thermodynamics lies in the need, in life, to transfer heat from one subject to another, to keep things warm, or as in refrigeration, to keep them cool. Many great physicists studied the problem of heat transfer and sought to establish general laws. The French physicist Sadi Carnot (1796–1832), was the first to study the question in detail. Subsequently William Thomson, Lord Kelvin (1824–1907) and the German physicist Rudolf Clausius (1822–1888) both reached, independently, the same conclusions, and these, as stated by Clausius, became accepted as the Second Principle of Thermodynamics. This law states that heat cannot be transferred from a colder to a hotter body without work being done – or energy being applied. This is, theoretically, a complex matter, but it is helpful to remember the simple

Top: thermal energy is conveyed through the air or in a vacuum by radiation; this is how an electric heater transmits its heat. Below: the visualization of convective movements produced by the mingling of hot and cold masses which were initially separated by a dividing wall.

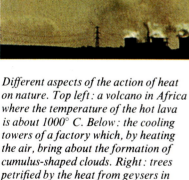

Different aspects of the action of heat on nature. Top left: a volcano in Africa where the temperature of the hot lava is about 1000° C. Below: the cooling towers of a factory which, by heating the air, bring about the formation of cumulus-shaped clouds. Right: trees petrified by the heat from geysers in Yellowstone National Park (USA).

example represented by a refrigerator: if cold food is left out on a warm day, it will become warmer from its surroundings; it will never become cooler by giving off some of its heat to its surroundings. If it is to be made cooler, it will have to be put in a refrigerator – a machine which, using energy from another source, will, through a complicated process, reduce the temperature of the food.

This second law of thermodynamics has a number of important implications, and severely limits the ways whereby heat which is naturally stored in many bodies can be used by man. If there is a substantial difference in temperature between two bodies, the hotter of the two can be employed without any additional energy source, to transfer heat to the colder body. Thus a block of ice put next to stones made hot in the sun will be warmed and melted by them; if the stones, however, are almost as cold as the ice, they will be ineffective as a melting agent. Thus many natural resources which are heat stores are nevertheless impossible to exploit economically – for example, sea-water which, although it contains heat, is not at such a temperature as will allow the heat to be extracted and used without the use of intermediary machines that would use more energy in extracting that heat than the heat

extracted would provide. On the other hand, sea-water can be used to cool steam produced by power stations.

Another consequence of this second law of thermodynamics is that it is never possible to convert into work all the heat contained in a body. In the course of extracting the heat some of it will inevitably be turned into some form of energy other than that which is needed. So because the useful end-product is always slightly smaller than the potential amount of energy contained in the body, perpetual motion machines are shown to be theoretically impossible, according to this law.

Thus, if we transform a certain quantity of work into heat, we can never turn it back and re-obtain all the initial work from the heat produced, without entailing some other secondary effect. Moreover, while both mechanical work and heat are forms of energy, there is (at least in every isolated system) a tendency towards an increase in the amount of energy present in the form of heat, at the expense of the energy present in the form of work.

The reason why this is so has been investigated by physicists working in the field of statistical mechanics. The theory is that heat contained in bodies is the result of the random movement of particles (molecules) of which the body is

Top left: the gap between two adjacent pieces of railway track, left to allow for the effects of heat expansion.
Below: a solar furnace in Israel.
Right: three infra-red photographs of the planet Jupiter.

Top left: an infra-red photograph of the Gulf of California. Below: the same picture deciphered, when the various colour-tones have been separated with special filters. Top right: a photograph of a forest; on one level one can see only leaves and trees. Below: with film which is sensitive to infra-red rays, the 'warm' body of a man is clearly visible behind the undergrowth.

made up; the higher the temperature of a body, the more intense this movement becomes. Let us now try to imagine what happens when, for example, we strike a hammer against a piece of metal attached to a surface. The initially orderly movement of the molecules of the hammer is transmitted at the moment of impact to the molecules of the metal being struck; but because the metal is fixed, its molecules cannot all move in the same direction, the direction that is, in which the sudden blow of the hammer would push them. Instead and in order to withstand the blow, they are forced to move in a disorderly manner in countless different directions. And the effect of this development of disorderly movement is a rise in temperature of the metal struck. The same applies to every transformation of work into heat. There is a transition from a situation of order (overall movement of all the molecules of a body in one and the same direction) to a situation of considerable disorder (disorderly movement of the molecules in every conceivable direction). This is usually expressed by saying that energy tends to lower naturally of its own accord, passing from a 'higher' form – work – to a 'lower' form – heat. The theory has in fact been put forward that the destiny of the universe is a state of total disorder. But this is still pure hypothesis, because of the scarcity of available

information about the behaviour of phenomena developing in parts of the Universe which are far removed from the planet Earth.

Top: successive transitions from an orderly distribution to an increasingly disorderly or random distribution. Below: the expansion of aniline dyes in water. This is a practical illustration of the situation shown schematically in the figure above; here too an initially orderly situation (with the anilines clearly separate and isolated) develops into a disorderly or random situation (with the anilines becoming mixed up as they spread through the water).

ELECTRICITY AND THE ELECTRON

Today we live in a world which in a way is dominated by electricity. Think what would happen if, for just a split second, all the appliances in the world which work by electric current, stopped working. We would find ourselves in a silent, death-like world of darkness. In fact not only would all those machines which derive their source of energy directly from electricity come to a halt (no more radio, television, trains, trams or light-bulbs). In addition, all those machines which need electricity indirectly would also grind to a halt. For example, we would be cold, because the thermostatic systems which regulate *oil*-fired central heating boilers nevertheless run on electricity; there would be no more vehicles in the streets because although cars run on petrol, they also use electricity in their ignition system.

And yet it was not very long ago that electricity became indispensable. In fact it was only towards the mid-eighteenth century that studies dealing with electrical phenomena began to be intensified and the various industrial applications of electricity did not occur until the nineteenth century.

There is no doubt that people knew about electrical phenomena in the very earliest times, and it is perhaps not as rash as it may seem to say that mankind's greatest discovery of all – fire – may well have been due to a flash of lightning striking a tree in a forest (due, in other words, to an electrical phenomenon). The name itself comes from a Greek word: legend has it that an unknown Greek shepherd wanted to polish a small piece of amber (in Greek: *elektron*) and so rubbed it against the fleece of one of the ewes in his flock; then he noticed to his surprise that the piece of amber he had rubbed had the curious property of attracting small chippings of wood. Leaving legend aside, we know for certain that the philosopher Thales of Miletus (who lived between 600 and 500 B.C.) described in considerable detail the electrification of amber, and it is also known that certain primitive peoples in Amazonia have the habit of electrifying, by rubbing, the seeds of certain plants.

Anyone can carry out such an experiment, and neither ewes nor amber are needed: it is enough simply to rub an ordinary fountain-pen with a piece of wool and then put the pen near some little bits of paper.

However, it was not until William Gilbert's work, *De Magnete*, published in 1600, more than 2000 years after the time of Thales, that the term *vis electrica* was coined, (the term *electricitas*, on the other hand, is due to Boyle). The electricity in question was still obtained by friction or rubbing, but Gilbert did start classifying the various groups of substances which gave rise to phenomena similar to that produced by amber, and he considered that glass, certain precious stones, sulphur, rock-salt (or halite), and sealing wax were bodies endowed with *vis electrica*; he also observed that atmospheric humidity was not favourable to the manifestation of the phenomenon of attraction.

After Gilbert the major figure to emerge in this field was the Magdeburg physicist, Otto von Guericke, already well-known for his research into the vacuum. Guericke was the first to build a machine, still based on friction, capable of producing electricity on a continuous basis, from a sphere of sulphur. Hence he was able to observe the effect of repulsion between electrified bodies, as well as the more easily observable phenomenon of attraction.

Towards the end of the seventeenth century there were already plenty of

This detail from the large panel by Raoul Dufy, Histoire de l'électricité *(Museum of Modern Art, Paris), shows some of the principal pioneers in the field of electricity.*

scientists who had repeated the experiments made by Gilbert and Guericke, with more and more improved versions of the friction machine in which, usually, sulphur was replaced by glass as the agent for friction.

Little by little, as invariably happens in the field of scientific knowledge, the countless and often *ad hoc* experiments (sometimes carried out simply to astound the public) started to be assembled, catalogued and interpreted. In about 1700 two famous scientists, Stephen Gray and Charles Du Fay, started, independently, to undertake a series of research studies into electrical phenomena which opened the way for subsequent understanding of their true nature. Gray discovered that electricity could be conveyed by some bodies and not by others, and thus made the first distinction between *conducting* bodies (like metals) and *insulating* bodies (like glass and ceramics); he also managed to transmit 'electric effects' along a cable two kilometres (1.24 miles) in length. Even more important, perhaps, was the fact that he demonstrated with accurate experimentation that electricity influenced other bodies at a distance, and thus laid the bases of electrostatic induction.

Du Fay's major contribution, on the other hand, was clearly to single out (even though he had had precursors) two different types of electricity, *resinous*

and *vitreous*, nowadays known as positive and negative electricity; and to recognize that bodies having the same type of electrification (for example, both resinous or both vitreous) repel one another, whereas bodies having different types of electrification attract one another. Following the Cartesian spirit of the times, Du Fay also tried to explain why electrified bodies attract or repel one another, and he himself supposed that when one rubbed a body in order to electrify it, vortices were formed which revolved in one direction for the vitreous bodies and in the other for the resinous bodies.

These theories were rejected at a later date, but what did remain intact were the precise and accurate experiments which, independently of their fanciful interpretations, formed the basis of much subsequent research.

Let us now briefly summarize how much was known about electrical phenomena at the beginning of the eighteenth century. It was known that certain bodies, when subjected to friction or rubbing, attracted or repelled other bodies; that these properties could be transmitted via even quite long wires, or even through the air; that bodies of the same type, once electrified, repelled one another; and that bodies of different types attracted one another. Along with all these experimental results, which were nevertheless to remain

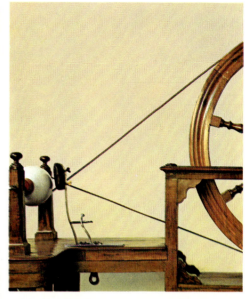

valid, went a large number of theories which sought to explain (by means of imponderable fluids and by means of vortices) *why* these phenomena came about. The next step was to try and understand not so much *why* but how.

A major step forward came with the invention of the piece of apparatus which soon became known throughout the world as the 'Leyden Jar' after the name of the Dutch town in which the physicist Peter van Musschenbroek carried out the first experiments in about 1745 (a similar experiment had been carried out in Germany by von Kleist). In trying to electrify the water contained in a glass jar (the famous 'Jar' in fact) by means of a long nail, van Musschenbroek received a powerful shock; quite by chance he had stumbled across the phenomenon of the accumulation of an electric charge between two condensers separated by an insulator.

News of the discovery of the 'shock' spread throughout Europe and numerous physicists, and amateurs, repeated the experiment, and varied it to explore the matter further.

Some killed small animals, such as birds and mice; some subjected rows of people holding hands to the shock (in France the shock was administered to all the monks in a monastery – who made a line which was almost three kilometres

Examples of electrostatic machines. Left: a modern Van de Graaf generator used as a particle accelerator. Top right: an electrostatic machine designed by Abbot J. Nollet (1700-1770) in which a sphere is electrified by friction by rubbing on a leather pad. Below: Wimshurst's machine, which is used in many training laboratories.

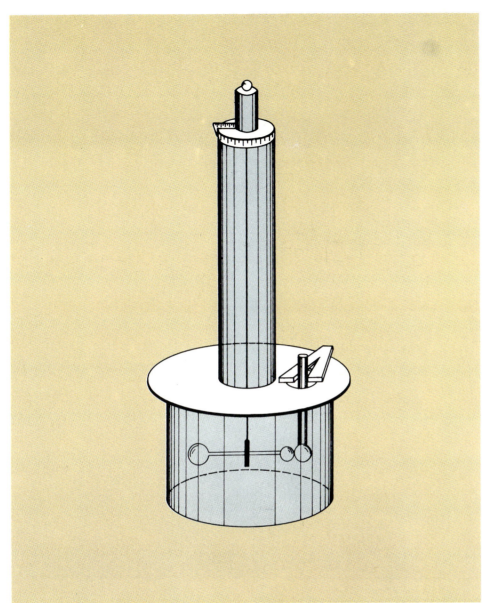

Left: the torsion balance with which Coulomb established the law of attraction and repulsion between electric charges. Top right: experiments made with a Leyden jar (an illustration dated 1700). Below: a typical Leyden jar, prototype of electric condensers.

long); others astounded audiences by setting fire to a spoonful of alcohol held by an onlooker. But there were some who tried to proceed in a more scientific manner, and during this period many improvements were made to instruments which could be used to show the presence of electrical charges, for example, the gold-leaf electroscope which is still in use, although its first version dates back to 1705, the year in which Francis Hawksbee devised it.

Naturally enough, medicine was quick to snap up these new discoveries, almost as soon as the first physiological effects were identified, but alongside serious research work many a charlatan and profiteer cashed in on the new fashion. As early as 1745, a professor Kratzenstein of St. Petersburg proposed activating the heart with this new commodity called electricity and at the same time many an unscrupulous scoundrel made his fortune selling 'electric water' which was meant to cure all ills.

It is worth noting that it was precisely with research into electrical phenomenon that the New World joined the scientific community; in the person of Benjamin Franklin. Although he did not begin to busy himself with physics until he was 40, Franklin carried out a great deal of interesting research work; he discovered that the plates of a condenser are charged with opposed

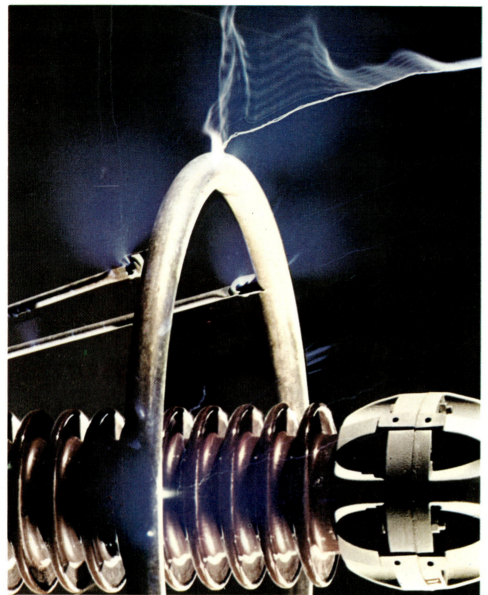

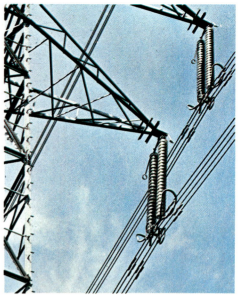

electricity (which he called 'positive' and 'negative' – both terms which have survived right down to the present day), but became best known of all for his experiments which had to do with atmospheric electricity and, in particular, for his invention of the lightning conductor. His observations were received initially with considerable scepticism, and in some cases even with derision, by the Royal Society in London, whose erudite members found it inconceivable that the centre of scientific knowledge was on the verge of shifting from England to a country which was then considered to be little better than primitive.

There were then no really important developments until the French engineer Charles Coulomb published the results of his experiments between 1784 and 1789.

With the *torsion balance*, which was subsequently used by Cavendish to measure gravitational attraction, Coulomb measured the strength with which opposite electrical charges attracted one another, and similar charges repelled one another. Having noted that the electricity within a conducting sphere is divided in half when the sphere is touched by another identical sphere, and having thus found the method of dividing an electric charge into equal parts, he was able to declare that the strengths with which two charges attract each other

Modern electric insulators with particular mechanical and electrical properties. Left: testing an insulator. Top right: glass insulators in a high tension line. Below: porcelain insulators; these are much more expensive than glass ones.

The length of an electric discharge is an indicator of the difference of potential, or 'electric tension', between the points where it occurs. In the air there are discharges of just a few millimetres for differences in potential of 1000 volts. Left: a discharge between monitoring equipment in the CGE-Ansaldo laboratories in San Giorgio. Top right: huge flashes of lightning over an astronomical observatory. Below: a discharge in a rarefied gas.

(if they are opposed) or repel each other (if they are the same) is in direct proportion to the square of the distance between them. This means that if the distance between two charges is doubled, the strength between them is one quarter of the initial strength; conversely, if the distance is halved, the strength increases fourfold, and so on. This law is identical to the law of gravitational attraction discovered by Newton, and all the deductions made by Newton himself concerning gravitational attraction could be transferred to the area of electrical attraction, bearing in mind, naturally enough, the fact that electrical attraction is vastly more intense than gravitational attraction. Just as the Newtonian law had stimulated important mathematical research, so Coulomb's law gave rise to the major work done by Poisson (1811), Green (1828) and Gauss (1839). What is more important, however, is that with Coulomb electricity was finally given solid quantitative bases, and with him the period of mere observation of sensational effects was closed for ever.

With the various friction machines, and even more so with the Leyden Jar, electric currents had been obtained which were very powerful, but short-lived. The decisive step forward came with the introduction by the Italian physicist Alessandro Volta of the 'pile' or cell, a device which made it possible to carry

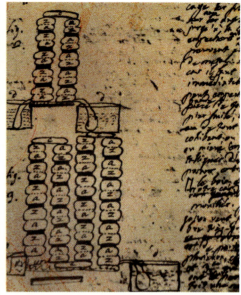

out research on a current of electricity that continued to flow. With the discovery of the pile or cell, physical research was directed for some twenty years (from 1800 to 1820) almost exclusively to the study of the phenomena presented by electric current.

Volta, who was a professor at Pavia and already very well known for certain inventions such as the endiometer and the inflammable air-pistol, but above all else for his discovery of 'marsh gas' (i.e. methane), had become involved in a scientific dispute with the distinguished anatomist Luigi Galvani. While carrying out some research into animal electricity, the latter had observed, in 1789, the contractions of a dead frog when the lumbar nerves and the thigh muscles were touched simultaneously with a metal arc. In Galvani's eyes the frog was to be considered as a kind of biological Leyden Jar, i.e. a small generator of electric energy, whereas, according to Volta, the frog was just the *detector*, and the metal arc was the actual *generator*. After carrying out increasingly painstaking and crucial experiments, Volta's view was shown to be correct and precisely this awareness of having gained a new type of current generator gave rise to research into bimetallic arcs which then culminated in the invention of the famous *pile*, when Volta noticed that the electric imbalance

Alessandro Volta (left), born in Como and professor of physics at Pavia, invented a new and powerful generator to assist in electrical research and study – the battery or pile. *Top right: the famous* Lettera sulla pila *in which Volta described the constructional and operative features of the battery. Below; one of the first electroscopes, which are instruments for detecting differences in potential, conceived by Volta.*

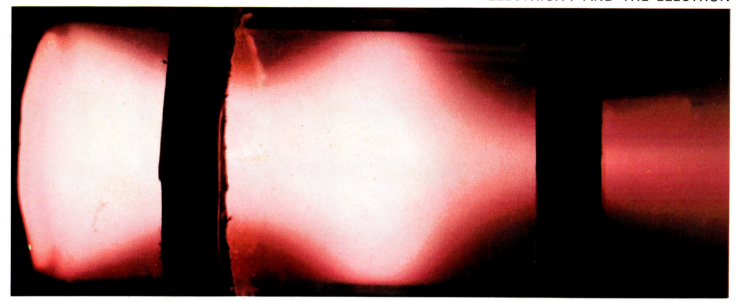

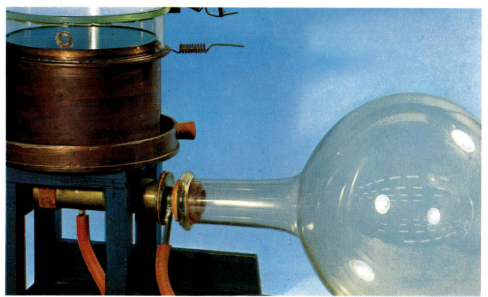

Top: a plasma, a strongly ionized gas (an ion is an atom to which an electron has been added, or from which one has been removed). Below left: the first cloud-chamber with which C.T.R. Wilson showed the traces of charged particles; with the addition of a magnetic field the charged particles, and in particular the electrons, make clearly visible circular orbits. Right: traces in a Wilson's cloud-chamber with a magnetic field of negative and positive (positrons) electrons. The radius of the orbit reduces as a result of the progressive loss of energy of the particles.

(today this would be called the potential difference) was increased if there was a chain of bimetallic conductors separated by an acid-based solution. This was the origin of *voltage*, which is the driving force which causes an electric current to flow. What Volta had discovered in fact was that an electric charge will flow from a point or body with a higher charge to a point or body with a lower charge (just as heat will flow from the hotter to the colder body). The difference between the high charge and the lower electric charge is the *potential difference*, and it is that which today is called voltage. Volta called his new instrument the *electromotive column machine*, or *pile* (in Latin 'pila' means column), and announced his discovery to the Royal Society in a letter dated March 20, 1800. With the pile, which was subsequently developed and improved by a host of other scientists, research into electric phenomena advanced by leaps and bounds, and a new era in the history of science was launched. Previously vague and imprecise concepts were clarified: this 'electric tension' (today called potential difference) was defined as the cause of the movement of electricity from one body to another, in the same way in which a difference in height causes the fall of a weight from a higher to a lower point, or a difference in temperature causes the passage of heat from a hotter to a colder body. In other

words a theory had been arrived at in which, as in the case of the 'heat-fluid', there was considered to be an 'electric fluid', which presented itself as current when it flowed from the conductor with the highest potential to the conductor with the lowest potential. In fact the analogy was not a perfect one, because whereas it was enough to consider just one single fluid to explain the various thermal phenomena, it was necessary to consider two fluids – one positive and one negative – to describe the electric phenomena, and it was impossible to say whether the current could be ascribed to the flow of one or other of the fluids, or to both of them. This fundamental question was not in fact cleared up until early in the twentieth century, with the discovery of the electron.

As in the case of numerous other basic discoveries in physics, the discovery of the electron cannot be put down to any one scientist, but was rather the fruit of countless and varied research projects, both experimental and technical, with which are associated the names of W. Crookes, J. J. Thomson, J. Perrin, H. A. Lorentz and R. Millikan. While observing the discharge in rarefied gases, a 'radiation' coming from the cathode (i.e. from the negative pole) had also been observed; this was called cathode radiation by Goldstein in 1876. The properties of this radiation, many of which were brought to light by Crookes,

Man's understanding of the electronic nature of current has enabled him to realize almost miraculous technical feats. In this photograph all the electrical appliances and the cables and contacts in the background, are replaced by the printed circuit in the foreground.

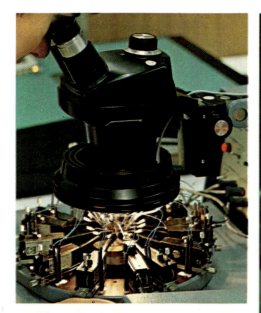

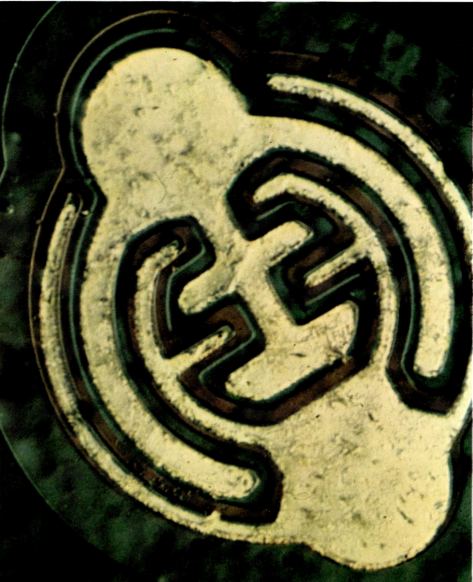

The use of transistors, the basic components of electronic circuits, has made it possible to miniaturize circuits and at the same time maintain and in some cases improve their efficiency.
Top left: detail of a micro-connection.
Below: a one-pole silicon transistor.
Right: a highly magnified transistor.
Transistors are now part and parcel of every type of electrical appliance, from radios to the ignition systems for internal combustion engines.

stirred up a lively scientific debate over the corpuscular or undulatory nature of the cathode rays, but in the end, and principally as a result of the work of Perrin and Thomson, the initial hypothesis won through. The rays were extremely light particles (about 2000 times lighter than the lightest atom), negatively charged, which moved within the cathode tubes at very high speeds indeed, in the region of tens of thousands of kilometres per second, and they were present in all matter. In other words they were 'atoms' of electricity.

For many historians of science, modern atomic or nuclear physics actually started with the discovery of the electron, or the first 'elementary particle'; it was followed by the discovery of vast numbers of other particles, starting with the proton, but we shall deal with this more fully in a separate chapter. Observing that the cathode rays could penetrate substances, J. J. Thomson also put forward various hypotheses about their dimensions, concluding that they must be extremely small, and much smaller than atoms themselves. Thomson's reasoning led to essentially correct conclusions, even if modern quantum mechanics has stripped the classical term 'dimension' of much of its meaning.

The discovery of the electron led, although not in the short term, to countless practical applications which have been so important that this elementary

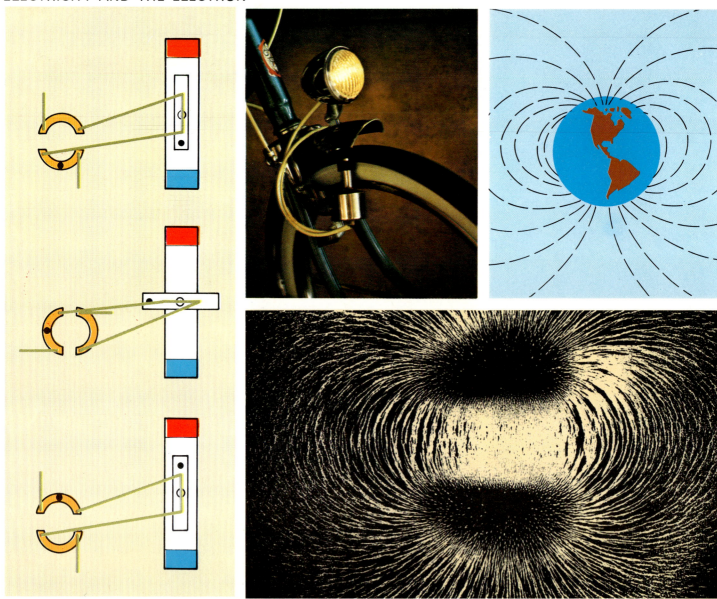

particle has christened a whole new branch of physics, i.e. electronics. Just as the nineteenth century can rightly be called the century of electricity, the twentieth century can just as rightly be called the century of electronics. Here we can only mention a few consequences of the discovery of the electron, which are now part and parcel of our day-to-day lives. They range from radio, which uses valves in which electrons are emitted, to television, which can be considered to be the latest development of the cathode ray tube with which Crookes carried out the original experiments, to the oscilloscope, which is now so familiar to all students of physics, and lastly to the transistor, which has made it possible to 'miniaturize' complex circuits to extremely small dimensions. Among other things, the process of miniaturization has been vital for the construction of those huge, extremely fast computers which can handle millions of operations per second. These computers are the fruit of the progress of science and at the same time the necessary condition of this very development. The fact that the real nature of electricity was still unknown did not pose any serious difficulties in these technological developments.

These developments were largely due to the realization of the close connection between electric and magnetic phenomena, both of which had been

The interdependence between electrical and magnetic phenomena was first brought to light by the Danish physicist H. C. Oersted. Left: the principle on which an electromagnet works. The winding receives a current which gives the bar a polarity which causes it to be repelled by the permanent magnet; when it is near the opposite pole, the switch sends current in the other direction and the bar continues to turn. Top centre: a bicycle dynamo. Top right: earth's magnetic field. Below: visualization of the lines of force of a magnetic field by means of iron filings.

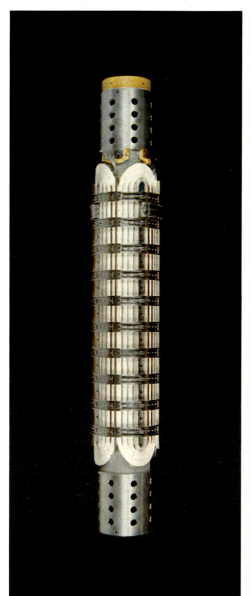

Left: parts of a superconductor electromagnet (Brookhaven National Laboratory) which works at a temperature of 4.8° (Kelvin). Cooled by liquid helium, it generates a magnetic field of about 7000 Gauss. Right: polar expansions of a magnet used in experiments with paramagnetic electronic resonance (Institute of Physics, Parma).

known about since time immemorial. This connection was first brought to light by the Danish physicist Hans Christian Oersted. In 1820 he showed that a wire carrying electric current rotated if placed close to a magnet, and that, likewise, a magnet rotated if placed close to a wire carrying electric current. The observation of these phenomena, which astounded the scientific world of the day, gave birth to electromagnetism (the name allocated by Oersted himself), the development of which was carried out by the famous physicist and mathematician James Clerk Maxwell. Progress in this field was consummated by Albert Einstein's theory of relativity, in which electric and magnetic phenomena are once and for all linked together. The discovery of the electromagnetic phenomena was to see industrial application as early as the end of the nineteenth century. Many of the appliances which we now use every day of our lives work on the basis of the principles established by Oersted. Suffice it to mention hydro-electric power stations, which transform the mechanical energy of waterfalls into electric energy, or on a smaller scale the dynamo of an ordinary bicycle by which pedal power is turned into electric energy. Similarly, electric power which comes to our homes is reconverted into mechanical energy when used for all those electrical appliances such as washing machines, mixers

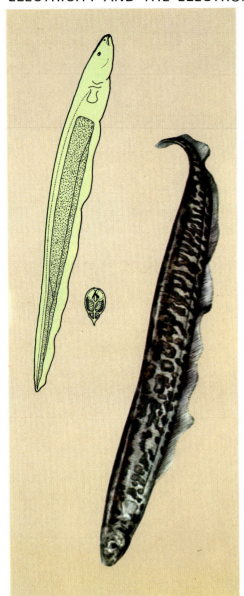

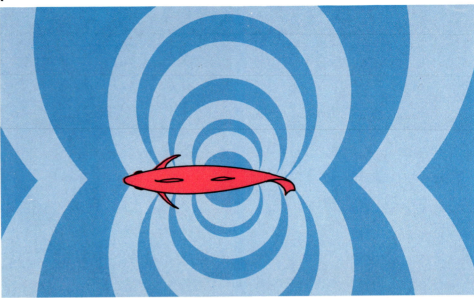

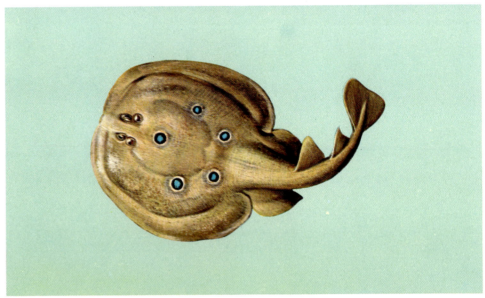

and cleaners which are now so much part of our daily life that we do not even realize how indispensable they have become. With the electron the first of a long series of elementary particles had been discovered. Electric current in conductors is therefore nothing more than the flow of negatively charged particles, the basic ingredients of all matter, with clearly defined charges and masses: the electrons. J. J. Thomson was certainly right when he asserted that the cathode rays were small hammers and not waves, but . . . (and in fact there is a 'but' as we shall see when we discuss the atom and the nucleus), the electron does not have such a simple structure as was first thought. In the light of modern physical theories, it has a dual nature, and is a particle but is also a wave. It is precisely this latter property which has led to new technical achievements of inconceivable importance, such as, for example, the electron microscope.

Animal electricity. Left, a South American electric eel (Gymnotus Carapo) *whose electric organs generate a dipolar type of field (top right). Below: a common electric ray.*

LIGHT: WAVES OR PARTICLES?

The relationship between man and the outside world is closely linked (as far as immediate experience is concerned) to man's senses, and of these faculties the one on which we rely perhaps most heavily is the sense of sight. It is therefore quite obvious that light, which is the principal means of seeing, should always have been the object of continuous study. Several centuries before the birth of Christ, people in China and Greece had already studied the reflection of light, the commonest of all the phenomena of light, using, in the first place, metal mirrors, and later, mirrors made of glass. The ancient Greeks used a concave mirror in order to concentrate the light of the sun and light the sacred fire of Vesta. The theory of sight propounded by the Greek philosophers was nevertheless somewhat naïve: they in fact believed that it was the eye which emitted light-rays, like a lighthouse.

Although the great Aristotle himself had objected to this theory, saying that if it were correct it should therefore be possible to see objects even on the darkest of nights, the theory of visual rays was even accepted by the great astronomer Ptolemy. He was the first to set forth the law of reflection, and he also carried out many experiments concerned with the refraction of light, that is, the way a diagonal or oblique light-ray bends when it passes, for example, from water into the air or *vice versa* (this is the phenomenon that makes a straight stick plunged into water appear to be bent at the point where it meets the water). Ptolemy also created a refraction table to correct the apparent altitude of the stars in the 'vault of heaven', having clearly sensed that the apparent altitude was influenced in a similar way by the refraction of light in the atmosphere. The more modern idea that it is light objects, not the eyes, which emit light occurs for the first time in the *Optician*, a work by the Arab scientist Al Hazen, who lived in the eleventh century.

Throughout the Middle Ages and during the sixteenth and seventeenth centuries considerable progress was made in the field of geometric optics, although this is more a mathematical than a physical science, and the great painters of the Renaissance such as Paolo Uccello and Piero della Francesca were interested in optical problems to improve the effects of perspective in their work. At the same time optical instruments were also developed. Yet no one formed a theory to explain the nature of light, and not even Galileo was successful with experiments to find out the speed of light. Only in 1676 did the Danish astronomer Ole Röme succeed in measuring the speed at which light travels, by measuring the times of the eclipses of Jupiter's satellites and comparing them with the theoretical times, based on Kepler's laws. Light, he established, travels at an incredible speed, somewhere in the region of 300,000 kilometres (186,000 miles) per second, but not at an infinite speed! It was during the seventeenth century that new ideas were put forward about the nature of light and further experiments led to the establishment of new laws. The exact law of refraction was discovered by Willebrord Snell: the phenomena of the diffraction and dispersion of light were observed by Francesco Grimaldi. The time was ripe to tackle the riddle of the nature of light, and two famous contemporary physicists, Isaac Newton and Christian Huygens, created a veritable storm in scientific circles with their contradictory theories on the nature of light: the corpuscular theory and the undulatory

The play of lights and colours obtained by the reflection and refraction of a ray of light on crystal prisms in a work by the artist Alberto Biasi.

theory. While studying the phenomenon of light dispersion, Newton had observed that a ray of sunlight (white light) passing through a glass prism, broke up into all the colours of the rainbow, from red to violet, and that it was then possible to turn this range of colours back into white light by passing the beam through a second prism. The colourful world in which we live is due to the fact that the white light of the sun is in effect a mixture of many colours.

Newton interpreted these experiments by putting forward the hypothesis that, in white light, there are as many corpuscles as there are colours. By this corpuscular theory, that light is made up of small particles, it was possible to explain a great deal that was observed about light – the way it travels in straight lines (rectilinear propagation) and the way it can cross a vacuum (if it did not, then we should not be able to see either the sun or the stars). This theory also explains the reflection of light, quite simply, according to the laws of mechanics: light particles would be *reflected* when, hitting a surface, they were bounced off it. The explanation of refraction was a little more complicated; it was necessary to introduce a new idea and suggest that when the light particles strike the surface of water or a plate of glass, they are subject to a force which makes them veer off, just as a charged particle veers off when it enters an electric field.

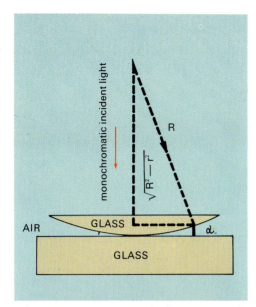

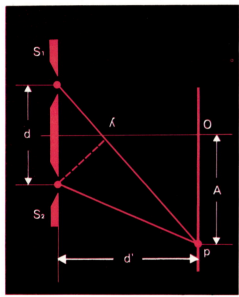

On the right we can see a series of concentric rings (Newton's rings), produced by light interference. In this case the fringes of interference are due to the thin layer of variable thickness between the flat-convex lens and the sheet of flat glass (top left). The phenomena of interference, explained by the wave- or undulatory theory of light, are due to the superimposition of light waves coming from the same source. A double slit placed in front of a source of light (below left) is a classic device for obtaining two images from the same light source.

Almost at the same time as Newton developed his corpuscular theory, Huygens was developing a theory that light was made up of waves (the undulatory theory). Such an idea had already been put forward by Descartes, but Huygens based his theory principally on the results of experiments by Grimaldi who had observed that the shadow of a very small object (for example the point of a needle) placed in front of a spot-like source of light, is slightly larger than it should be, according to the principle of rectilinear propagation and that, in addition, there appear on the rim of the shadow faintly coloured lines. This is the phenomenon of light diffraction.

Grimaldi had carried out many other experiments, observing that the light-rays reflected by a finely-lined pane of glass turn out to be coloured; this is also an effect of diffraction which, for example, gives rise to some of the iridescent colours in the wings of butterflies.

Before presenting the physical concepts which lie at the root of the undulatory theory, it is worth noting that for almost 100 years the Newtonian corpuscular theory was accepted and preferred by most physicists, even though, in principle, there should have been no difficulty whatsoever in carrying out experiments which would have established conclusively that light

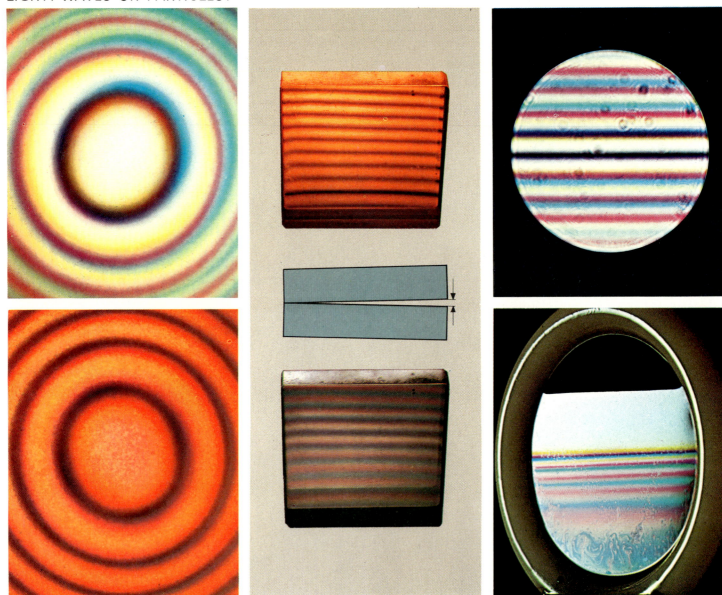

is an undulatory phenomenon. In some sense the power and prestige attaching to Newton delayed a clear understanding of the nature of light, and it was not until the early nineteenth century that an English doctor, Thomas Young, revived the undulatory theory. Earlier in his career Young had been concerned with physiological optics and had shown that the curvature of the lens in the human eye changed according to the distance between a person and what he is looking at. He had then switched his interest to the problems of acoustics and the human voice. It was probably this, the study of sound which brought the idea to mind that light was also made up of waves.

What is a wave? A wave is very difficult to explain accurately in simple non-technical language, but it is something that anyone who has ever watched a pond or dropped a stone into a lake will instantly recognize. It is defined as the flow of a periodic disturbance: the up and down movement of the surface of water, or the alternate compression and rarefaction of a solid, liquid or gas. The particles or molecules of the water or gas do not move along as the wave motion is observed, they merely move up and down.

To return to the stone dropped in the pond. If one stone is dropped one wave will spread out in an ever widening circle, from the point where the stone hit the

Left: Newton's rings obtained from white light (top left) and from monochromatic light from a sodium lamp (below left). Centre: a wedge-shaped interferometer; the variable thickness of the wedge makes it possible to obtain interference fringes in white light (top) and in sodium radiation (below). Interference fringes can be obtained with laser light (below right), but can also be commonly seen in everyday objects as, for example, in a thin strip of soap (top right) in sunlight.

The 'white light' passing through a prism separates into its coloured components (the visible spectrum). For each colour there is a given frequency. Instruments commonly used to break up light into its different colours are called spectrometers.

water. If a number of stones are dropped into the water, one after the other, a series of waves will be created. These will move outwards at different intervals, depending on the time that passes between each stone being dropped in the water – the speed at which each wave moves will depend on whether the water is clear or thick and muddy. The speed is also different if the same stones are dropped, for example, into oil – the speed depends on the nature of the liquid. The speed at which the waves spread out is the *velocity of propagation*; the distance between the highest point, or crest of each wave is the *wavelength*, the height of each crest is the *amplitude* of the wave.

It is the *wavelength* that is the most important characteristic of a wave for the understanding of light. The wavelength, the distance between each crest, is different for every colour, although in fact the wavelengths are very very small. White light is made up of different colours, which as described above, are all to be seen when white light is broken down through a glass prism into the spectrum of colours. In the spectrum, red is at one end, violet is at the other end, but the wavelength of violet light is 4/1,000,000ths of a millimetre, whereas that of red light is 8/1,000,000ths of a millimetre.

Let us now look at a classical experiment of interference: let us imagine that

we have placed in front of a source of light a sheet of paper in which two very small holes have been made close together. On a screen placed in front of the two holes through which the two identical spots of light are thrown, we can observe a series of light and dark fringes, known as the *fringes of interference*, which are like the diffraction fringes seen around the image of a needle point mentioned earlier. Basing his ideas on the analogy with the sound waves, the English physicist Thomas Young (1773-1829) produced an exhaustive explanation of this phenomenon, which is the same as the explanation of Newtonian rings, the concentric rings formed when a lens placed on a flat sheet of glass is lit from above. Newton himself had not explained this effect with the corpuscular theory of light. The undulatory theory can be shown mathematically to explain this phenomenon completely. Young himself explained it non-mathematically, 'The spring and neap tides which are produced by the combination of ordinary solar-lunar tides provide a fine example of the interference of two immense waves; the spring tide is the outcome of this combination when the waves coincide in both time and place; a neap tide occurs when the waves follow one another at the distance of a half-interval in such a way that the only thing perceptible is the effect of their difference.' A

The total reflection of light coming from a source placed in a small water tank. When light rays fall on a surface which separates an optically denser substance (such as water) from a less dense one (such as air) the phenomenon of total reflection *can be seen: beyond a certain angle of incidence there is no refracted light and all the light is completely reflected.*

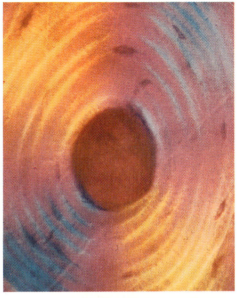

The phenomenon of polarization, which was initially studied to investigate the nature of light, today has many applications, both in scientific research and in industry. Shapes and sizes of biological structures may be deduced from the polarization of the light diffused by them. Top right: a microphotograph using polarized light of a cross-section of bone tissue. Below: thin section of a mineral (olivine) in polarized light. Left: a microscopic photo, using polarized light, of a hailstone; the various concentric layers correspond to the various stages of growth of the stone as a result of water freezing.

similar explanation applies to the formation of light and dark fringes, depending on whether the crests of two light waves combine or not.

There still remained the problem of the medium in which light is carried. Everyone knows that we can hear sounds and talk to each other because we are surrounded by air. Where there is no atmosphere, as on the Moon for example, there are no sounds. Hence nineteenth-century physicists suggested the existence of a material medium, which they called 'ether', in which light could be transmitted.

In 1810, Etienne-Louis Malus, of the École Polytechnique discovered the polarization of light by reflection, a phenomenon which, on the one hand, was to clarify still further the nature of light vibrations but made the problem of ether, on the other hand, even more complicated. As early as Newton's and Huygens's day, there was knowledge of the existence of specific crystals known as birefractive crystals, such as Iceland spar (or crystal), which had the property of splitting a light-ray in two. During his experiments Malus realized that when one looked at the light reflected from a window with a birefractive crystal, one did not, as usual, see the two images, but just one of the two, depending on the angle of the crystal. To begin with, being unable to explain

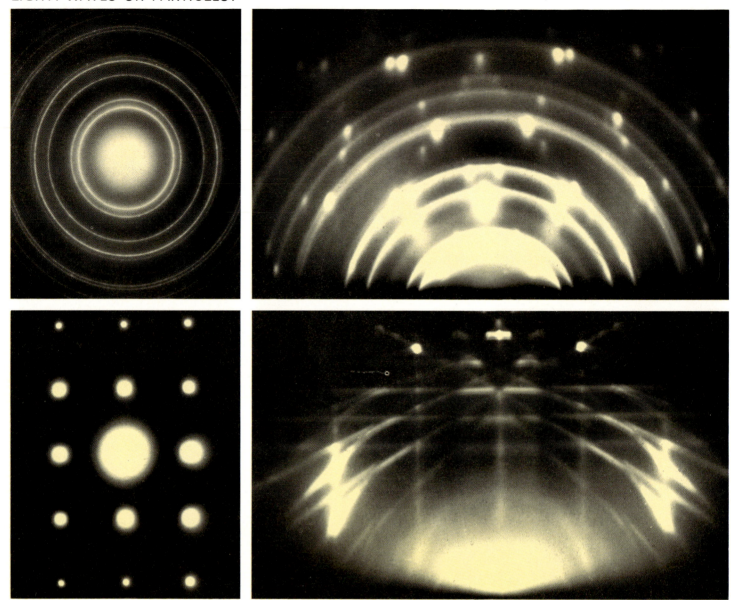

this theory by means of the undulatory theory, Malus took up Newton's ideas. Precisely to explain the phenomenon of double refraction (or birefraction), Newton had supposed that light particles had sides or 'poles'; hence the term 'polarized light' which Malus gave to reflected light. In every reflection and refraction polarization of the light coming at an angle occurs and François Arago showed that even the light reflected obliquely by the Moon is polarized, as is the light given off by incandescent or white-hot solids. The difficulties which seemed to stand in the way of the wave theory in explaining polarization were overcome when Young, and then Augustin Jean Fresnel, showed that this effect could also be explained, although only on the assumption that waves of light are transversal. That is, the particles of ether vibrate in every direction perpendicular to the direction in which the wave travelled. Light-waves are *transversal* waves, unlike sound-waves which are *longitudinal*, because air molecules vibrate in the same direction as the propagation of the wave. The polarisation of light is a difficult matter to explain in non-technical terms, but the effect may be perhaps clearer if the popular polaroid sunglasses are considered. These are welcome because they cut out glare from sunlight which sparkles or glitters. What the sunglasses do in fact is to act as a filter, only

Diffraction of electrons obtained with various types of crystals. These photos show how electrons (clearly defined mass particles) like light behave both as waves and corpuscles, for the rings seen at the edges of the pictures are a phenomenon which only occurs when a wave motion is present.

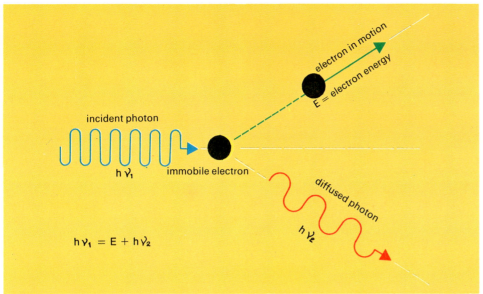

incident photon

$h\gamma_1$

immobile electron

electron in motion

E = electron energy

diffused photon

$h\gamma_2$

$h\gamma_1 = E + h\gamma_2$

Top right: the photo-electric cell in a modern exposure meter allows the intensity of the light falling on the lens to be measured. Top left: a diagram showing the Compton effect: a photon throws an electron from the atom with a wave motion. Below: a bundle of thousands of optic fibres placed on a photograph transmits the image of the eye. In the future it may be possible to use optic fibres as transmission 'channels'.

letting through light coming at a particular angle and cutting out the light hitting them at other angles. If two polaroid lenses are put together so that the filter in each is not aligned with that in the other, then no light at all will pass through. This demonstrates the transversal nature of light-waves.

But in polaroid glass, for example, the transversal waves have a medium through which they pass which has all the features of a rigid body. It is hard to think of an 'ether' with those characteristics of rigidity, which would not at the same time interfere with the movement of heavenly bodies. Several years later the conclusive experiments of Michelson and Morley (1887) were carried out, leading to the concept of ether being finally rejected.

The wave theory of light was fully developed in 1862 when J. Clerk Maxwell, using purely mathematical means, showed that light is an electromagnetic phenomenon.

Light thus becomes simply a particular example of an electromagnetic wave. Maxwell, however, was not unduly concerned with the verification and experimental development of his theory, and it was not until 1887 that Heinrich Hertz obtained electromagnetic waves with wavelengths of almost 100 metres (110 yards), which had all the common features of light. Hertz's

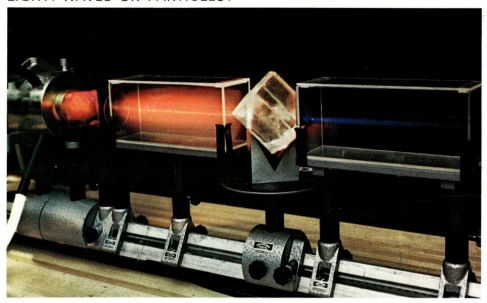

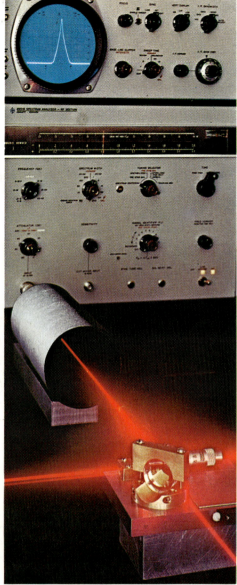

experiments opened the way to modern radio- and radar-technology. Light, or rather visible radiation, only occupies a small place in the spectrum of electromagnetic radiations. As things stand today, radio broadcasting entails wavelengths of dozens of kilometres for normal television communications, right down to micro-waves less than one centimetre in length for use in radar equipment. Thermal radiation, likewise – i.e. infra-red rays – is electromagnetic radiation with a wavelength which is slightly longer than the wavelength of red light. Beyond violet light we find other forms of electromagnetic radiation: ultra-violet light with a wavelength of 1/1,000,000th of a metre, X-rays (observed for the first time in 1885 by Röntgen and well known for their medical application) and lastly gamma rays which have wavelengths equal to 1/1,000,000th of the wavelength of visible light.

Hertz also discovered that a sheet of metal lit with monochromatic light (light of one colour in the spectrum, for example, red light) emits electrons which *all* have the same speed. By varying the frequency of the light (that is changing its colour) – and using violet light by way of example – the speed of the expelled electrons increased. This effect, which is known as the *photoelectric* effect, cannot be explained by the wave theory of light. In 1905, the same year in

Some types of Lasers, *devices which generate and amplify electromagnetic waves in the wavelengths of visible light. Top left: a ruby laser. Below: green light issuing from lithium hydrate on which an infra-red ray is being projected. Right: criss-crossing lights of varying wavelengths generated by laser and projected on to a crystal.*

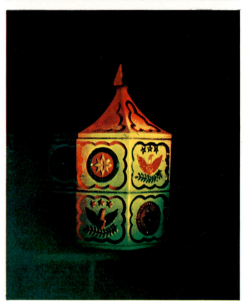

The use of the laser makes it possible to reproduce and record images by using the phenomenon of interference in a much more precise manner than with traditional optical sources. Top left: two laser beams interfering, produce a hologram. By projecting on to the hologram monochromatic light, (top right) an image of the object is produced which is three-dimensional. Below left: a hologram obtained with two lasers, using red and blue light. Below right: a device for recording a hologram.

which his two fundamental articles dealing with relativity were published, Einstein put forward a theory which could explain the photoelectric effect, a theory which won him the Nobel Prize in 1921.

Einstein's idea was to substitute Newton's material corpuscles with 'packets' of energy; in other words, light would be composed of quantities of energy called *photons* (from the Greek word *fotos* which means light). The energy of each photon is proportional to the frequency of the light; thus the violet light photons have twice the energy of red light photons. A quantity of violet light striking a metal electron yields it its energy; it is thus clear that an electron expelled as a result of the photoelectric effect from violet light will have a higher speed than the electron emitted by red light. Einstein's intuitions were confirmed by the experiments dealing with the diffusion of photons carried out in 1924 in Mexico by Arthur H. Compton.

Compton verified that a photon striking a stationary electron behaves precisely like a particle; the same laws apply as those applying to the collision of two billiard balls, and consequently the diffused photon has less energy, a lower frequency, than the incident photon.

At this point it is obvious that one should ask oneself what exactly light is.

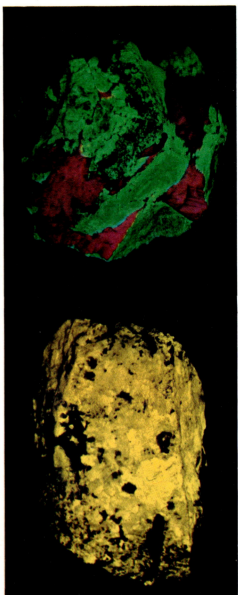

The validity of the electromagnetic theory of light in explaining interference, diffraction and polarization is undeniable; it could be asked whether light has a dual character and so behaves sometimes as a wave and sometimes as a particle? In a certain sense this question is meaningless; quantum mechanics shows that wave and corpuscle are only *models* which we have made for ourselves to explain and describe a reality which is different. According to modern quantum mechanics, the instrument with which one makes a measurement and the object in question interact in such a way that wave or corpuscular behaviour will result according to the type of experiment carried out. Thus when electrons (which are corpuscles with a clearly defined mass) strike a crystal, they produce diffraction figures which are identical to those of X-rays; so that they are behaving like a wave.

This greater understanding of the quantum basis of light phenomena led in 1960 to the construction of one of the most astonishing and useful pieces of apparatus both for fundamental and applied research, the LASER, which is an abbreviation for 'Light Amplification by Stimulated Emission of Radiation'.

With these new sources of light one can produce extremely intense and extremely thin beams with high monochromatic properties which can keep a

*Various animals give off light. Top left: an abyssal fish (*Argyro pelecus*) which converts the chemical energy in its food into light energy by means of specialized organs (below). Right: phenomena of fluorescence in minerals.*

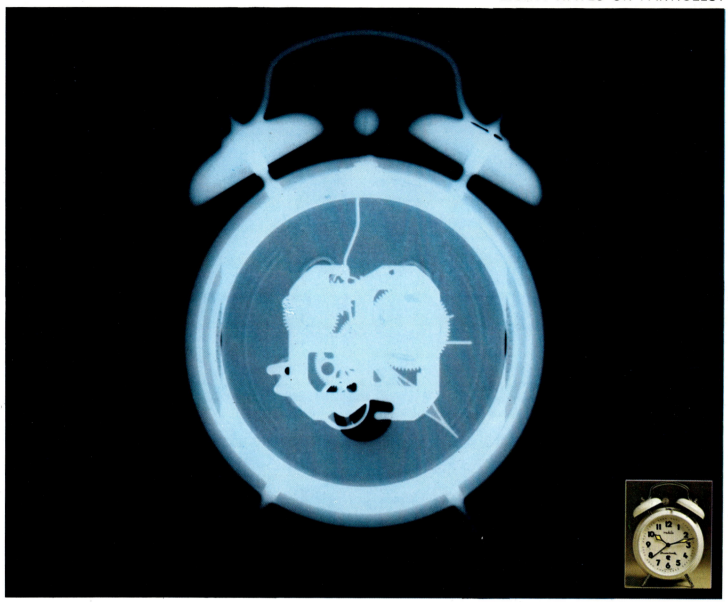

X-rays are high-frequency electromagnetic waves, invisible to the human eye, but capable of making an impression on photographic plates. They have a high power of penetration and are vital to medical diagnosis. Above: an alarm-clock and its X-ray image in which one can make out details of its inner mechanism.

very small diameter for many kilometres. This has made it possible to use them not only for measuring distances (for example the distance between Earth and the Moon) with hitherto inconceivable accuracy, but also for concentrating on very small surfaces large and easily measurable amounts of energy. As an outcome of the flexibility of the laser we can mention, by way of example, certain highly successful micro-surgical operations (like that to cure a detached retina).

Recently there have been remarkable developments in the use of lasers for military purposes, and in the United States there are already systems of aerial bombardment in which the laser beam focussed on the target guides a missile when released from an aircraft with a quite extraordinary degree of precision. A bomb released in the traditional way has a margin of error in the region of hundreds of metres, whereas the use of lasers makes it possible to reduce this margin to slightly more than 45 centimetres (18 inches).

THE ATOM AND THE NUCLEUS

Atomos is a Greek word which means 'indivisible', and the origins of atomic theory go back to ancient Greece. The atomic theory says that all matter is made up of small particles which themselves cannot be broken down further. These particles are the atoms, and it is the countless variety of combinations of atoms that leads to the virtually infinite variety of the things around us. The first atomists were Leucippus and Democritus, who lived some 2500 years ago. According to Democritus, atoms alone had properties of extension, form and movement; reality was made up, quite simply, of atoms and the void.

It was, however, only the discoveries made by chemists in the nineteenth century which definitively established the atomic nature of matter. The names associated with the atomic theory are those of Antoine Lavoisier, Joseph Proust and John Dalton. With the publication of Lavoisier's *Elements of Chemistry* in 1789, science broke away completely from the ancient art of alchemy, and took on a fundamentally quantitative, and so 'modern' character. The principal contribution made by Lavoisier was his authoritative statement that in every reaction the total mass of the reagents is strictly equal to the mass of the products; and his suggestion that primary substances, the chemical elements, were relatively few in number (today there are in fact just over 100 known), whereas compounds produced by these elements, are many.

Proust is credited, above all, with having set forth the law of defined proportions. This law asserted that two or more simple elements would only combine together, to make a chemical compound, if the elements were mixed together in particular proportions according to weight. Proust defended this law against the opposition of one of the greatest and most authoritative chemists of the day, Claude Berthollet, who maintained that elements could be combined with one another in any proportion or ratio whatsoever. Proust was a convinced upholder of the atomic theory, but the one who clearly realized that the truth of the law of defined proportions implied the existence of atoms was John Dalton who, in 1808, set forth the law of multiple proportions. Dalton asserted that the same volumes of different gases contained different numbers of particles, or atoms. He measured these differences according to the weight of the number of atoms, thus establishing the concept of 'atomic weight'. He also asserted that these atoms would link to each other, to give compounds, in simple proportions.

Another decisive step forward was made by researches of the French chemist Joseph Louis Gay-Lussac and the Italian Count Amadeo Avogadro. The former discovered that two gases would react with each other and so form a third gas, in particular proportions, those being whole numbers, and usually small ratios, that is, two gases would make a third if mixed in the proportion two to one, or one to three. Avogadro defended the terms 'atom' and 'molecule', and also linked the laws of defined proportions to the law on the volumes of gases; he put forward the fundamental hypothesis that equal volumes of gas (at the same pressure and temperature) contain the same number of molecules. This hypothesis, which was subsequently confirmed but not actually understood for a further 50 years, allowed the straightforward establishment of the ratios between the weights of the individual molecules in the various gases. If it were possible to know how many molecules there were in

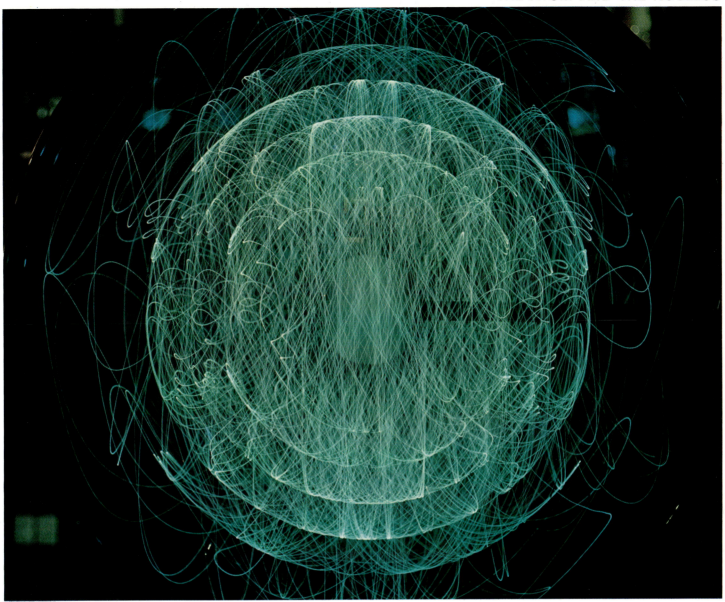

The orbits of electrons around the nucleus in a model of the uranium atom.

a given volume, it would then be possible to know the actual weights of the molecules themselves.

The number of molecules contained in the same volume of any element is different; and thus elements are distinguished by measuring the volume in which is contained $6,022 \times 10^{23}$ molecules (this is an extremely large number – almost one million milliard milliards or 1,000,000,000,000,000,000,000,000!). This number, known as *Avogadro's number* (or *Avogadro's constant*) is one of the main constants in physics, and is usually indicated by the symbol N. The calculation of Avogadro's number, which is essential for the determination of the atomic weight of an element, can be carried out in many different ways. One of the most accurate is that devised by the French physicist Jean Perrin. In 1908, while studying the *Brownian movement* of smoke particles, and interpreting it with the help of the theory set forth three years earlier by Albert Einstein, he determined N_0 by finding a value which agreed perfectly with that obtained by totally different methods. The determination of N_0 also allowed an approximate estimate to be made of the dimensions of molecules.

But all this knowledge about atoms did not provide incontestable proof that the atomistic view of matter was correct. And it was not until the early

twentieth century that the last doubts were dispersed as the result of two fundamental discoveries: the electron, and radio-activity. It was then quickly realized that electrons were basic ingredients of matter and could easily be 'removed' from the atoms themselves. As a result the atom ceased to be something *indivisible* and some kind of model of its structure was called for. In 1900 it had already been established that electrons have a mass 1837 times smaller than the mass of the hydrogen atom and were negatively charged with electricity. Because matter is electrically neutral, and the atom contains granules of negative electricity (electrons), matter must also contain granules of positive electricity, otherwise without that balance it would not be neutral. The problem was to identify the positive charged granule in the atom. J. J. Thomson, one of the main architects behind the discovery of the electron, then conceived the atom as a positively charged sphere containing negative electrons. Thus the atom could be compared to a currant bun, the bun itself being the positively charged sphere, the individual currants being the negatively charged electrons. This model did not stand up, however, to experiments carried out in the early twentieth century by Ernest Rutherford. To make his experiments, Rutherford used the new radiation (what are now

Pierre Curie and Marie Sklodowska Curie (top right) with their daughter Irene and (below) at their home in Sceaux. They were among the pioneers in nuclear physics. In 1903 they were awarded the Nobel prize for their discovery of the two new radioactive elements polonium and radium. Below left: instruments for measuring the ionized rays emitted by radioactive substances. Top: two types of radioactive minerals: labradonite (left) and Congolese curite (centre).

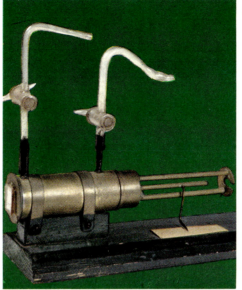

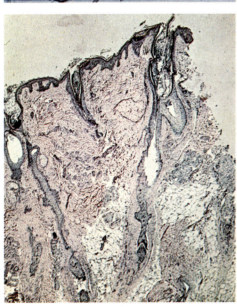

The nuclear model of the atom, i.e. of a small and positively charged nucleus surrounded by negatively charged electrons, is largely the work of the New Zealand physicist Ernest Rutherford (top left). He used alpha particles (helium ions) emitted by radioactive elements to 'bombard' nitrogen nuclei and thus investigate atomic structure. Top centre: the apparatus used by Rutherford. Right: traces of alpha particles in a Wilson cloud-chamber. Below left: the effects on human skin of radiation caused by the atomic bomb which was dropped on Hiroshima. Below right: a technician in the laboratories at Ispra handles contaminated plant with automatic arms.

known as alpha rays, beta rays and gamma rays) emitted by radioactive bodies.

The discovery of this natural radioactivity had come about, almost by chance, in 1896, when the French physicist Henri Becquerel, interested in the discovery of X-rays by Röntgen, tried to see if, likewise, fluorescent substances (i.e. substances which glow after being exposed to light) emitted rays of the same type. Becquerel worked with crystals of a uranium salt; first he exposed them to sunlight then placed them on a photographic slide which, when developed, revealed a faint mark where the crystal had been placed. Becquerel maintained that the sunlight in some way 'excited' the uranium salt which then gave off the radiation it had absorbed from the light. But when he forgot about a crystal placed on a photographic plate in a box (thus well removed from any source of light), he discovered a very distinct mark on the plate. This could only mean that the uranium crystal emitted rays continuously and naturally, and did not need any form of preliminary excitement by the sun. After being subjected to every conceivable form of chemical 'torture' (heat-treatment, grinding, acid treatment) the crystals (or rather all that remained of them) continued to emit energy, apparently tirelessly, in the form of rays. Radioactivity, therefore, it was concluded was a property typical of atoms, and

Becquerel thought that the uranium atoms were in effect responsible for this.

Becquerel's discovery stimulated more systematic research, particularly that of the Polish scientist Maria Sklodowska and her husband Pierre Curie.

After having examined tons of matter containing uranium salts, Madame Curie managed to isolate two substances, *polonium* and *radium*, which were much more radioactive than uranium, but she was also the first scientist to fall the victim of radiation sickness. She died of leukaemia caused by prolonged exposure to the penetrating rays, the first of many researchers to die of cancer.

But to return to the model of the atom. Rutherford had discovered that the alpha radiation emitted by various radioactive substances was in fact made up of positively charged helium ions travelling at very high speed; he had the idea of using these to bombard atoms with the aim of getting to know something more about the make-up of the atom itself. In 1911 he put small and carefully aligned strips of alpha particle onto a very thin piece of gold-leaf and was able to determine the angle at which the particles themselves were deflected by the impact. If Thomson's model of the currant bun had been correct, the deflections should have been very small, but Rutherford noticed that some particles were deflected by several degrees, and that others were even bounced

One of the foremost figures in nuclear physics was the Italian Enrico Fermi (right) who made vital contributions to theory by experimental investigation. Top left: Fermi receiving the Nobel prize for physics at the age of 37. Below: Fermi standing near the experimental cyclotron in the University of Chicago. Research into nuclear structure requires ever larger and more advanced machines which are capable of accelerating nuclear projectiles to extremely high energies. Opposite: the large linear electron accelerator at Stanford (California), which is more than three kilometres (nearly two miles) in length.

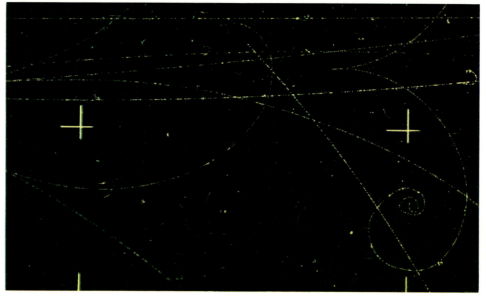

back. These results disagreed considerably with a 'cake-like' atomic model, but they could be interpreted quite straightforwardly if one imagined the atom as made up of a tiny central *nucleus*, in which almost all the mass was concentrated, surrounded at a distance, by a swarm of orbiting electrons – in other words a kind of solar system in miniature in which the Sun was the nucleus and the planets the electrons. In this way the present-day model of the atom was born: the model of the *nuclear atom*. It should be remembered (and Rutherford made this point himself) that a similar model had already been proposed by the Japanese physicist Nagaoka, but it was Rutherford who carried out the crucial experiments to prove the truth of the idea.

Keen young scientists worked alongside Rutherford at the Cavendish laboratory, Cambridge, which became the world's most important centre for atomic research. One of the most important discoveries was made in 1932 by James Chadwick who showed that many experimental properties could be explained if one bore in mind that the atomic nucleus contained not only positive charges (which it had to contain to neutralize the negative charges of the electrons) but also neutral particles: the *neutrons*. The atom thus turned out to be formed of a central nucleus, formed by particles with a positive charge (the

Top left: the SIN cyclotron at Villigen. Below: a nuclear occurrence in a bubble chamber. Top right: to check the alignment of an accelerator tube a neon laser is used. Below: a liquid hydrogen bubble chamber belonging to CERN in Geneva.

Right: various stages of the Russian Serpukhov accelerator which was the first to produce protons with a speed of 0.999915 times the speed of light. At the present time the record is held by the Batavia protosynchroton in the United States. Left: two details of the 6.28 kilometre (nearly four miles) long tunnel which contains the ring in which the protons of the Batavia accelerator circulate.

protons) and neutral particles (the neutrons), around which, and at a distance equivalent to some 10,000 times the diameter of the nucleus itself, the electrons orbit – these being negatively charged. The chemical features of an element are given by the number of protons (or electrons), which is called the *atomic number*, whereas the number of neutrons may vary, thus giving rise to *isotopes*.

The atomic model proposed by Rutherford did, it is true, explain the deflections of the alpha particles and many other phenomena, but it was incompatible with the ideas of traditional electromagnetism. In fact, if the electrons – i. e. electric charges – revolve around the nucleus, they should, according to Maxwell's laws, continuously lose energy in the form of electromagnetic waves, which would involve a very rapid 'drop' by the electron in the nucleus, and this 'drop' is totally incompatible with the obvious stability of the atoms which form the world about us. Moreover, according to the classical laws of electromagnetism, an atom should have a life of not more than one hundred-millionth of a second!

The only solution was to propose that, at an atomic level, classical electromagnetism was no longer valid. This was in fact done by the Danish physicist Niels Bohr (Rutherford's close friend). In Bohr's view, the electrons

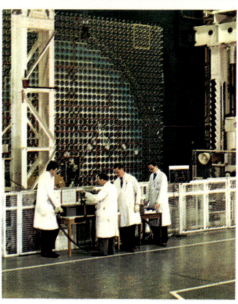

in an atom can continue revolving without any loss of energy, but do give off or absorb energy when they 'hop' from one orbit to another. Bohr's atomic model, which was later improved by Sommerfeld, managed to explain many features of the spectra of the various elements, which were already known to spectroscopists and obeyed rather mysterious laws. Today it can be said that this is undeniably an outdated model, the results of which form part of a more general type of 'quantum mechanics' with much greater natural consistency: but there can be no doubt that it is still extremely convenient, as a model, in many applications of pure physics, such as chemistry or solid state physics. The laws of the spectral lines or line spectra of the various elements can be easily interpreted in this way: each orbit has its own energy, when an electron 'hops' from an orbit of energy (E2) to another, with less energy (E1) an amount of light (i.e. a *photon*) is emitted whose energy is equal to the difference in energy between the two levels (E2 – E1). Bearing in mind Planck's ratio between energy and frequency (i.e. $E2 - E1 = hf$, where f is the frequency and h is a universal constant), a clear explanation emerged both of the bright lines observed in the emission spectra and in the dark lines of the absorption spectra (in this case the electron hops from the lower-energy orbit to the higher-energy

The splitting or fission *of a uranium nucleus by the absorption of neutrons into two lighter fragments is the principle at the basis of the controlled chain reaction in nuclear reactors. Left: the 'swimming pool' reactor at the Saluggia centre. Top right: A Boiling Water Reactor at the Garigliano power station. Below: the natural uranium reactor at Harwell in England.*

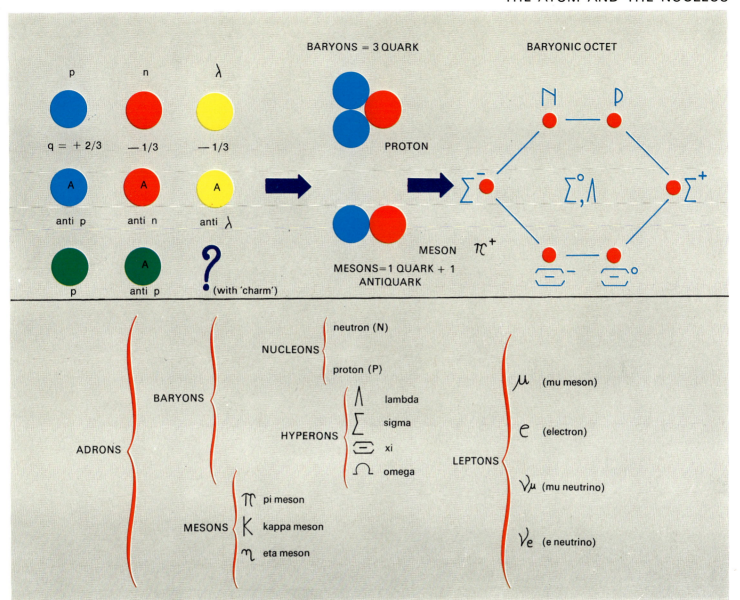

The structure of 'elementary' particles
has been shown to be considerably
more complicated than at first thought,
by the information derived from large
accelerators. Only recently have
cataloguing principles been
established, using the so-called 'quark
model', which is not entirely
satisfactory but does at least reduce
the truly 'elementary' particles to just
three or four. These new particles have
the odd feature of having an electric
charge which is a fraction of that of the
electron. All known particles (which
now number somewhere in the region
of 100) are made up of quarks.

orbit by absorbing a photon in which the energy is equal to the difference in energy between the two levels). Another major success scored by Bohr's theories was the rational explanation of the periodic table of elements, due largely to the German physicist Wolfgang Pauli who introduced the principle of *exclusion* (the *Pauli exclusion principle*), whereby in each orbit there can be no more than two electrons. When one orbit is completely full, the electrons must arrange themselves on another orbit, when this other orbit is also full, they must find another, and so on. It was then necessary to put forward the hypothesis that as well as revolving around the nucleus the electron also revolved on its own axis like a spinning top, and could not, therefore, be considered to be a 'granule' of electricity, but rather as 'something' equipped with a structure.

Rutherford's 'nuclear' model and Bohr's reasonably satisfactory interpretation were based on experiments and not the result of a comprehensive theory. The real advance was made by Erwin Schrodinger, taking up the idea of the Frenchman Louis de Broglie, and asserting that a wave was associated with each electron; the behaviour of this wave was described by an equation similar to that which described the behaviour of waves in traditional physics.

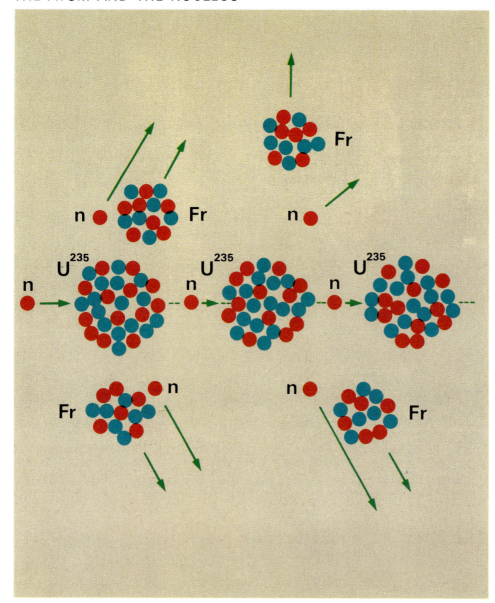

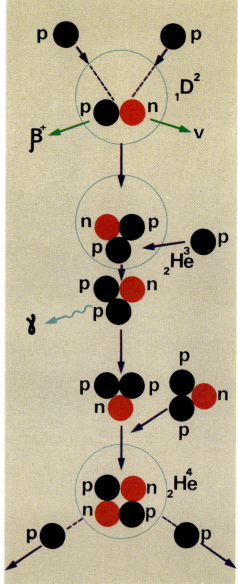

Just as light has a dual nature – undulatory and corpuscular – so does the electron; it was not long before this was confirmed by experiment. Schrödinger's 'material waves' seemed to explain everything. In 1930 the English physicist P. A. M. Dirac showed, purely theoretically, that if one also took into account the theory of relativity in the description of the movement of electrons, the electron *had to* behave like a top. But Dirac's equations involved an astonishing consequence: as well as the electron there had to be another particle with the same mass but with the opposite charge – the *positron*. The positron, which was discovered not long after, was the first in a long series of 'anti-particles', and the existence of the particle/anti-particle pairs is currently one of the fundamental principles of physics.

The atom was far from being a single and 'indivisible' entity, and when research focused on a more detailed study of its small central section, the nucleus, it became even more complicated! Nuclear physics study programmes which entailed the need to 'bombard' the atoms themselves with more and more powerful projectiles, brought to light a whole collection of particles. Not only protons, electrons and neutrons, but, as more and more powerful machines were developed to 'launch' these projectiles, a vast number of

Left: a diagram of controlled chain-reaction fission; a uranium 235 nucleus splits by the absorption of a neutron into two fragments (Fr), producing neutrons; one of these triggers off another fission while the others leave the system. Right: a diagram of fusion, in which two light nuclei join up to form a strongly linked, energy-releasing nucleus. While the process of fission can be controlled, and thus produce usable energy, the process of fusion does not yet have any practical applications, despite the intense research currently under way.

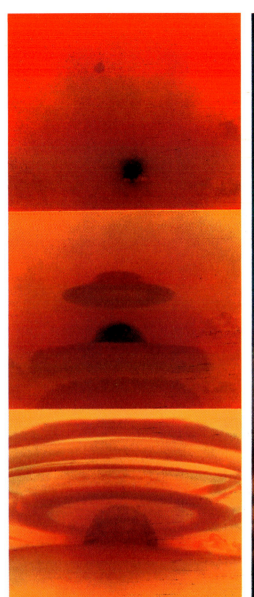

The utilization of the vast energy of atomic nuclei has, lamentably, been more oriented to war than to peaceful purposes. On 6 August 1945 the first fission bomb (right) was dropped on the city of Hiroshima, killing 80,000 people: the after-effects of the bomb, often unforeseen, lasted for several years. Left: a sequence showing the development of the explosion, using a film which was sensitive to the different temperatures.

'elementary' particles: mesons, neutrinos, hyperons ... with masses of every conceivable value but with a charge invariably the same as the charge of the electron. This collection is still being classified, and numerous laws have been discovered which seem to make it possible to achieve some sort of order. One of the most interesting schemes is the so-called *quark model*; on the basis of this model, which has managed to explain many properties of the known particles, all the elementary particles are formed by the combination of just three, the *quarks* and their anti-particles, the *anti-quarks*. The energy which keeps protons and neutrons together in the atomic nucleus (i.e. the nuclear energy) is a greater energy than its chemical counterpart. This theoretical discovery was followed, not many years later, by 'practical' applications, made by the German Otto Hahn and the Italian Enrico Fermi. Energy is obtained when a heavy nucleus is shattered (i.e. by fission) or when light nuclei are merged (i.e. by fusion).

MOLECULAR PHYSICS

CHEMISTRY

Man goes in search of knowledge. Knowledge includes the investigation, to even greater depths, of the matter about us, and the environment in which we live.

Scientific knowledge implies, further, an investigation carried out with mathematical rigour, in such a way, that is, that the deductions or recorded results of investigation are exact. It is nevertheless hard to give an exact definition of any science other than pure mathematics, and accordingly more accurate to talk in terms of experimental sciences. Chemistry, to take a case in point, is precisely one such experimental discipline; everything that we know or shall know in this field has or will be demonstrated by means of experiment. The investigation into the inner structure of matter is the specific task of physics. It is the task of chemistry, on the other hand, to investigate the transformations of matter, whether these transformations or changes happen naturally or are caused by external agencies.

In chemistry, as in all the other experimental sciences, there is a tendency to turn into general rules, as far as is possible, the results obtained from a specific series of experiments. When analyzing the various experimental evidence he has gathered, the chemist tends to deduce from it a general law: on that basis it is possible to give a rational interpretation of all the experiments he has made to predict the outcome of experiments in the future. This implies a continuous process of checking results, if necessary revising the general laws. The general law set forth will be revised if just a single phenomenon is discovered which cannot be explained according to it.

Chemistry cannot be an independent science on its own. It is a discipline that uses principles and data drawn from physics, biology, engineering and other scientific disciplines. It is a very practical subject, although it often encounters a certain hostility on the part of readers and students because it is imagined to be a sterile list of formulae and reactions which are apparently divorced from how things really are.

This is as mistaken and limited a view as could be imagined. It is impossible not to appreciate the logic and coherence with which so many things that at first glance seemed mysterious and abstract can be explained on the basis of the laws of nature that science and chemistry have established. How many times has each one of us twizzled a rod, or some other kind of metal object, between our fingers, and pondered, with scepticism and disbelief, that that small rod contains several billion billions of particles, some of which travel at speeds approaching the speed of light.

It seems impossible, but if one goes further and asks oneself why, for example, when the rod is heated, the heat is transmitted quickly from one end to the other, why it conducts current, and why a carbon rod behaves differently; then what is impossible at first sight seems less so, for these phenomena are all explicable on the basis of those same general laws. The atomic theory supplies exhaustive explanations for all these phenomena, and this helps to dispel initial scepticism about the validity of such and such a theory, taken as a general principle. The atomic theory has at last transformed into reality man's ancient dream of one unitary explanation for all phenomena. The unification of matter, irrespective of its different physical states, in as much as it is always and only formed of atoms or molecules, represents the basic concept and the point of departure of all the modern sciences. Only a few decades have passed since the moment when courageous scientists put forward the precise statement of atomic theories. And yet, despite the short lapse of time, chemistry and physics have as a result made huge advances in every quarter, which have in turn constantly confirmed the validity of the basic theory. This daily increase in confidence in the theory has, inevitably, led science to the most testing problem of all: the origin of life, and its evolution.

It is inherent in man that he should question his origins: and also that the availability of more scientific knowledge should lead to fields of enquiry concerning the human body itself. Chemistry and Bio-chemistry are studies immediately empirical in their research and pragmatic in their application. Also, more than physics, they have taken on the role of the mediaeval alchemist. D. N. A. could be said to have replaced the Philosopher's Stone. In advanced studies of brain chemistry one cannot but doubt the existence of the 'spirit', when all the ills of man are traceable to chemical reactions and interchanges. The chemist has open to him the tools by which to measure the very nature of thought. This quest is obviously one both of great self confidence – leading to the current eulogy of 'The Marvels of Science'; and also one of great fear, as people feel themselves reduced to nothing more than a chain of chemical compounds. However, it is perhaps arrogant to think this crisis any greater for us than was the parallel crisis for the alchemist, who felt himself also trembling on the brink of the 'answer to Life'. For alchemists, the ultimate end of life was held to be the synthesis of gold; gold was a symbol of permanent beauty, incorruptible splendour and everlastingness. Robbing nature of the secret behind the preparation of gold meant finding the key to all the mysteries.

The Philosopher's Stone, which would make it possible to transform any base metal into gold, would thus have made a permanent transition from the ephemeral to the immortal and unveiled the very essence of life. Such ideas, in all their refreshing naïvety, still manage to fire people's minds today, although they take different forms. What counts for man is the search, over and above any possible results; in this way the various paths of knowledge seem to lead towards one single principle. Various sciences have joined forces with one another and developed awesome powers; together they strive to wring from nature what, hitherto, has defied all explanation.

The laws of chemistry are not always as exact as those of physics, even though the borderlines between these two sciences are anything but clear-cut. This makes chemical studies more difficult but also more challenging: the greater the margin of uncertainty in a science, the more that science will engage the intellect and develop its capacities.

Nowadays, of course, physicists themselves find they have to grapple with problems which are increasingly indefinite, and with matter which appears to be less and less material and seems to reduce itself to nothing more than energy or electromagnetic radiation.

The architecture of the atom makes our whole civilization seem to hang together by no more than a thread, and the progressive whittling down of the dividing-lines between matter and nothingness poses greater and greater scientific and philosophical problems.

The cold materialism which lies behind the boldest scientific research does not, it is true, exclude a faith: faith in the boundless potential of the human intellect, and hence in whoever equipped man with this intellect – whoever it may have been, and if, indeed, there ever was such a figure. The elements with which chemists work are less intangible, more real and as a result more reassuring; but the basic principles guiding them are the same. We are thus left with a dramatic question: experiment has suggested a principle; this has been formulated and reconfirmed by other experiments, and upon it a science has been built. But what would happen if this principle were to be wrong?

This is the nature of scientific discovery. The refutation and re-assertion of theories and principles often uses a Hegelian principle of dialectic process to break new boundaries; for it is only by querying 'established' principles that new discoveries are made.

ATOMS AND REACTIONS

The definition of the atom now was that 'the atom of an element is the smallest part of it which still retains all the physico-chemical characteristics of the element'. Because the most important property of atoms is the bond between them, the union of two or more atoms produces complex particles to which the name *molecules* was given; the features of these depend solely in the type of association or pattern which binds the ingredient atoms together.

Specific problems emerged when it was necessary to quantify the atom in real terms. It was evident that because of its extremely tiny size an atom must have a weight which could not be determined experimentally. This difficulty certainly did not discourage the pioneers of the microcosm. If in fact it was impossible to establish the real weight of individual atoms, the same could not be said of their *relative weights*. Thus, taking an element as a standard of reference, and giving the atom of this element an arbitrary value for its atomic weight, the weights of the other elements in relation to the standard could be easily deduced by examining various compounds. The adoption of a universal standard of reference was not easy, in fact it only occurred twenty years ago when chemists and physicists decided to choose as the standard the carbon atom, with an atomic weight of twelve. This is the *unit of atomic mass* (u. a. m.) which is defined as a twelfth part of the carbon atom, and the atomic weight of each element is the number indicating how many times heavier than this twelfth part is the element in question. The molecule is a complex of atoms, and *molecular weight* is simply the sum of the atomic weights of all the atoms forming the molecule. *Gram-atoms* and *moles* are the weights in grammes of an element or compound, and are equal, repectively, to the atomic weight of the element and the molecular weight of the compound.

The French chemist Antoine Lavoisier was responsible for the principle of the conservation of mass in chemical reactions. By means of a painstaking investigation carried out on hundreds of reactions, he was able to state with absolute certainty that a chemical reaction occurs without any appreciable variations in mass, or that the weight of the reagents before the process is identical to the weight of the products once the reaction has taken place. Lavoisier's principle removed chemistry from the sphere of magic and witchcraft; the constant nature of the mass in the experiments showed that a chemical reaction was a phenomenon which could be interpreted rationally. Another major figure in the history of chemistry was Michael Faraday. As a pioneer in the field of research into the structure of the atom, Faraday devoted much of his work to the chemical phenomena caused by electric current. Starting from the experimental observation that current was capable of causing chemical reactions or, at least of inducing changes in the substances involved, he deduced that matter and electricity must be somehow linked, or rather that matter must be made up of electrically charged particles. Further study and experiments followed which led to the discovery of negative particles. In particular, and based on the examination of phenomena observed in 'exhaust' pipes filled with gas at low pressure, it was discovered that particles with a negative charge were present in all matter; these were then called *electrons*. Again with the use of 'exhaust' pipes, particles were discovered which

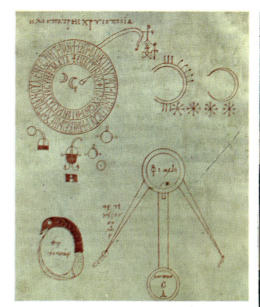

The contribution of alchemy to the development of science should not be underestimated; in the Middle Ages, when the climate was unfavourable for science, it represented the only move in a scientific direction. Top left: the formula for turning a base metal into gold. Below: the frontispiece to Giabir's Treatise on Alchemy. *Right:* Alchemists at Work, *from an eighteenth-century painting by G. Stradano. Alchemy did not, however, develop a true scientific method because it did not attempt to relate individual experimental results to any one general principle. Its association with magic and witchcraft was also unfortunate and helped delay the development of modern chemistry.*

had an electric charge equal to that of the electrons, but a positive charge. The mass of these particles turned out to be approximately 1840 times greater than that of the electrons; these particles were called *protons*.

Once the discoveries of these basic units of charge and mass had been made, the first theories on the structure of the atom were formulated. Thomson thought that the atom was formed by a positively charged cloud, which contained the electrons, in numbers such as to render the whole complex electrically neutral. Not long afterwards, Rutherford showed this hypothesis to be unfounded by discovering the *atomic nucleus*, that tiny spherical region inside the atom in which all the protons (thus all the positive electric charge) are concentrated: it is in the nucleus that the entire mass is concentrated, for the mass is due almost exclusively to the protons (the electrons have a mass equal to only 1/1840 of that of the protons). In addition, the existence of particles with a mass equal to that of the protons but having no electric charge was also postulated. The function of these particles was, supposedly, to reduce the strong electrostatic repulsion between similarly charged protons within an extremely small space (the radius of the atomic nucleus is in fact about 10^{-13} cm.) Neurons were then discovered by Chadwick.

Protons and neutrons have an almost identical mass and are situated in the nucleus; the protons have a positive charge and the neutrons do not have a charge at all. The electrons have a negligible mass and a negative charge, equal in intensity to that of the protons.

For the atom to be electrically neutral, the number of protons must be the same as the number of electrons. Later experiments have shown that the electrons do not obey the laws of traditional physics, but, rather, the principles of a new science which applies only to subatomic particles, known as *Quantum Mechanics*. In particular, the electrons cannot absorb energy continuously, but only in *quanta*, that is, in the form of very small, clearly defined and non-divisible amounts.

The first atomic model based on the quantum theory was proposed by the physicist Niels Bohr in about 1913. Bohr postulated that the electrons in the atom travelled along closed orbits, called *energy levels*, and that there could be neither emission nor absorption of energy by the electrons as long as they remained in these orbits. The theory was right, in so far as it managed to explain many phenomena and, in particular, solved the problem of *spectral lines* or *line spectra*. It was not in fact clear why, by supplying energy to a compound (e.g. a

The emphasis laid on experimentation led alchemists to question the Aristotelian theories; in this respect alchemy can be considered as the forerunner of modern chemistry. The last shaky alchemistic theories, such as the phlogiston theory, were decisively demolished by A. Lavoisier. Top left: the microscope used by Lavoisier. Below: the scientist in his laboratory; and (right) with his wife.

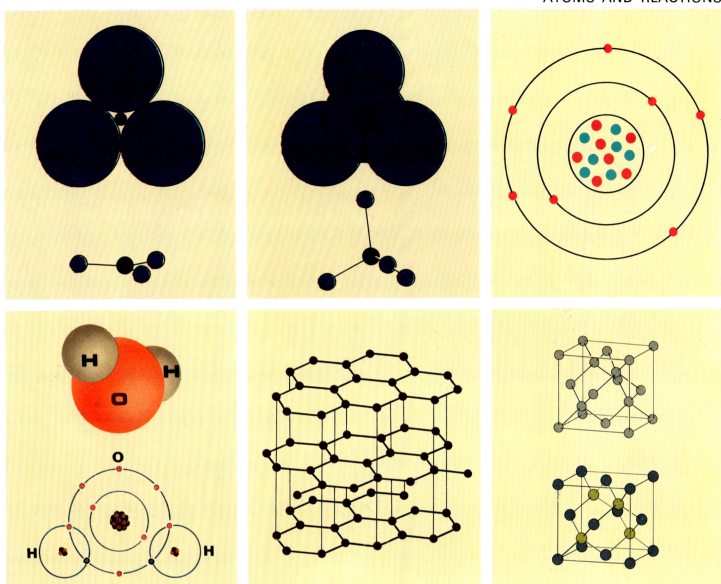

Representation of the atomic structure of various elements. Such representations seek to show the composition of the atoms, and the relationship of the particles of each other. Top right: a diagrammatic drawing of the nitrogen atom, with seven protons and seven neutrons in the nucleus. Top left and top centre: the nitrate ion with its triangular structure and the phosphate ion with its tetrahedric structure. Below left: the structure of water; centre and right: structures of two different forms of carbon: graphite and diamond.

salt) by means of heating on a flame, a coloration was obtained which, when observed with the appropriate instruments, turned out to be a spectrum formed by a large number of distinct lines, rather than by various colours which merged, by degrees, into one another, as in the case of a ray of light crossing through a prism. Bohr considered that these lines could be put down to precise energy jumps; in other words, the salt atoms absorbed energy from the flame, and passed on to higher energy levels. Once the stress had ceased, they returned to the basic level, by emitting the energy absorbed in the form of light; this emission was equivalent to the difference between the energy at the level to which they had been raised and the level to which they returned. This is the phenomena previously described as the 'hop' made by the atom.

Bohr's model has remained right to the present day but it has undergone a substantial and important modification as a result of work carried out by Erwin Schrödinger. Basing himself on Heisenberg's hypothesis (the latter being a physicist who had shown that it was impossible to know, simultaneously, the position and the amount of movement of electrons), he elaborated a new type of physics for the subatomic particles, known as *Wave Mechanics*.

This is a very complex subject but briefly the changes in Bohr's hypotheses

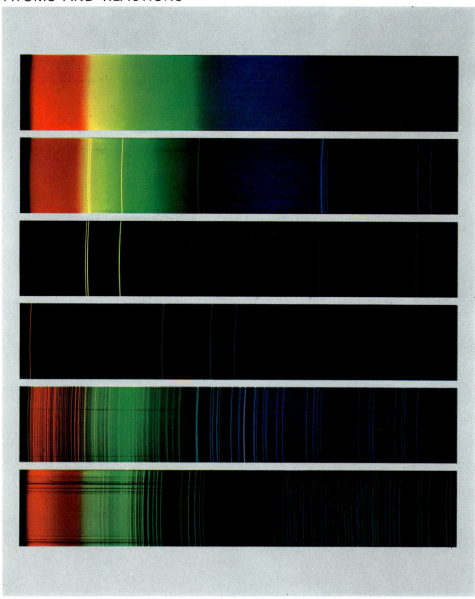

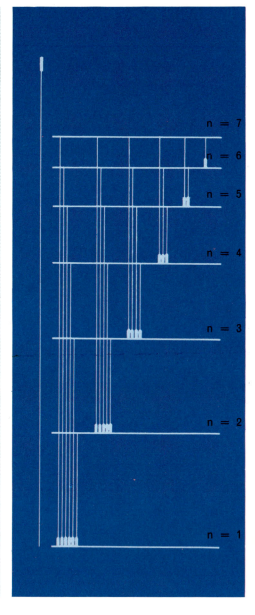

proposed by Schrödinger were these: Bohr had supposed that it was possible to know exactly the areas of the space around the nucleus in which the electrons orbit in the atom and that it was possible to calculate their shape and size. Schrödinger suggested that it was more correct to talk of the *probability* that electrons moved in given areas, he did not think these areas could be known with certainty. By solving complicated equations, he managed to calculate the areas in which the electrons *were most likely* to be found, and the energy associated with them. In about 1870 the Russian chemist D. Mendeleev analyzed all the elements known at that time and discovered that these showed surprising chemical and physical analogies at regular intervals. As a result the first table was brought out in which the elements were arranged on the basis of their recurrent properties, and this table is very similar to modern periodic tables. The greatest success of the atomic theories referred to was the rational explanation of the recurrence of properties, based on the examination of their atomic structure, in the various elements. The demonstration that the chemical properties depend solely on the electronic configuration gave the atomic theory added credibility. Among the various elements in the periodic table, those with the greatest stability and thus with the highest chemical inertia are the *noble* (or

The colour spectrums of different lights. Left, from top to bottom: the spectrum light from an incandescent lamp; from a fluorescent mercury lamp; of mercury and cadmium discharge; from a xenon lamp; the arc spectrum of iron. Right: diagram of the transition of matter is shown up by a spectrum: at each level the photons have a different speed. Each level contains various sub-levels: two for the second level: three for the third and four for the others: each one corresponding to a specific line on the spectrum.

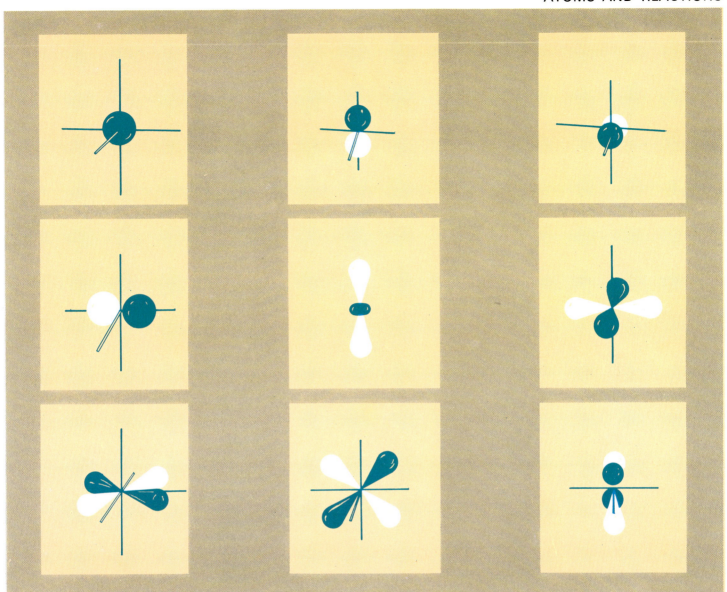

Illustrations showing the principles of Wave Mechanics. Bohr had suggested that it should be possible to trace exactly the area in which electrons orbit round the nucleus in an atom. Schrödinger suggested that it was more correct to talk of the probability of electrons orbiting in a particular area. These probable areas can be spherical, or the relatively simple S-types, illustrated top left. Other orbits, the P types, seem to be situated opposite to the centre of the axes on which the nucleus lies: others, illustrated below, the D-types, are even more complex.

rare) gases, at the extreme right of the table itself. The nucleus of the atom contains the entire positive charge and the electrons have a negative charge. It follows from this that in an atom with several electrons, the outermost ones will be less affected by the nuclear force of attraction at the centre, because this force is shielded and reduced by the electrons nearer to the central nucleus. The capacity of each atom to join together with others depends solely on the outer electrons and it is thus clear that all the factors which determine how closely these electrons are joined together (total number of electrons, dimensions of the atom etc.) have a direct influence on the formation of the unions. The most important chemical bonds are of two types: *ionic* and *covalent*. A molecule whose atoms are joined by an ionic type of bond is formed by one part with an excessive positive charge (positive ion) and one with an excessive negative charge (negative ion). Ionic bonds are formed, preferably, between atoms which have little tendency to retain their own outer electrons and atoms which, on the contrary, tend to attract electrons. A classic example of this is given by the compound CsF – cesium (caesium) fluoride.

Cesium loses its own outer electron very easily, but fluorine conversely, has just as strong a tendency to acquire an extra electron. By losing an electron,

MENDELEEV'S LONG-INTERVAL PERIODIC TABLE

I A	II A	III A	IV A	V A	VI A	VII A	VIII			I B	II B	III B	IV B	V B	VI B	VII B	O
1 H																1 H	2 He
3 Li	4 Be											5 B	6 C	7 N	8 O	9 F	10 Ne
11 Na	12 Mg											13 Al	14 Si	15 P	16 S	17 Cl	18 Ar
19 K	20 Ca	21 Sc	22 Ti	23 V	24 Cr	25 Mn	26 Fe	27 Co	28 Ni	29 Cu	30 Zn	31 Ga	32 Ge	33 As	34 Se	35 Br	36 Kr
7 Rb	38 Sr	39 Y	40 Zr	41 Nb	42 Mo	43 Tc	44 Ru	45 Rh	46 Pd	47 Ag	48 Cd	49 In	50 Sn	51 Sb	52 Te	53 I	54 Xe
55 Cs	56 Ba	57* La	72 Hf	73 Ta	74 W	75 Re	76 Os	77 Ir	78 Pt	79 Au	80 Hg	81 Tl	82 Pb	83 Bi	84 Po	85 At	86 Rn
87 Fr	88 Ra	89** Ac															

lanthanides* 4f	58 Ce	59 Pr	60 Nd	61 Pm	62 Sa	63 Eu	64 Gd	65 Tb	66 Dy	67 Ho	68 Er	69 Tm	70 Yb	71 Lu
actinides** 5f	90 Th	91 Pa	92 U	93 Np	94 Pu	95 Am	96 Cm	97 Bk	98 Cf	99 Es	100 Fm	101 Mv	102 No	103 Lw

cesium acquires the electronic configuration of the rare gas xenon, and fluorine that of the rare gas neon. The compound formed is thus extremely stable; this stability is increased by the fact the opposite ions formed, $Cs+$ and $F-$, are held together by a considerable electrostatic force of attraction. Clearly ionic bonds occur in the union of elements in group 1 with those in group 7 in the periodic table. The typical example of a pure covalent bond is the union of two atoms of the same type. In this case, with identical atoms, the bond is not due to the greater or lesser tendency to acquire electrons, but to other factors. If we take the case of the hydrogen H_2 molecule, it can be shown that two interacting H atoms at a certain distance form a more stable system than the two separate atoms. It has been found that the distance which permits this greater stability is less than the sum of the radii of the electronic clouds of the two atoms. This means that these clouds have interacted and penetrated each other, giving rise to a single *molecular orbital* in which the electrons are mixed in together. Pure covalent bonds occur only in the union of two atoms of the same sort, but the commonest type of bond is the so-called polar covalent bond. This bond is formed whenever different atoms unite, when atoms are involved which have features which exclude the formation of pure ionic bonds. In this case, we find

The periodic system: the horizontal lines are called periods *and the vertical lines,* groups. *H is the archetype, groups IA and IIA contain, respectively, the alkaline metals and the alkaline-earth metals, while all the elements from IIIA to IIB are called transitional metals. Proceeding towards the right, the metallic characteristics gradually diminish (the demarcation line). Group O, on the extreme right, contains six rare gases. The number above each element is the atomic number, that is the total number of electrons and thus the number of nuclear protons in the atom in question.*

Molecular movement depends exclusively on temperature. Left: water in the solid state; centre: in the liquid state; right: as steam. The intermolecular forces of attraction are considerable in the solid state (such that they keep the molecules tightly packaged in the crystals), slightly less in the liquid state (the molecules can move over one another) and, lastly, markedly weak in the gaseous state (the molecules here being free to move and occupying all the space available to them)

distortions in the bonding electronic clouds by the atom which has the greater tendency to attract electrons. The unit of this distortion constitutes the so-called *ionic percentage* of a polar covalent bond. One last important phenomenon which must be mentioned is the *hybridization of orbitals*. It often happens that an atom forms a greater number of bonds than those forseeable on the basis of examination of the electronic configuration of its outermost layers. This is due to the fact that one or more outer electrons can pass over into empty, adjacent orbits, in an 'energetic' way. As a result of such transitions, one can observe a rearrangement of the outermost orbitals which intermix and produce new electronic distributions, the geometry of which differs from the initial geometry. These are called *hybrid orbitals*. In the chapter devoted to organic chemistry this concept will be clarified and we shall examine one of the most important instances of hybridization: that concerning carbon.

It is no easy task to deal fully with the topic of chemical reactions in just a few lines; the term *chemical reaction* can, in general, be applied to a process in which one or more compounds are transformed, in appropriate conditions, into other compounds with different features than the original ones. Chemical reactions may be natural or provoked. Because every system tends invariably

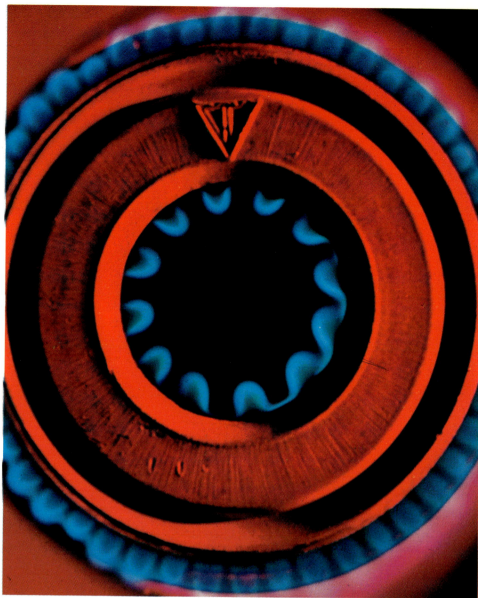

to attain a more stable configuration, that is, with lower energy, it follows that a reaction is natural if it leads to products with a lower energy content than the reagents. But this does not always apply: in fact, many reactions which give rise to lower-energy products, can be impeded by kinetic factors, that is, they may proceed at very, very low speeds. In practical terms, for a reaction to occur, it is necessary that the reagents go beyond a certain potential barrier, after which the process towards the formation of the products gets under way. This barrier is known as the *activation energy* of the reaction, and may have both very low and remarkably high values, independently of the fact the reaction may then absorb or release energy. This situation is comparable to that of a mass – let us say a spherical mass – situated on a base near a slope. The mass represents the system formed by the reagents; it will have a marked tendency to roll down the slope (the development of the reaction towards the products) but to do so it will have to be nudged (activation energy). The force required to shift it will be small if the base is flat (reactions with low activation energy which occur easily), but fairly strong if the base is, for example, concave (reactions with high activation energy which are hard to achieve). Each reaction is hallmarked by a precise difference in energy-level between reagents and products If this

Gases contain many fewer particles than solids and liquids of the same volume. The volume of a gas is inversely proportional to the pressure on it, whereas it is directly proportional to the temperature (although at a constant pressure). The pressure is likewise directly proportional to the temperature, at a constant volume. Left: pressurized gas pipelines. Right: a butane gas domestic oven.

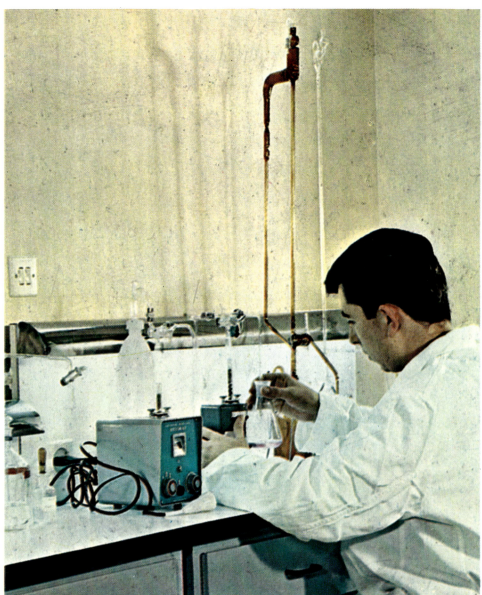

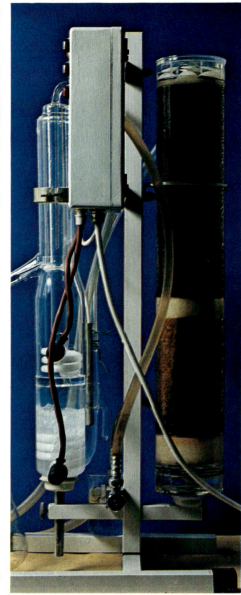

From theoretical principles to day-to-day practice. Left: an analyst at work at a titration bench: by adding one chemical to another, and recording the reaction, the composition of the substances can be analyzed. Right: modern apparatus for the double distillation of water. Water coming from an ion-exchanger resin column passes into the first heating chamber. Here it is brought to the boil and the steam is released, condensed and collected in the upper chamber, where the process is repeated and the condensed steam is collected for use.

difference is positive, that is, if the products have less energy than the reagents, once the activation energy has been obtained the reaction will develop by restoring this excess energy in the form of heat. This gives rise to what we call *exothermic* reactions. When, conversely, the difference is negative – the products having more energy than the reagents – heat must be constantly supplied for the reaction to proceed. In this case it is as if the original mass were initially at a lower height than the level to which it will roll having negotiated the slope. The energy required to take it to the top of the slope will thus be higher than the energy which the mass will develop as it rolls downhill from the other side; the reaction in this case is *endothermic*. A reaction may be influenced by a very large number of variables: temperature, pressure, concentration, solubility of the products, acidity of the medium etc. Marked effects are produced by the catalysts – substances which can accelerate (or slow down) a reaction to a considerable degree, and yet remain unchanged at the conclusion of the process. Their active mechanism is not always evident, but they certainly act by lowering (or increasing) the activation energy of the reaction. All chemical reactions are processes which develop towards a situation of dynamic equilibrium. In other words, a reaction never proceeds

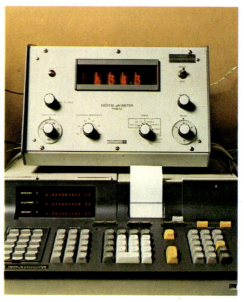

completely towards the formation of the products, and although the transformation may often be very forced, a point is always reached at which the speed at which the products are formed equals the speed with which the products react to re-produce the reagents.

There is an important empirical principle which governs a chemical reaction in the state of equilibrium: this is the *Le Chatelier principle* or law of mobile equilibrium.

This principle states that every system that is in a state of equilibrium moves, when subjected to an external stress, in such a way as to annul this stress and restore the conditions of equilibrium, thus minimizing the effects of the stress itself. The application of this principle is particularly useful in the case of reactions which are only partly proceeding, in so far as it permits intervention with the equilibrium, to move it towards the formation of the desired products. Let us take a typical case in point: the formation of ammonia from nitrogen and hydrogen. The balanced reaction is:

$$N_2 + 3H_2 \longrightarrow 2NH_3 + 22 \; Kcal/mole.$$

It can be seen that from four moles of reagent gases we pass to two moles of

Engineering and electronics have brought about revolutionary innovations in chemical apparatus, allowing complicated analyses to be carried out both very quickly and extremely accurately. Left: a general view, and (right) details of equipment for completely automatic computer-aided electrochemical analyses. The electronic computer has stored in it all the data necessary for instant temperature control, instant checking of the progress of the analysis and for the addition of the appropriate reagents.

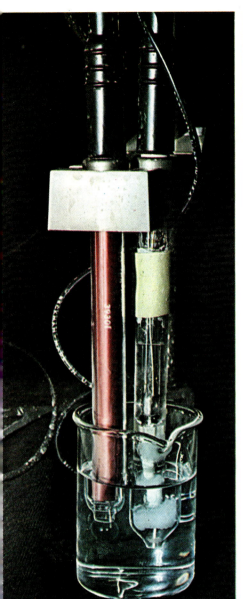

The acidity of a solution is expressed by the pH *factor (which technically is the concentration of hydrogen ions) with the negative logarithm. The* pH *scale can range from 0 to 14. Pure neutral water, neither acid nor alkaline, has a* pH *value of 7: solutions with a* pH *value 0–6.9 are acid, those with a* pH *value 7.1–14 are alkaline. Left: the measuring electrodes of a* pH-meter. *Top: the scale. Below: (left) bromocresol blue in an acid solution and an alkaline solution; (right) showing the different colour obtained, according to the nature of the solution.*

gaseous ammonia and as a result, at a constant volume, we shall have a reduction of pressure. In addition, the reaction is exothermic, that is, it occurs with the development of heat. To obtain the desired product in the most complete way possible, we must, on the basis of the Le Chatelier principle, work at very high pressure and low temperature. In fact by greatly raising the pressure, the system will move in such a way as to reduce it, in order to annul the stress. Because the formation of NH_3 from nitrogen and hydrogen occurs with reduction in pressure, the reaction will develop towards the formation of ammonia. If, on the other hand, we reduce the pressure a great deal, the system will react in such a way as to minimize the effect of this reduction. Raising the temperature is again obtained by the formation of ammonia, which occurs with the development of heat.

Of the large number of balances which can be observed in watery solutions, particular interest attaches to that of the dissociation of water. Pure water gives rise to the following balance of ionic dissociation:

$$H_2O \rightleftarrows H+ + OH-$$

This reaction develops towards the right to a minimal degree; in fact the

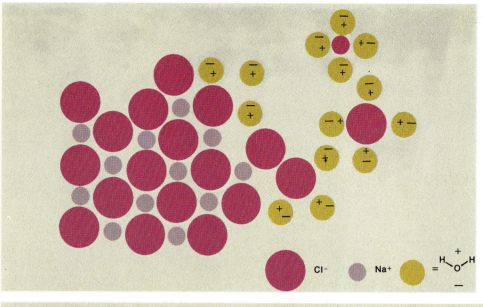

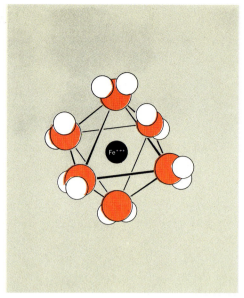

Cl⁻   Na⁺   = H—O—H

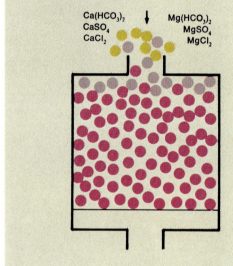

Ca(HCO₃)₂
CaSO₄
CaCl₂   Mg(HCO₃)₂
MgSO₄
MgCl₂

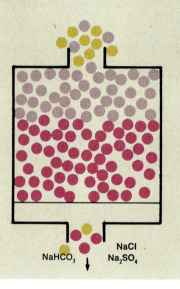

NaHCO₃   NaCl
Na₂SO₄

NaCl

MgCl₂

concentrations of $H+$ and $OH-$ are both 10^{-7} moles per litre of water at an ambient temperature. The predominance of the $H+$ ions gives the solution features of *acidity*; on the other hand, if the $OH-$ ions are more numerous than the solution turns out to be *basic* (or alkaline). Water owes its extraordinary solvent capacities to the fact that it has a high *dielectric constant*, i.e. a strong tendency to reduce the electrostatic attraction between ions of the opposite signs. All the ionic compounds and many of the polar covalent compounds are thus water-soluble and as they dissolve they give rise to the phenomenon known as electrolytic dissociation. Many of these soluble compounds have the property of interacting with water, and altering its dissociation balance.

In this, as in any other form of equilibrium, the product of the concentration of the ions that have formed must remain constant; for water, the constant has a value of $10^{-7}.10^{-7} = 10^{-14}$. It follows that every substance capable of supplying a large quantity of $H+$ hydrogen ions (e.g. HCl, $_2SO_4$, HNO_3, etc.) enormously increases the number of these ions in solution, gives them acidity characteristics and is thus called an *acid*. As a result of the dissolution, in water, of an acid compound, there is a corresponding reduction of the $OH-$ ions,

Top left: diagram of the process of dissociation of a salt in water. The solvent directs the negative part towards the sodium ions, and the positive part towards the chlorine ions, reducing the forces of attraction. Top right: diagram of a ferric hydrate ion, with the water molecules arranged octahedrically. Below: diagram of a water-softening plant in the phases: preparation, execution and regeneration. The calcium and magnesium ions present in the water are replaced by the corresponding sodium ions.

The production and processing of salt, sodium chloride (Na Cl), a basic raw material. Top left: saline rocks. Below: an aerial view of salt-works in France. Centre: the salt market in Mexico. Top right: the Margerita salt-works in Savoy. Below: detail of a processing plant. The marked solubility of NaCl in water (approx. 350 g/litre) makes it possible to prepare very concentrated solutions with great ease, and these are particularly useful for electrolytic processes.

which maintains the above-mentioned value of 10^{-14} for the product.

So if, for example, the $H+$ ions assume the value of 10^{-3}, the $OH-$ ions are reduced to 10^{-11} moles per litre. Vice versa, the substances which give rise, by dissociation, to $OH-$ ions ($NaOH$, KOH, $Ca(OH)_2$, etc) give the solution alkaline characteristics and cause a corresponding reduction of the $H+$ ions. Likewise, many salts may influence the dissociation balance of water, because of the tendency of some ions to interact with the $H+$ ions or the $OH-$ ions. If, for example, we dissolve sodium cyanide $NaCN$ in water, this produces $CN-$ and $Na+$ ions. These latter do not have any tendency to react, whereas the $CN-$ ions unite with the $H+$ ions and give the undissociated stable compound HCN. The solution is then low in $H+$ ions and rich in $OH-$ ions, as it becomes alkaline. An opposite pattern of behaviour is shown by the salt NH_4Cl, ammonium chloride. The $Cl-$ ion is inert, while the NH_4+ has a strong tendency to assume $OH-$ to give the undissociated compound NH_4OH. The solution is low in $OH-$ ions and rich in $H+$ ions and becomes acid. The phenomenon described here is known as the *hydrolysis* of salts.

Pure water is a very poor electricity conductor, i.e. it has a huge resistance to the movement of current. But if water contains a dissolved salt and we put two

electrodes into this solution which are connected at the poles to a battery with continuous current (or direct current), we can see that conduction is now possible, and that in fact the current passes through the solution very easily. Electric conductivity is due, for this reason, to the ions present in the solution, deriving from the electrolytic dissociation of the salt. At the same time as the movement of the current, interesting phenomena occur at the electrodes. The positive ions (or cations) head towards the negative electrode which is thus known as the *cathode*, while the negative ions (anions) head for the positive electrode or *anode*. At the cathode the positive ions neutralize their charge by taking on electrons and thus undergoing a process known as *reduction*, while, at the anode, the negative ions relinquish their excess electrons, giving rise to the process known as *oxidation*. Let us take a simple case: let us suppose we are dissolving, in water, the salt $NaCl$, sodium chloride, and applying electrodes. $Na+$, $H+$, $Cl-$ and $OH-$ ions are all present in the solution. The $H+$ ions head towards the cathode, which take on electrons on the basis of the reaction $2H+ + 2e \longrightarrow H_2$, and at this electrode we find the formation of molecular hydrogen which is released as gas. The $Cl-$ ions head towards the anode; these relinquish their electrons on the basis of the reaction $2Cl- \rightarrow Cl_2 + 2e-$ and we

A battery of mercury electrolytic cells for the electrolysis of salt, sodium chloride. The cathode is formed by metallic mercury on which the $H+$ ions do not discharge, because of the phenomenon of overvoltage, based on the nature of the cathode itself. Discharge is produced by the $Na+$ ions, which form a metallic amalgam with the mercury, and are reduced to elementary sodium. The amalgam is made to react with water and the sodium contained in it produces hydrogen and $NaOH$.

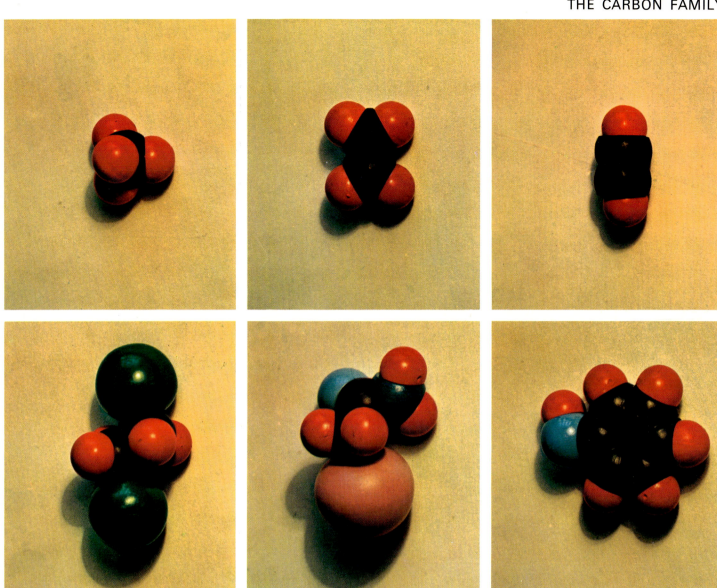

Atomic models of simple organic molecules. Top, left to right: methane, *which is the simplest organic molecule,* ethylene, *the simplest olefine, and* acetylene, *the simplest hydrocarbon with a triple bond. Below left:* dichloroethane; *centre:* iodoacetamide; *right:* phenol *formed by a benzene ring substituted by an* OH *group.*

atoms are arranged differently. On the basis of precise physical measurements and theoretical considerations relating to a large number of organic compounds, it has been deduced that the four simple covalent bonds made by carbon, and hence its substituents, are not all on the same plane, but rather oriented in space like the segments which unite the centre of a tetrahedron with its vertices. At the centre of the tetrahedron we thus find the *C*, with the four *H* substituents at its vertices; the angle formed between any two of these four bonds is approximately 109°, like the angle at the centre of regular tetrahedron, if the four substituents are equal.

Thus, whenever a carbon forms four simple covalent bonds, its substituents are situated in space in relation to one another like the vertices of a tetrahedron. But this is not always how things are: in many compounds the four carbon bonds are on a plane or in a line, rather than tetrahedric. This occurs when there are less than four substituents around the *C*. In these cases, and in order to achieve the stable configuration of rare gas, the carbons find themselves as it were constrained to reinforce their union and bring four or six electrons into interplay, rather than just two as in the case of a simple covalent bond. Because these bonds are no longer simple they are called double or triple

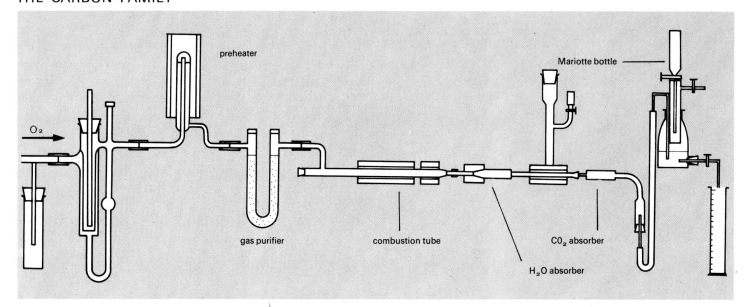

preheater

Mariotte bottle

O_2

gas purifier

combustion tube

CO_2 absorber

H_2O absorber

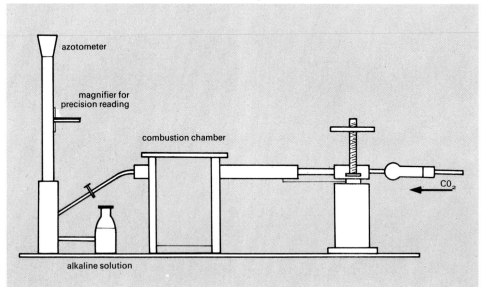

azotometer

magnifier for precision reading

combustion chamber

CO_2

alkaline solution

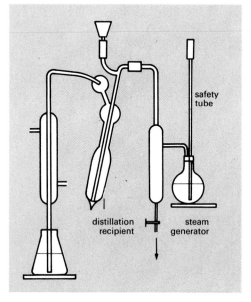

safety tube

distillation recipient

steam generator

bonds, and accordingly change the geometric form of the compounds in which they occur. Ethylene, for example, with the empirical formula C_2H_4, is a plane-type molecule with the following structure: $\begin{smallmatrix} H & & H \\ & C{=}C & \\ H & & H \end{smallmatrix}$. As can be seen, each of the two Cs forms a further four bonds, two of which are simple with the H and one of which is double, between them. The atoms are no longer in the vertices of a tetrahedron, but in a plane; around each carbon the bonds are oriented like the bisecting lines of an isosceles triangle, at the centre of which is the carbon. Acetylene, with the empirical formula C_2H_2, has an even smaller number of substituents than ethylene, and contains a triple bond between the two carbons and the molecule, as a consequence of this, is linear in form; i.e. its structural formula is: $H{-}C{-}C{-}H$. As well as being used to clarify the geometry of the carbon compounds, these compounds are the archetypes for three series of compounds known as hydrocarbons, in as much as they are formed by just carbon and hydrogen. Methane is the first of the series of the

Top: diagram of apparatus used to determine the quantity of carbon and hydrogen present in organic compounds. Below: two diagrams of apparatus for the regulation of organic nitrogen. In the first, the nitrogen released during combustion is measured for volume in the azotometer; in the second, the substance is first decomposed with sulphuric acid and then distilled in KOH; the nitrogen present is released as ammonia which is titrated with boric acid.

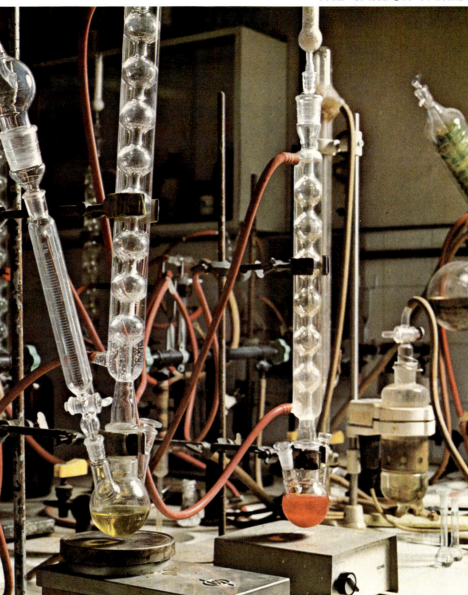

Left: an analyst producing the combustion of an organic compound for the determination of carbon and hydrogen, with apparatus similar to that already described. Right: a laboratory bench with various reaction apparatus. Depending on the reaction, appropriate glass recipients are generally used in which the substance, dissolved in a suitable solvent, is kept in constant motion. The bubble-coolers mounted on the glass flasks are used to retain the steam from the solvent.

saturated hydrocarbons or paraffins or alkanes; ethylene is the first of the unsaturated hydrocarbons with a double bond, or olefines or alkenes; and acetylene is the first of the unsaturated hydrocarbons with a triple bond, or acetylinics or alkines.

These three series of compounds show fairly similar physical properties, but differ in their chemical properties, i.e. in their reactions when brought into contact with other substances. The alkanes show no reaction to most of the common reagents and are not attacked at normal temperatures or by acids, but they do react with the molecular oxygen O_2 which is present in the air, at high temperatures, by burning. Carbon dioxide (CO_2) and water (H_2O) as well as large amounts of heat are developed from their complete combustion; it is precisely for this reason that petroleum is important, because it is formed mainly of hydrocarbons and not only of the paraffin type. Unlike the alkanes, the alkenes and alkines are very reactive with many substances, such as, for example, acids, halogens and oxidizing agents. This behaviour is caused by the ease with which the double bond (or the triple bond) can, for example, add one molecule (or two molecules) of chlorine Cl_2, and in this way be transformed into two (or four) simple bonds. In fact, by reacting with chlorine, ethylene

gives rise to dichloroethane:

$$\begin{array}{c} H \\ \diagdown \\ C \end{array} = \begin{array}{c} H \\ \diagup \\ C \end{array} \ + \ Cl\text{--}Cl \longrightarrow \ H\text{--}\overset{\displaystyle H}{\underset{\displaystyle Cl}{C}}\text{--}\overset{\displaystyle H}{\underset{\displaystyle Cl}{C}}\text{--}H$$

Once a reaction has been carried out, it is often necessary to remove the solvent, by distillation, before analyzing the products. In this illustration can be seen a modern piece of apparatus called a revolving evaporator, with which it is possible to carry out this operation in a vacuum.

Each of the two bonds of the double bond is less strong than a single bond, and for this reason more reactive; in addition, the electrons of the double bond are more exposed to the reagent than are those of a single bond.

It is interesting to note how the higher the number of carbon atoms in a molecule, the greater the possibilities of arranging these atoms in different ways in the space. Let us give some examples, and take a look at the first six of the series of saturated hydrocarbons, namely: methane CH_4, ethane C_2H_6, propane C_3H_8, butane C_4H_{10}, pentane C_5H_{12} and hexane C_6H_{14} (from these formulae it can be seen that the general formula of the alkanes is C_nH_{2n+2}, where n is the number of carbon atoms which constitute the molecule in question, and n can reach a very high value. Whereas for the first three there is just one possibility of setting out their structural formula, for butane there are two, for pentane three and for hexane five.

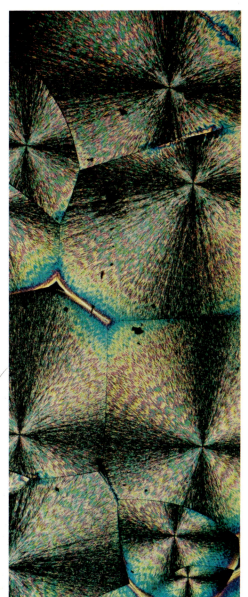

Many solid organic substances can be obtained in a pure, crystalline form. Crystallization, where possible, is a worthwhile operation because it makes it possible greatly to increase the purity of the substance. Left: ascorbic acid crystals. Right: crystals of nicotinamide, photographed under polarized light. These two substances play a very important part, as vitamins, in our bodies.

$$CH_3-CH_2-CH_2-CH_3$$

$$CH_3-CH-CH_3$$
$$|$$
$$CH_3$$

$$CH_3-CH_2-CH_2-CH_2-CH_3$$

$$CH_3-CH-CH_2-CH_3$$
$$|$$
$$CH_3$$
$$CH_3$$
$$|$$
$$CH_3-C-CH_3$$
$$|$$
$$CH_3$$

$$CH_3-CH_2-CH_2-CH_2-CH_2-CH_3$$

$$CH_3-CH-CH_2-CH_2-CH_3$$
$$|$$
$$CH_3$$
$$CH_3-C-CH_2-CH_3$$
$$|$$
$$CH_3$$
$$CH_3$$
$$|$$
$$CH_3-CH_2-CH-CH_2-CH_3$$

$$CH_3 \quad CH_3$$
$$| \quad |$$
$$CH_3-CH-CH-CH_3$$

These structures are in effect so many distinct chemical compounds and are called structural *isomers* – compounds with the same empirical formula but with a different structure. There is another type of isomer called a geometric

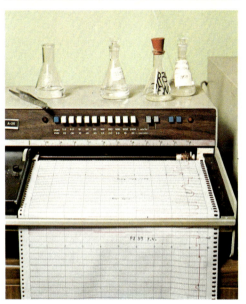

isomer which occurs in compounds containing one or more double bonds. A simple example of a geometric isomer is dichloroethylene, $C_2H_2Cl_2$. If the two chlorine atoms are on the same carbon atom, then only structure I is possible, but if they are joined to different carbon atoms, then structures II and III are possible:

I:
$$
\begin{array}{ccc}
Cl & & H \\
& C=C & \\
Cl & & H
\end{array}
$$
II:
$$
\begin{array}{ccc}
Cl & & Cl \\
& C=C & \\
H & & H
\end{array}
$$
III:
$$
\begin{array}{ccc}
Cl & & H \\
& C=C & \\
H & & Cl
\end{array}
$$

The last two compounds are isomers in themselves, the first *cis*-form, the second *trans*-form (cis- if they are on the same side, and trans- if they are on the opposite side in relation to the line of the double bond). The general empirical formula of the alkenes is C_nH_{2n}, but it should be noted that this corresponds with yet another series of compounds: the cyclic alkanes. In fact four alkenes and two cycloalkanes – six isomers in all – correspond to the empirical formula C_4H_8.

There is another series of unsaturated hydrocarbons, the chemical inertia of

It is often possible to determine the purity or composition of an organic sample by using the technique of steam-phase chromatography. A minute quantity of a volatile substance is introduced into a tube filled with some absorbent matter, and moved by a nitrogen-type inert gas to the tube outlet, where it is collected. Left: the technician is carrying out a micro-injection in the gas-chromatograph. Right: part of the measuring and recording programming.

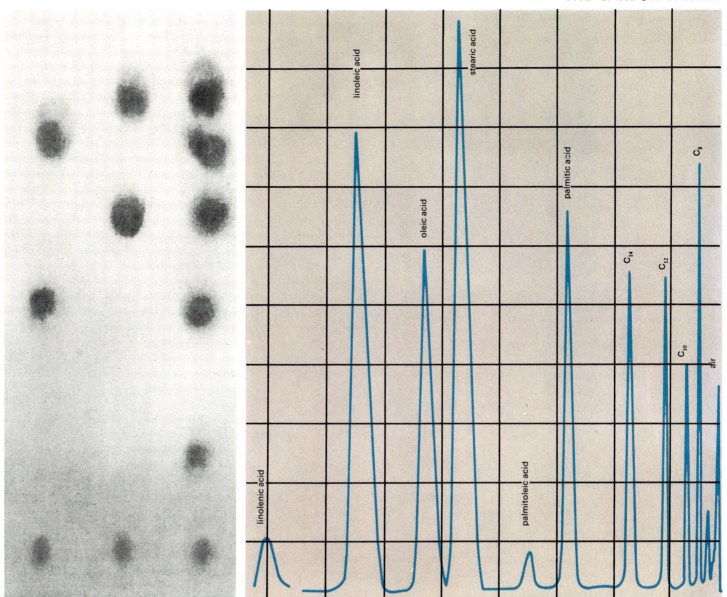

The results of gas-chromatographic analyses are usually recorded on paper, for greater accuracy, and convenience. Left: we see the separation of the fatty acids from eight up to eighteen carbon atoms. Right: certain volatile organic acids have been separated by the technique of chromatography on paper. The paper containing the samples is placed with its lower edge in a dish, so that it touches a mixture of solvents which slowly spread by capillarity as far as the other edge of the paper, conveying the various components of each sample at different speeds.

which, however, makes them more akin to the alkanes than to the alkines: these are called aromatic hydrocarbons because of the smell they give off, or also benzene–hydrocarbons, after their more simple name, benzene, with the empirical formula C_6H_6, and a flat hexagonal structure. For this compound the two equivalent structures I and II below were proposed more than a century ago by the chemist Kekulé, who also proposed the tetravalency (or ability to join with four hydrogen atoms or their equivalent) of carbon:

Today we know that these structures are inaccurate; for example, they do not explain why benzene does not react rapidly with bromine Br_2. This was recently understood due to the development of the theory of electronic resonance. As far as we are concerned it is enough to know that in benzene the electrons do not form three double bonds but are equally distributed among all the carbon atoms of the molecule as in structure III. As well as the hydrocarbons there are very many other series of organic compounds with distinctly different properties and structures, but which, for the sake of simplicity, can be considered as deriving from the hydrocarbon structures by the substitution for one or more of the hydrogen or carbon atoms, of other elements or atomic groupings. In effect, organic compounds have been classified on the basis of the varying chemical nature of these substitutions, known as functional groups, which hallmark the chemical and physical behaviour. In general, only the functional groups take part in the chemical reactions, whereas the skeleton of the molecule remains unaltered. The most important functional groups of organic chemistry are: hydroxyl–OH, carbonyl $C\!\!=\!\!O$, carboxyl $-\overset{\displaystyle\,}{\underset{\displaystyle O}{C}}\!\!-OH$, and the amino-group $-NH_2$. For example,

The apparatus on the left, known as the Soxhlet extraction apparatus, is usually used for the continuous extraction of a solid compound by a hot solvent. The solid, which is placed in a porous thimble made of filter-paper, is introduced into the glass recipient. The solvent, which evaporates from the heated flask, condenses on the sides of the cooler placed above the thimble and falls into it, gradually dissolving the solid. When the thimble is full of solvent, this is automatically siphoned into the flask and thus takes with it that part of the solvent which has been dissolved. Right: a viscometer, for measuring the speed at which a solution flows.

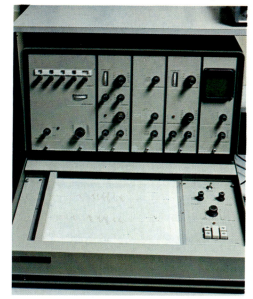

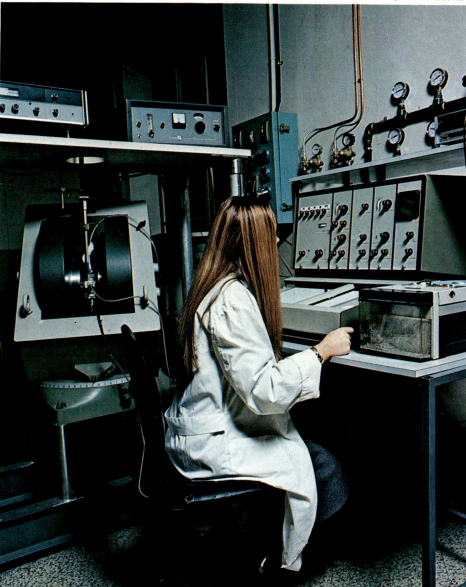

Electronic spin resonance (ESR) is a physical technique which makes it possible to single out and sometimes distinguish molecules containing free radicals and compounds containing transition metals. The free radicals, usually typified by a marked chemical reactivity, have been shown to be transitory chemical intermediaries in many reactions with ESR. The quartz cell containing the sample is placed between two large magnets, cooled by liquid air and analyzed. Above we see details (left) and a general view (right) of apparatus for the measurement of ESR.

the compound CH_3-CH_2-OH is ethyl alcohol, whereby the functional group $-OH$ isolates the series of alcohols. The compounds containing a carbonyl are called: aldehydes, if a carbon and a hydrogen are associated with this group, and ketones if two carbons are associated with it: two examples are acetaldehyde $CH_3-\overset{\displaystyle |}{\underset{\displaystyle H}{C}}=O$ and acetone $CH_3-\overset{\displaystyle \|}{\underset{\displaystyle O}{C}}-CH_3$. If the OH group is directly linked to a carbonyl group, a carboxyl is obtained, i.e. an organic acid such as, for example, acetic acid $CH_3-\overset{\displaystyle \|}{\underset{\displaystyle O}{C}}-OH$. The acid properties derive from the ease with which the carboxyl group can dissociate the H in the form of proton $H+$. Vice versa, the amino-group or amines are substances with alkaline properties, given that nitrogen can acquire a proton and thus increase the pH of the solution:

$$CH_3-CH_2-NH_2 + H+ \longrightarrow CH_3-CH_2-NH_3+$$

FROM TEST-TUBE TO CONTAINER

It would be wrong to think that industrial chemistry consists simply in the large-scale application of processes and reactions already discovered and tested in the laboratory.

In fact there are hundreds of syntheses which, although of great interest, cannot be applied on an industrial scale because they involve compounds which are either very costly or difficult to come by. Others, although based on simple compounds, require reaction conditions that pose major technological problems to solve which would make the finished product prohibitively expensive. Only what is commercially economic is of interest to the chemical industry, as distinct from the research chemist. Thus industrial chemistry concentrates on the manufacture of cheap compounds with a good return.

Of course both technological development and other factors, including political events that may affect the availability of raw materials, may mean that a process which seems commercially viable today may also, before long, prove unacceptable.

The development of industrial chemistry brings with its blessings its own problems. As each process is established, it becomes an integral part of the economic and industrial structure, so that if events cause raw materials to be suddenly unavailable, the repercussions can be serious. Ideally of course, by way of insurance, more than one process ought to be available, so that alternative raw materials could be used, to obtain the same industrial product, but this is not always possible. Sometimes, when it is possible, short-term commercial considerations inhibit its development. The oil crisis, which in recent years has affected the West, illustrates this point well. During the 1939–1945 war, with a sudden shortage of fuels and the urgent and mounting needs of the war effort, the Germans developed the Fisher-Tropsch process and synthesis, whereby, starting with coal and water, they could obtain remarkable quantities of paraffin-based and olefinic fuels (hydrocarbons). This process, which represented a valid alternative to oil, was subsequently abandoned, when the precious 'black gold' became available once more.

Industrial chemistry is divided into the two major sectors: organic and inorganic. The former is far and away the more far-reaching and important; proof of this is the fact that many of the main inorganic compounds are prepared to be used as reagents in major organic industrial syntheses, the inorganic industry thus serving the organic industry. The pattern of the industry can be seen from considering the production of some typical products of the industry, such as chlorine, caustic soda and sulphuric acid. Very considerable importance attaches to the production of chlorine, because of its many uses in organic chemical industries as a chlorinating agent for the hydrocarbons, and also because of certain applications in the inorganic field. It is prepared industrially by $NaCl$ (salt) electrolysis. Almost saturated with chloride (brine), the solution is continuously fed into diaphragm cells where chlorine is developed by means of anodic oxidation, while, in the cathode, there is a reduction of H_2O with the formation of gaseous hydrogen and $OH-$ ions. The cathode compartment thus becomes rich in $NaOH$ (deriving from the combination of the $Na+$ ions with the $OH-$ ions which have formed) which is another important product. In the mercury cells it is this metal that acts as a

In many cases the supply of basic raw materials is the responsibility of the mining industry; the chemical industry, on the other hand, supplies semi-processed and finished products. Top left: a pile of bauxite, a mineral containing aluminium, which is extracted both chemically and electrochemically. Below: bars of pure copper, obtained by the electrolytic purification of the crude metal. Right: a plant for the production and purification of ammonia, obtained by direct synthesis from the elements.

cathode, and because a H_2 discharge in it is not possible, there is a discharge of the sodium ion which forms an amalgam with the mercury, while the chlorine is released as a gas at the anode. The amalgam is then made to react with water (only the sodium reacts in fact) with the formation of $NaOH$ and H_2. Caustic soda is recovered by the evaporation of the cathode liquid; its most important uses are as a strong base in the chemical and petrochemical industries, and in the soap- and fibre-manufacturing sectors. Sulphuric acid accounts for about 90 per cent of the production of the sulphur industry. Its uses range from the phosphate and ammonium sulphate industries to the titanium pigment industry, and from the metallurgical to the chemical and petrochemical industries. Now that the lead-chamber process has been abandoned, production is based on the catalytic process. This consists in the oxidation, with air, of sulphur dioxide (obtained by roasting sulphur minerals or elementary sulphur) to sulphur trioxide SO_3 by means of platinum or vanadium catalysts. The subsequent hydration of the SO_3 produces sulphuric acid at the various concentrations desired. Ammonia is the principal product in the nitrogen industry. Production is based on the process of direct synthesis between H_2 and N_2, now that all the other systems of recovery from coke gases

etc. have been dropped. The reaction between H_2 and N_2 is fairly exothermic and occurs, in addition, with a reduction in volume. The Le Chatelier principle thus proposes the use of high pressures and low temperatures. In practice, however, one can see that below 500° C (932° F) the reaction does not proceed at a useful speed, which is why temperatures within the range 450–600° C (842°–1112° F) are used on the whole. Pressures range from a minimum of 300 atm to 1000 atm. These drastic conditions created considerable problems, which is why, although known about since the early nineteenth century, the process only became operative in about 1920 in the BASF Haber-Bosch plant, the first of its kind. The catalysts used in the synthesis are iron-based (oxide or metal) with the addition of activators with a base of alkaline oxides and aluminium. The enormous production of NH_3 (more than two million tons in Italy alone) is destined for the nitrogenous fertilizer industry (70 per cent) and for the urea industry, as well as for a variety of uses in the pharmaceutical industry and in organic chemistry. In addition it is the basic compound for the synthesis of another inorganic product, nitric acid.

Petroleum

We shall leave aside the various theories put forward about the origin of

Top left: graphite, an allotropic form of carbon. Graphite has good electric conductivity and high chemical inertia. One of its commonest uses is therefore for the manufacture of electrodes for electrochemical processes. Below: graphite electrodes obtained from amorphous carbon just out of the furnace. Right: a plant for the production of small diamonds from graphite, subjected to a pressure of more than 200,000 atm and a temperature of 2600° C (4712° F).

Of the various olefines, ethylene is certainly one of the most important. Above: a plant for the production of ethylene by steam-cracking. The feed-load, generally petrols and light fuel oils, is vaporized at 900° C (1652° F) with superheated steam, which acts as a diluent, a heat-conveyor and prevents the carbonization of the substances. The ethylene yield from the process is quite high (40 per cent).

petroleum, and confine ourselves to saying that the most important oilfields are situated in various Middle Eastern countries, in the United States (Kansas, Ohio, Virginia and Pennsylvania), in Venezuela, in the Caspian region, and in the Sahara (Libya). Estimates relating to known fields make it possible to say that the supply of crude oil will pose no problems for at least the next 30 years.

Petroleum is a mixture of several hundred liquid hydrocarbons, in which certain solid and gaseous hydrocarbons are dissolved. The principal constituents are the linear and ramified paraffins, and the cyclic, saturated and aromatic hydrocarbons. The place of origin or source of the crude oil hallmarks its composition, there being some forms which are extremely poor in aromatic hydrocarbons, and others particularly rich in them. In addition petroleum contains (but in fairly small amounts) sulphur compounds, in particular H_2S and mercaptans or thio-alcohols (compounds containing $-SH$ groups) and their derivatives, as well as small quantities of oxygenated compounds (fatty acids and phenols) and nitrogenous compounds; there are, however, no olefines whatsoever.

In the refinery crude petroleum undergoes an initial operation of fractional distillation, carried out in a single multi-stage column or in several columns in

series; this separates the major 'fractions' listed below:
– *gases which cannot be condensed* such as: hydrogen, hydrogen sulphide, methane, ethane and so on.
– *low-boiling fractions*, i.e. all the hydrocarbons which distil within the 40–170° C (104°–338° F) range, formed by petroleum ethers, ligroin and benzine (first distillation).
– *kerosene*, i.e. hydrocarbon fractions boiling within the 170–270° C (338°–518° F) range.
– *gas oils*, crude oil fractions which distil within the 270–360° C (518°–680° F) range.

Next come the heavy distillates, which can in turn be divided into lubricating oils and waxes. We should note that, in this case, distillation is carried out in vacuum plant, so that temperatures of 550° C (1022° F) or lower can be used: at atmospheric pressure higher temperatures would be needed and fires or explosions would occur. The residue from all these processes is made up of asphalts, resins and bitumen, as well as heavy fuel oils and coke, used principally in metallurgy.

The processing of petroleum cannot, however, end with these distillation

From the remote desert sites from which it is pumped (below, left and centre: drilling tower and drill-bit), petroleum starts its journey through large and extremely long pipelines; then it is loaded on board oil-tankers until it reaches the large refineries, where it undergoes initial processing (right: a tangle of pipes in a refinery), and the petrochemical plant where it is transformed into countless products. Top: an aerial view of the petrochemical works at Gela.

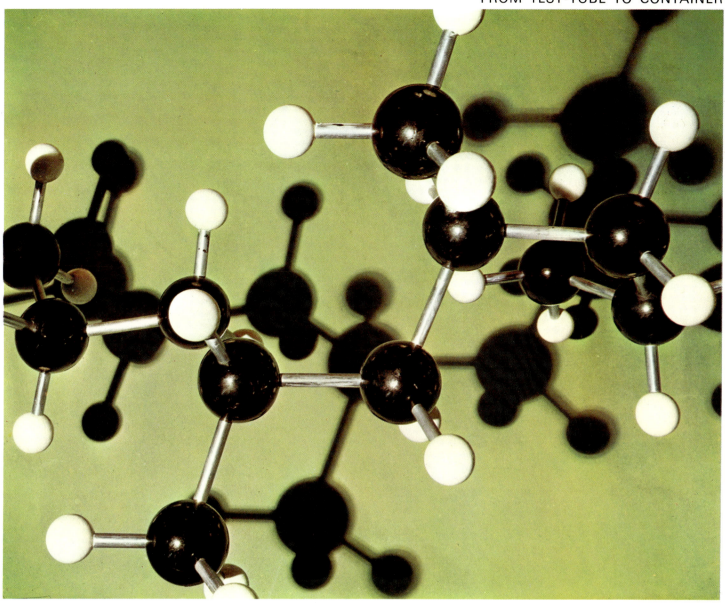

Stereochemistry, which has been mentioned in connection with organic chemistry, is of great practical importance in industrial organic syntheses.
The different spatial arrangement that the various monomers can assume in the formation of a polymer bring about very considerable differences in the mechanical and chemical properties of the polymer. Above: schematic illustration of a section of an isotactic polypropylene chain (moplen).

processes. There are two main reasons for this: firstly to obtain the maximum amount of the most sought-after product (benzine) and at the same time to produce large amounts of olefines from the hydrocarbons extracted from the crude oil: the olefines when subjected to various processes of halogenation, hydration, oxidation, alkylation and polymerization in turn yield the greatest possible variety of chemical compounds.

This is why the various processes of heat- and catalytic cracking have been developed. In the heat-cracking process olefines are formed by breaking the paraffin chemical chain, leading to the formation of an alkane with a shorter chain, and complementary olefines. In addition, many olefines can be provided by the dehydrogenization at high temperature, of the corresponding alkanes.

Because it is based on an ionic process, catalytic cracking is only possible if the feed-load contains olefines, because only olefines undergo the addition of the $H+$ ions with the formation of carbon ions. If the feed load does not contain olefines, it is possible for these to be formed first of all by heat-cracking and then subsequently to undergo attack from the acid-type catalyst, thus giving rise to the chain of reactions peculiar to the catalytic process. The feed-loads for these processes are generally formed by gas oils and the heavier

fractions derived from distillation in a vacuum. The principal products from oil result from alkylation reactions, and are made up of about 60 per cent high octane petrol, twenty per cent light gas oils, butenes, butanes, propylene and light gases: the residue is usually coke (eight per cent). The petrols thus produced then undergo other processes designed to increase their anti-knock qualities: they are subjected to processes known as aromatization and isomerization (reforming), which produce many varied concentrated substances. Other processes are designed to eliminate the sulphur compounds (blending) and the possible residual olefines (hydrogenation) which would tend to polymerize in engines at high temperatures, and give rise to harmful rubbery compounds.

Petrochemistry

The production of large quantities of petrols is not however the only important process in oil-refining. In fact, many (or most) of the products obtainable by distillation are of concern to the petrochemical processing industry, which provides through its various processes, raw materials and finished goods used in every sphere of industrial activity. The lighter paraffins are usually used for the production of olefines, the importance of which has already been

Plastics have found countless uses. Left: a dome made with sheets of Perspex, an acrylic polymer that has proved a fine substitute for glass, with many advantages over it. Top right: seats made of ABS, another polymer that has fired the imagination of designers. Below: a polypropylene colander. Polymers have also found their way into our kitchens. The colours are made with the use of inert organic pigments.

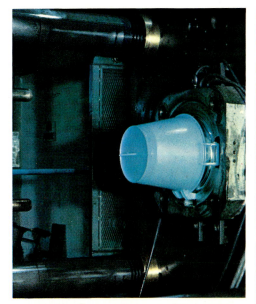

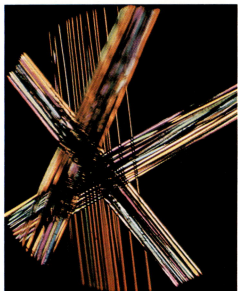

Polymers can be processed in various ways, the most important of which are hot-pressing (top left), extrusion, vacuum moulding, and calendering. Polypropylene (a fibre of which is shown at bottom left) is one of the most widely used polymers for conventional purposes. Right: flexibility testing a strip of polymer).

underlined, whereas the paraffins with more than five carbon atoms are used for oxidation reactions (the production of alcohols, acids, aldehydes etc) and chlorination reactions (to produce chlorinated solvents).

The olefines produced by cracking and dehydrogenation are used as raw material for a vast number of chemical syntheses. All the most important products can be derived from these by means of the reactions just mentioned.

Of these, the processes of polymerization require special attention, given the increasingly far-reaching importance and use of products so obtained.

Acetylene, obtained from methane and ethylene, is the point of departure for the production of very important intermediaries and useful monomers.

The aromatic hydrocarbons are also very important. Of these, benzene is the most exploited, especially for alkylation reactions and subsequent dehydrogenation reactions (for the production of styrene), for phenol synthesis and for dodecylbenzene alkylation, a compound which, after sulphonation, supplies the base product for the preparation of detergents. Naphthenes, which can also be produced from oil, are less important, with the exception of cyclohexane, a solvent, which is sometimes extracted likewise by the dehydrogenation of benzene.

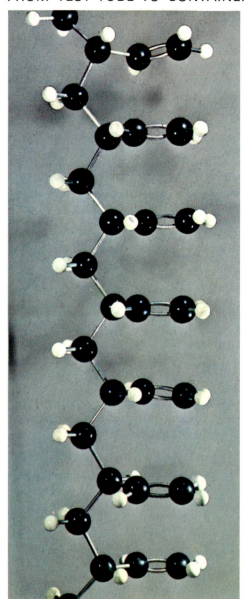
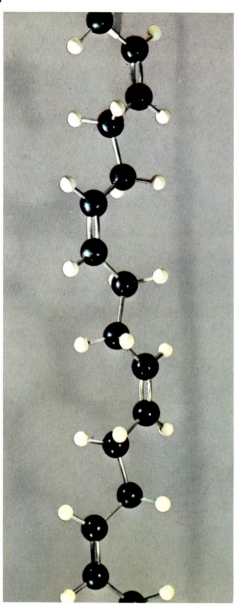
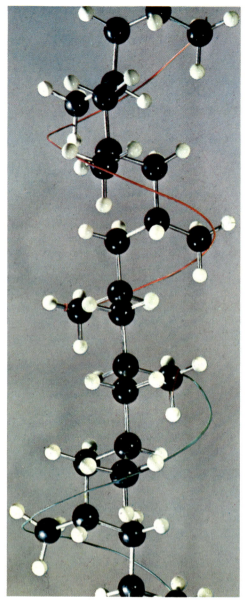

Too much space would be needed to examine in detail the reactions here described; for this reason we shall just list the various transformations undergone by certain basically important products and the reader will be able to gain a rough idea of the vast number of compounds derived from petroleum. – *Methane and light hydrocarbons.* Methane can be transformed into acetylene, and from this it is possible to obtain acetaldehyde, acetic acid, acetic anhydride and acrylic esters. The chlorination of and addition of *HCl* to acetylene bring about the formation of trichloroethylene and vinylidene chloride, which is essential for polymerization; in addition, tetrachloroethane and tetra-chloroethylene. The light hydrocarbons are used to produce olefines such as: ethylene, propylene, butene and butadiene.
– *Ethylene and homologous higher olefines.* Ethylene undergoes a large number of reactions, among which we should mention the addition of water to give ethyl alcohol, and subsequently acetaldehyde, acetic acid and acetic anhydride, unsaturated aldehydes and the higher alcohols. The addition of oxygen gives ethylene oxide from which it is possible to obtain acrylonitrile and acrylic esters (very important for producing fibres, synthetic rubber and plastics); the halogen reaction brings about the formation of various derivatives, most

The growing demand for synthetic rubbers and the possibility of producing them by means of butadiene copolymers have made this monomer one of the most important. The butadiene polymer gives rise to a chain containing double bonds. Left: a small model of a 1.2-sindiotactic polybutadiene chain. Centre: 1.4-cis polybutadiene. Right: another section of an isotactic polypropylene chain.

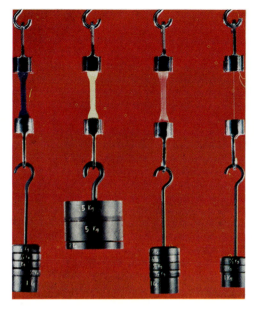

Top left: traction test on strips of moplen of varying thicknesses. Right: stretching a strip of rayon fibre (polyamide). Below: a plant for the production of propylene glycol and homologous higher products, basic monomers for the preparation of polyesters. Right: another detail of the olefine production plant.

important of which is vinylidene chloride, and lastly the addition of benzene gives styrene from which polymers and copolymers of importance are derived.

Propylene undergoes similar treatments and among the products that result are phenol, detergents, isopropyl alcohol and acetone, acrylic esters and glycerin, as well as the products of polymerization. Benzene is almost exclusively converted into styrene and phenol, as well as undergoing other reactions, including alkylation.

Polymers

In this brief review frequent mention has been made of products obtainable by polymerization. The enormous technological importance of these products deserves more attention. Polymerization is a process in which simple molecules (called monomers) unite, giving rise to compounds with a very high molecular weight (polymers). The process may occur in two ways, with or without the elimination of a sub-product. In the first case we talk of polycondensation, in the second of polyaddition. To obtain a polycondensation process it is necessary that the two reagent monomers have at least two functional groups each, e.g. a diacid and a dialcohol. The first step in the polymerization process consists in the reaction between one acid group and one alcohol group to give a

Plastics are supplied by the petrochemical industry in the form of granules. The final processing industries then handle their remelting, dyeing and the manufacture of the widest possible variety of objects (right: plastic granules). Left: collecting the latex from a rubber-tree. The production of natural rubber has now given way to synthetic products, such as the butadiene–styrene polymers, polyisoprene, 1.4-cis polybutadiene and many others.

long-chain ester, leaving two more functional groups free at the ends of this new molecule. To illustrate this, let us give the reaction between ethylene glycol and a bifunctional acid:

$$HOCH_2CH_2OH \ + \ HOOCRCOOH \xrightarrow[-H_2O]{} HOCH_2CH_2OCORCOOH$$

$$\xrightarrow[-H_2O]{+ \ glycol} HOCH_2CH_2O\text{–}OC\text{–}R\text{–}CO\text{–}O\text{–}CH_2CH_2OH$$

At this point there is another attack from another acid molecule, then another glycol molecule, and so on. The chain continues to lengthen and at each stage there is elimination of a sub-product (water). In polyaddition this is not the case, and we find a reaction initiator which gives rise to radicals or ions (radical, cationic, anionic polymerization). Propylene is an example of cationic polymerization with an acid-type initiator, there will be an $H+$ attack on the double bond with the formation of a carbon ion. This will be attacked by a new molecule whose double bond has, in the meantime, polarized, and so forth. Schematically, we shall thus have:

In addition to its high mechanical and chemical resistance, rubber is easy to work with a wide variety of techniques (left: the extrusion process). Top right: view of a rubber factory. Below: an enormous synthetic rubber dome, incorporating dacron, with a diameter of 65 metres (213 feet).

$$CH_3-CH{=}CH_2 \xrightarrow{H^+} CH_3-\underset{\underset{CH_3}{|}}{CH}(+) \xrightarrow[\underset{(+)\ (-)}{CH_3-CH=CH_2}]{} CH_3-\underset{\underset{CH_3}{|}}{CH}-CH_2-\underset{\underset{CH_3}{|}}{CH}(+) \longrightarrow$$

and the process will continue, giving rise to a linear chain. In the first case, because a reaction between alcohol and acid gives an ester, the product obtained will be a polyester; in the second case it will be a polyolefine, or, more precisely, polypropylene. The controlled growth of a polymer is important because very long chains lead to products which are sticky or rubbery. The molecular weight of a macro-molecule (polymer) gives fairly exact indications of the degree of polymerization reached. It is also important to know, when these processes are set up, the way in which the monomers attack each other during the formation of the polymer. In fact, for polypropylene, the attack can be described in the following way:

$$\underset{\underset{(+)}{}}{CH_3}-CH-CH_3 \xrightarrow[\underset{CH_3-CH=CH_2}{}]{(-)\ (+)} CH_3-\underset{\underset{CH_3}{|}}{CH}-\underset{\underset{CH_3}{|}}{CH}-CH_2(+) \longrightarrow$$

  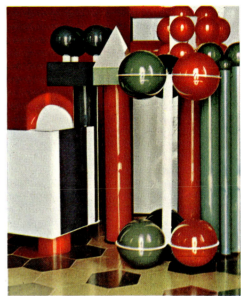

and it will then continue along one path or the other. In addition, in the case of the polyaddition of diene monomers – those, that is, containing two double bonds (and some of them are extremely important), one of the double bonds remains in the polymeric chain, and it is thus possible to obtain cis- or trans-polymers. The macro-molecules in which the monomers are always arranged in the same way are called isotactic, but when they are alternating at each stage, they are called syndiotactic. If the arrangement is random, we have atactic polymers. The isotactic polymers are much the most important because of their superior mechanical and chemical properties and much study has been carried out to control the polymerization processes, to achieve stereo regularity. Research has culminated in the discovery of stereospecific catalysts, which make it possible to achieve this end very well. Molecular weight and stereoregularity of the chains: by keeping these variables under control and by acting at the right moment, it is possible to obtain very wide varieties of characteristics, even from one and the same polymer.

Moreover there are many important copolymers, deriving from the union of monomers capable of polymerizing individually. It is perhaps unnecessary to mention the importance of this sector of industrial chemistry, because it is part

Manufacture and uses of organic dyes; the production of these compounds must meet strict quality standards. Top left: plant for coupling reaction, for the production of azoic dyes. Below, left and top centre: test on the finished product to try out the dye-strength and any possible impurities. Below centre: synthetic dyed fibres. Top right: plastic shapes designed by Bolla. Below: cosmetics.

Petroleum is the dominant ingredient of modern life. Right: switching plant in a large oil pipeline. There is also a use for the very last residue after processing (top left: laying a surface of tarmac, bitumen and asphalt). Below left: modern detergents; likewise derived from petroleum, and designed to clean things more thoroughly: their indiscriminate use may well end up by polluting everything, beyond repair.

of our day-to-day practical experience. It is such that the extent to which polymers are industrially exploited can be taken as an indicator of the technological development of a society. One need only mention plastics, rubber products, fibres, resins, adhesives, etc. All these extremely useful products are formed by a wide variety of polymers; plastics are formed by simple and substitute poly-olefines, acrylic polymers and various copolymers; rubbers by isoprene polymers and butadiene-styrene copolymers, fibres by acrylic, acrylic and phthalic esters, and resins for paints by polyesters, acrylic, vinylic and variously modified polymers.

All these products, which industrial chemistry has made and continues to make available, should contribute to the improvement of our daily well-being.

THE VITAL MOLECULES

The smallest unit of matter which has all the typical properties of life is the cell; in fact all living organisms consist of one or more cells. What, however, do we mean by the term 'living organism'? What are the properties which make a living organism distinct from a non-living organism?

Although, today, there is still discussion about whether certain structures – such as the viruses for example – should be considered as living or not, certain characteristics have nevertheless been singled out which, if all present at the same time, appear peculiar to living organisms. The first of these is the high degree of structural organization found in them. Even the smallest cells, like the bacteria, contain extremely complicated and highly functional internal structures. This is true not only at the microscopic level but also for the various cellular organs, such as the mitochondriae, for example, and also for the various types of molecules present in cells, such as the proteins and nucleic acids.

Another characteristic of living organisms is their chemical composition, which is significantly different from that of the earth and its atmosphere. In fact only sixteen chemical elements have been consistently found in all living organisms, and living cells are largely made up of hydrogen, oxygen, carbon and nitrogen, while the most common chemical elements in the lithosphere (or Earth's crust) are oxygen, silicon, aluminium and sodium.

It is therefore evident that living organisms have, first and foremost, made a selection at the chemical level; this is so because only a few of the hundred or so chemical elements present in the Earth's crust have the property of forming the molecules best suited to carrying on the vital processes within a cell. Within the cells there is a whole range of molecules with increasingly complex structures, the most important of which are macro-molecules, with a molecular weight ranging from about 1000 to about 1000 million, formed by the cells themselves, starting from very simple compounds present in the environment, such as carbon dioxide, water and atmospheric nitrogen.

At this juncture we may well ask ourselves the following question: why do so many different particles co-exist within an organism, be it pluricellular or unicellular? The answer introduces into the discussion another characteristic of living matter. Every part of a cell is, in fact, useful in carrying out one or more biological functions and is, for this very reason, vital if the cell is to be kept in the living state. This applies not only to cellular organs, such as, for example, the chloroplasts in plants, which are responsible for the photosynthetic, or the endoplasmatic reticulum in which the ribosomes appear to be the sites of protein synthesis, but also to the individual biomolecules. The phospholipids, for example, are essential for the construction of the cellular membrane, also known as the plasmatic membrane; the nucleic acids are nowadays recognized to be the molecules responsible for storing and transferring the *genetic code*, those characteristics, in other words, which, taken together, hallmark a particular species; lastly we have the proteins, a direct product of the nucleic acids, which are essential not only for structural upkeep but also for the normal metabolic functioning of the cell. It can, today, be said with certainty that everything present in a given cell, be it animal, vegetable or bacterial, is the fruit of biological evolution, in other words of that

A schematic drawing of the structure of a cell. The various cellular organs (organellae) are as follows: (1) the plasmatic membrane, (2) granules of cellular secretion, (3) micro-bodies, (4) centrioles, (5) mitochondria, (6) the Golgi apparatus. Further towards the centre we can see the nucleus of the cell, surrounded by a porous membrane, containing the genetic material and the nucleolus. At certain points the cellular surface may fold, giving rise to a complex system of vesicles known as the endoplasmatic reticulum, to which the small particles called ribosomes are attached.

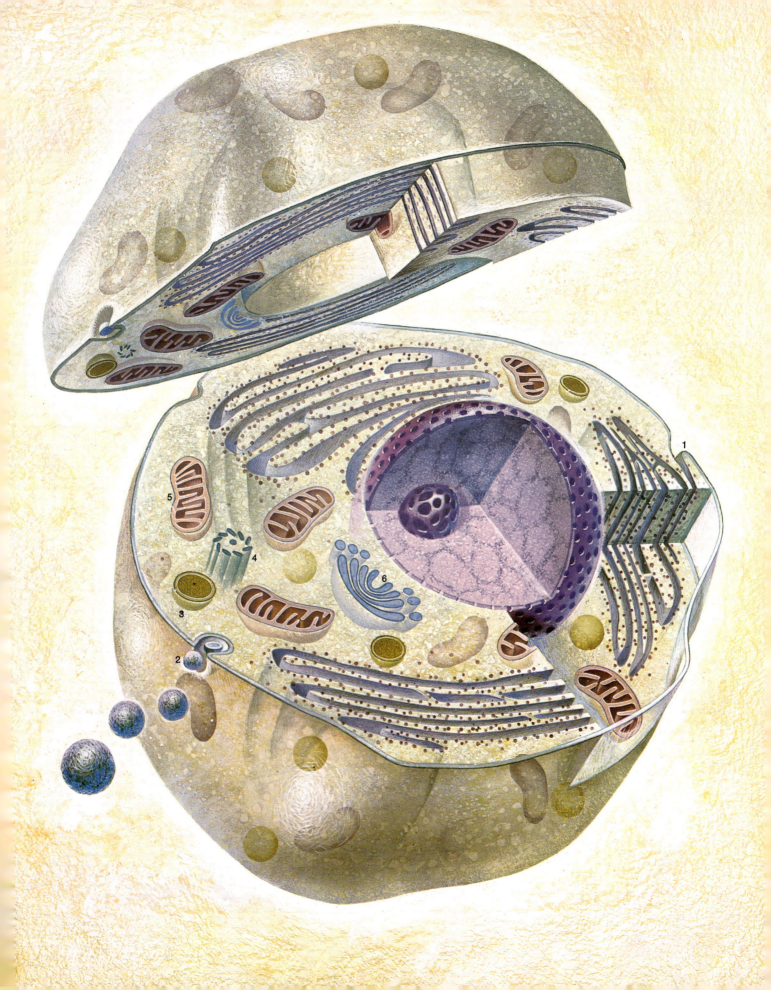

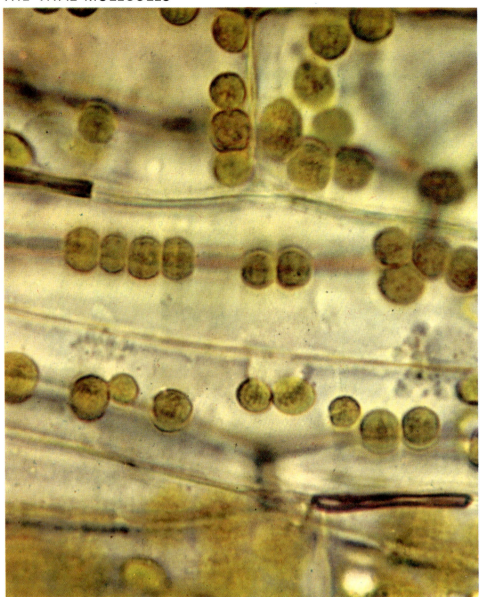

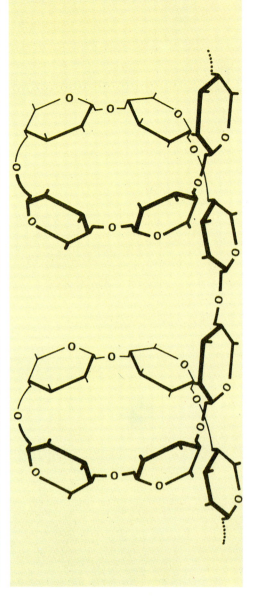

complex of physical and chemical processes which has, over a period of several thousand million years, brought about structures which are constantly updated so as to be able to survive in the specific environmental conditions which have gradually established themselves on our planet.

Whatever they may be, living beings need food in order to survive, whereas pebbles, for example, do not have this problem. From food, cells extract energy which is then used to maintain their highly organized internal structures. Glucose, for example, as present in foods, is transformed inside cells into carbon dioxide and water, in other words, into compounds which are very simple as compared with the initial molecule, and have less energy. The energy released during this process of biological oxidation is conserved for a while in the form of chemical energy contained in the phosphoric bond of the ATP (adenosine triphosphate) molecules. By means of these molecules which are sufficiently small to be able to spread rapidly to all parts of the cell, living organisms can have, at their fingertips as it were, all the energy necessary for carrying out the various types of cellular work, such as, for example, that associated with their movement or with the synthesis of their proteins.

A further property of living organisms is the capacity to reproduce

Left: chloroplasts of a leaf of a higher plant photographed microscopically × 2000 approximately. The biological role of these organs is extremely important: the process of chlorophyllic photosynthesis, with the formation of glucose and the development of oxygen, takes place in them. Right: an amide chain, consisting of a very large number of glucose molecules which are chemically inter-linked. It is in this form that the biosynthesized sugar in the chloroplasts is conserved by the plant.

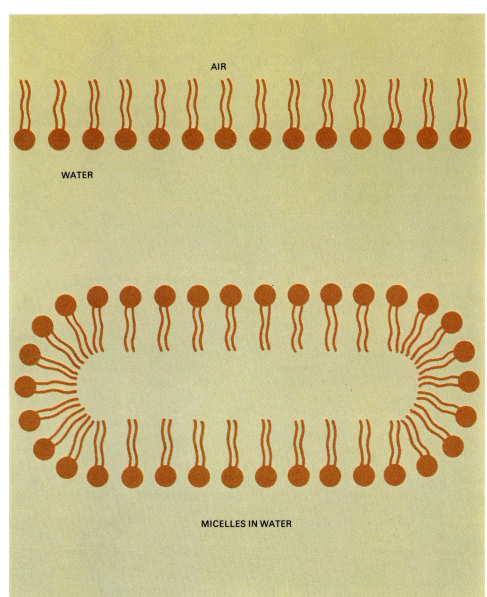

AIR

WATER

MICELLES IN WATER

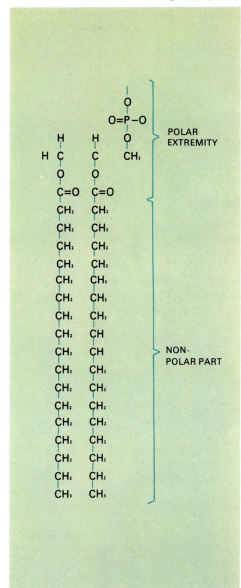

POLAR EXTREMITY

NON-POLAR PART

Above : because of the particular chemical composition the phospholipids tend to be arranged on the water-surface in monomolecular layers, or, alternatively, form micelles which are similar in structure to what we find in the cellular membranes. Right : the chemical formula of a phospholipid, in which we can see the head, formed by glycerin and water-soluble phosphate, and the tail, formed by two long hydrocarbon chains of stearic and oleic acids, which are apolar and thus not water-soluble.

themselves accurately; this means that individuals of a certain species will give birth, invariably, to individuals endowed with very similar characteristics, if not characteristics which are identical to those of their parents. This process of faithful reproduction has been repeated on Earth millions of times for millions of years and undoubtedly represents the fundamental mechanism of life. Today, we have detailed knowledge of the patterns according to which cellular reproduction takes place, but we do not know why it takes place. In other words, we still do not know which structures instruct a fertilized egg-cell not only to start multiplying itself, but also to grow in different ways at a suitable speed, in such a manner that it will have a normal development. This problem is currently the object of a considerable amount of research. But we do know which cellular structures contain the genetic data, i.e. that set of instructions necessary for programming the development of a given individual, starting from the fertilized egg-cell. These structures, called genes, in turn form long DNA molecules (deoxyribonucleic acid) which are normally found in the nucleus of cells, closely associated with nuclear proteins called histones, where they form complex structures known as chromosomes, which can be easily seen under a microscope during the first phase of cellular divisions, called *mitosis*.

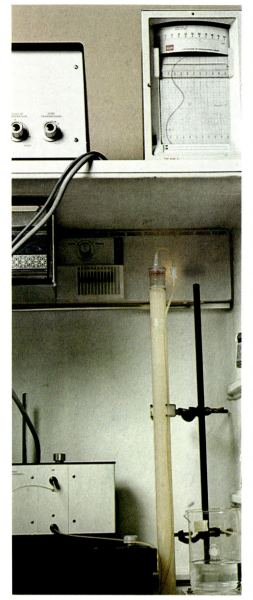

TABLE OF THE AMINO ACIDS FOUND IN PROTEINS

Type and name	Symbol
NON POLAR	
Glycine	Gly
Alanine	Ala
Valine	Val
Leucine	Leu
Isoleucine	Ileu
Proline	Pro
Phenylalanine	Phe
Tryptophan	Trp
Methionine	Met
NEUTRAL POLAR	
Serine	Ser
Threonine	The
Cysteine	Cys
Tyrosine	Tyr
Asparagine	Asn
Glutamine	Gln
ACID POLAR	
Aspartic acid	Asp
Glutamic acid	Glu
BASIC POLAR	
Histidine	His
Arginine	Arg
Lysine	Lys

Let us now describe the most important molecules of cells and their properties. One of the principal components of cells is a substance with which we are all familiar – water. In fact, on average, this represents about 70 per cent of the total weight of a cell, or, given that all organisms are formed by one or more cells, about 70 per cent of the total weight of all living matter.

But it should not be thought that, despite its predominance, water is just something akin to a 'filler', i.e. relatively inert. On the contrary, water is a liquid strongly reactive with almost all organic and inorganic compounds. For example, it is capable of dissolving salts, that is, of separating the various constituent ions of water by surrounding them with its various molecules; in this way it enables these salts to pass through the cellular membrane.

Another reason why living beings have selected water rather than some other liquid such as ammonia, for example, is because it can easily absorb or 'digest' heat without causing major temperature changes, or temperature changes similar to those that other liquids would produce, if the temperature of the surroundings or of the body of which the cell forms part, were to change. This is because water has a very high specific heat, both as a liquid and when changing from liquid to steam. This property of water simplifies the life of the

The amino acids are among the most important molecules of living organisms. Their union in the form of long chains gives rise to the proteins. One of the major problems to be solved is that of their separation from the complex mixture of compounds in which they are normally found. Column (left) or paper (right) chromatography is the simplest and most effective technique of doing so. The components of the mixture 'migrate' at different speeds across the column or paper, being pulled in different ways, and are in the end separated.

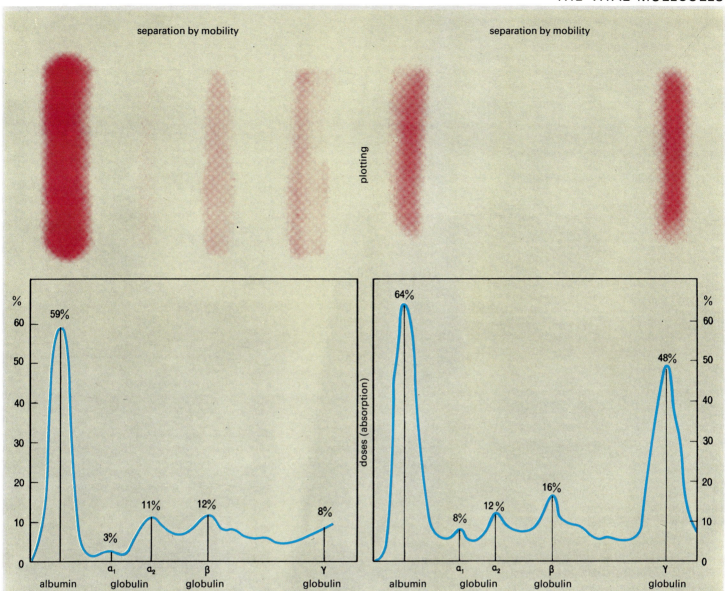

The various proteins contained in human serum can be separated by means of the technique of electrophoresis, based on their different migratory speeds in an electric field. The relative distribution and intensity of the various fractions, obtained after electrophoresis on paper of a sample of normal human serum (left) and a sample of pathological serum (right), makes it possible to study the various components separately.

cell, because reactions which take place within it are affected considerably by fluctuations in temperature. If this becomes too high, the cell can no longer control its growth or upkeep, in so far as the large molecules which govern the metabolic reactions are only stable within a very narrow temperature range. Outside this range they lose their capacity to function properly.

Furthermore, water is directly involved in many other vital processes; the cells of plants owe their growth, in the main, to a process of osmosis, i.e. to the absorption of water, which creeps up the plant from the roots. This can occur because the concentration of salts and other chemical compounds in the plant cell is greater than the concentration in the surrounding environment; and, because water can pass freely through the cell walls, it tends to move inside the cells to reach this heavier concentration of salts. This leads to an increase in the size of the cells which then reach a critical size, at which they divide; and the plant thus grows. Of course, water alone is not enough to turn a seedling into a tree. In reality the process is more complicated, and what has been described is only one aspect, but it is still a vital one.

The other substances in the cells which carry out the most important vital functions can be grouped into four major categories: *carbohydrates, lipids,*

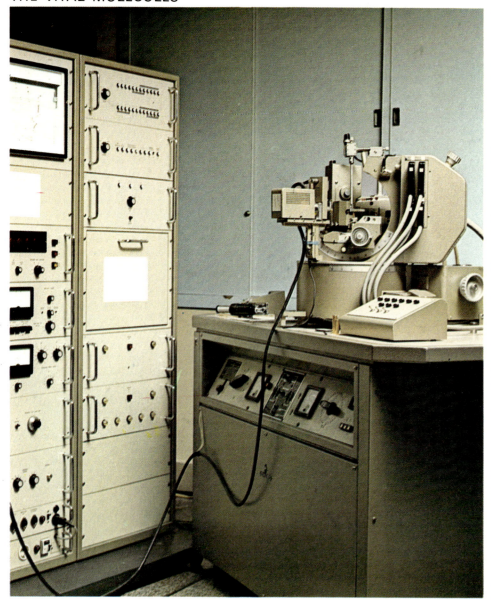

proteins and *nucleic acids*. Each of these categories embraces a vast number of different molecules, endowed with particular chemical and biological properties. Despite this huge variety, chemical analysis of these biomolecules has revealed the presence in them of just six elements: carbon, hydrogen, oxygen, nitrogen, sulphur and phosphorus.

The carbohydrates, so-called because their empirical chemical formula is $(CH_2O)_n$, where n may assume values equal to, or greater than, three, are commonly known as sugars. The simplest compounds in this group are solid, white in colour, very water-soluble and sweet-smelling. After water, the carbohydrates are the compounds found in the greatest abundance in the biosphere (or ecosphere) and they are used by cells principally as 'fuel' because cells derive from them most of the energy necessary for the various cellular processes. They also play the part of structural components of the cell-walls of bacteria, plants, and animal cellular membranes.

The most plentiful sugar found in nature is glucose, containing six carbon atoms, from which it is thought that many other sugars present in cells are derived. During the day, plants use the light-energy coming from the sun to transform carbon dioxide and water into glucose, which they then retain, and

Diffractometry with X-rays is, today, the most accurate technique for the determination of the structure of molecules in the crystalline state. The X-rays which strike the sample are diffracted by it in various directions and with various intensities. Left: a general view of a Siemens automatic diffractometer. Top right: the various concentric guides by which it is possible to make the crystalline sample assume every possible orientation during radiation. Below: a detail of the place on which the crystalline sample to be analyzed is mounted.

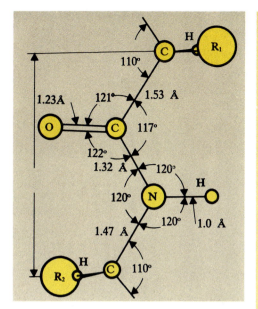

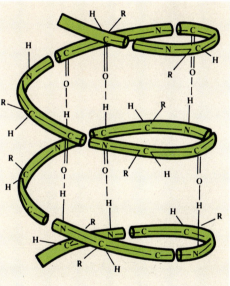

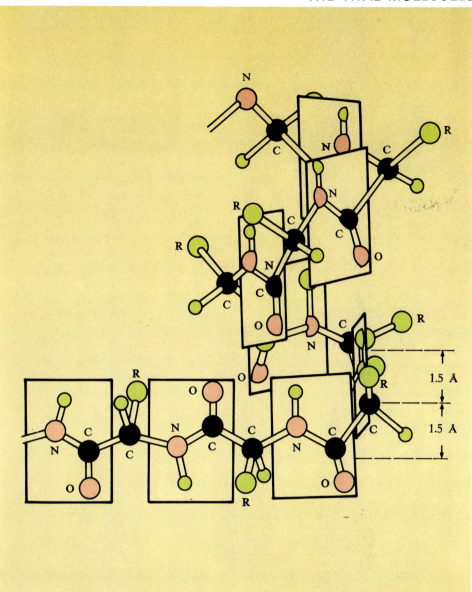

Top left: the structure of a peptide linkage, obtained by the X-ray analysis of simple crystalline peptides. Below: a schematic drawing of the skeleton of a polypeptide chain in the form of a helix. Right: a more detailed drawing of a polypeptide chain in a partly helical form, and partly in the form of an extended plane. The planar peptide linkages are inside the rectangular figures.

into oxygen, which they return to the atmosphere. This process, known as photosynthesis, takes place in specific cell organs called chloroplasts, which are present in leaf-cells. Given that the products of photosynthesis – glucose and oxygen – are in turn used by animals and plants themselves, and converted once again into carbon dioxide and water, it is quite legitimate to conclude that all life on this planet relies, in practical terms, on the flow of solar rays which reaches the Earth each and every day. This cycle of reactions could not, however, last very long if, in addition to plants and animals, there were no micro-organisms, called bacteria. It is in fact these bacteria which ensure the recycling of the organic waste products of animals and plants, and thus carry out an extremely important task in the survival of living forms on Earth. micro-organisms, called bacteria. It is in fact these bacteria which ensure the

A large number of simple sugar molecules can unite with each other, thus giving rise to compounds with a high molecular weight, known as polysaccharides. These represent the form in which the chemical energy contained in the glucose molecule and in the molecules of other simple sugars is conserved in living organisms. In fact, in plants we find amide present, and in animals glycogen, two polysaccharides with a protective function, known as

the cellulose function.

Unlike the carbohydrates, the lipids are not very water-soluble. This property derives from the fact that most of their chemical bonds are non-polar, being almost exclusively formed by carbon and hydrogen.

There is a low percentage of oxygen present in the lipids in the form of ester bonds, whereas in the carbohydrates it exceeds the 50 per cent mark and is mainly present in the form of hydroxyl, a functional group very akin to water. Because of the high content of extremely energetic hydrocarbon bonds, the lipids release, by means of biological oxidation, large amounts of energy which is used by the cell in a similar way to the energy derived by the oxidation of carbohydrates. Living organisms in effect, under normal circumstances, hold large quantities of lipids by way of a reserve. There are two main classes of lipids: the fats and the phospholipids. The former, also known as triglycerides, are formed by the union of a glycerine molecule with three organic acid molecules, such as stearic acid, oleic acid or palmitic acid. Examples of vegetables and animal fats are, respectively, olive oil and lard. The phospholipids are likewise formed from a glycerine molecule, but this time it is esterified with two fatty acid molecules and one phosphoric acid molecule.

The structure and function of certain enzymes. Top left: the atomic model of ribonuclease, an enzyme which breaks down the molecules of ribonucleic acid (RNA). Top centre: the active site of the ribonuclease enzyme. Below (left and centre): a stereographic model of lysozyme, an enzyme which attacks the cell walls of bacteria and disintegrates them. Top right: a drop of liquid containing catalase-rich white corpuscles; catalase is an enzyme which decomposes oxygenated water into water and oxygen. Below: the addition of a drop of oxygenated water causes the development of oxygen.

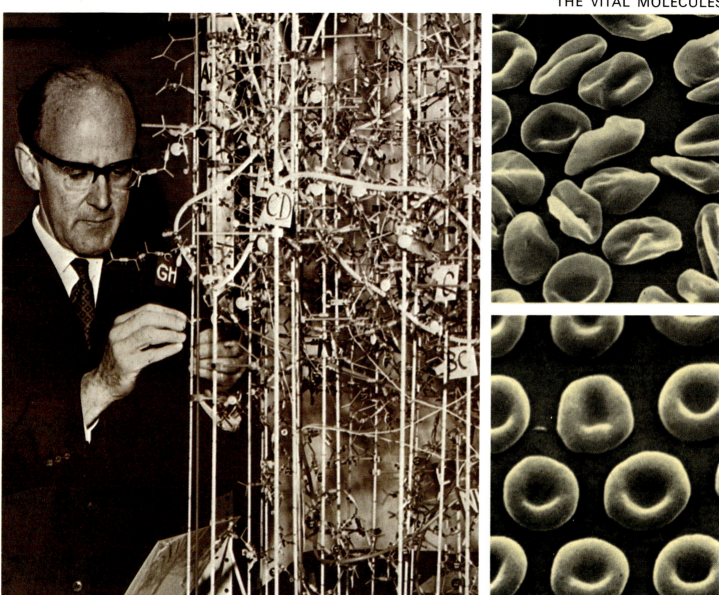

The molecular biologist Max Perutz, photographed constructing an atomic model of haemoglobin. In 1962 he was awarded the Nobel prize for his important discoveries about the three-dimensional structure of proteins, obtained by means of X-ray crystallography. Right: microphotographs of sickle-shaped (top) and normal (below) red corpuscles. Sickle-cell anaemia is a pathological genetic-type alteration which makes the transport of oxygen from the lungs to the tissues inefficient and is due to the presence of abnormal haemoglobin molecules.

Given the presence of the ionizable phosphate group, their behaviour in water is of particular biological interest. In fact they tend to unite in water by forming structures similar to those which can be observed in the cellular membranes with their phosphoric part polar, turned towards the water molecule, and the remaining part non-polar, formed by the long hydrocarbon chains of the fatty acids, well away from the solvent and inside these structures. The phospholipids have been found to be the structural components of the membranes, associated with proteins, which are called, specifically, membrane proteins, and not soluble in distilled water. It is largely for this reason that most of the research currently under way is directed towards deeper and deeper investigation of the chemical and physical properties of the lipids.

The last two categories to be looked at are the proteins and the nucleic acids. These molecules have in fact attracted much scientific attention to date, and rightly so, because they are the most interesting. The proteins are very plentiful in the cells, accounting for some 15 per cent of their weight. Chemically speaking they are polymers, i.e. long chains of amino acids inter-linked by covalent bonds. The average elementary composition of the proteins is 50 per cent carbon, about 25 per cent oxygen, about 15 per cent nitrogen, about 7 per

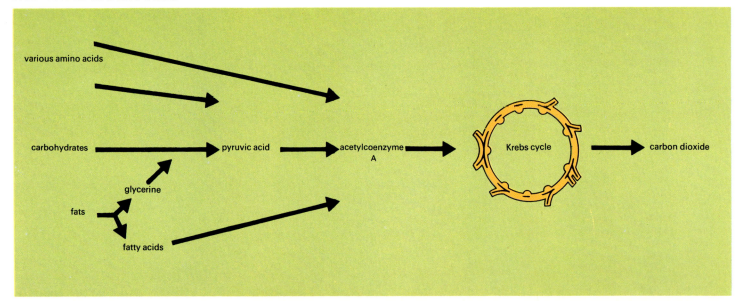

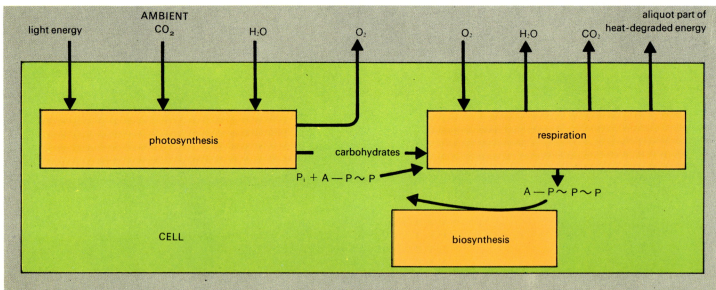

cent hydrogen, and 0–3 per cent sulphur. Comparing this qualitative composition with that of the carbohydrates and the lipids, we find a new light element: nitrogen. This derives from the molecular nitrogen present in the atmosphere. It is transformed by certain micro-organisms into nitrites and nitrates, forms in which it can be absorbed by plants and used for the synthesis of amino acids. The amino acids are fairly simple organic compounds; in fact their general formula is: $NH_2–CH–COOH$. The structure of the radical R, also

$$| \atop R$$

known as the lateral chain, varies with the variations of the amino acid in question. For example, in glycine, R is simply a hydrogen H atom, in alanine it is a methyl group CH_3 and in cysteine it is group containing sulphur CH_2SH. While some chemico-physical properties of a protein, such as its molecular weight or its water-solubility, depend essentially on the number and type of amino acids which make it up, other properties, such as for example its form and biological functionality, depend on the way in which the various constituent amino acids are arranged. The determination of the three-dimensional structure of proteins is a very taxing undertaking, and has been

The citric acid cycle or Krebs cycle, named after the man who discovered it, is present in the metabolism of all the aerobic cells. It consists of a sequence of cyclic reactions catalyzed by appropriate enzymes situated in the mitochondria, and allows a complete utilization of the energy contained within the nutritive molecules of sugars, amino acids and lipids, by being coupled with the conveyance of electrons and oxidative phosphorylation, from which adenosine triphosphate or ATP is formed. Below: a diagram of the energy cycle in living organisms.

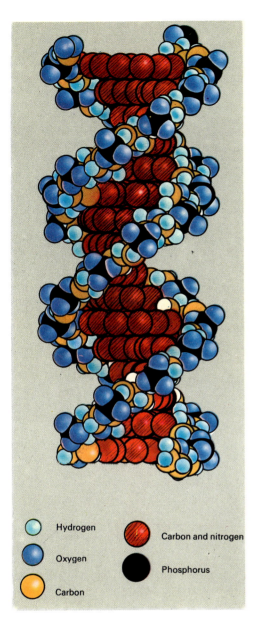

<table>
<tr><td>Hydrogen</td></tr>
<tr><td>Oxygen</td></tr>
<tr><td>Carbon</td></tr>
<tr><td>Carbon and nitrogen</td></tr>
<tr><td>Phosphorus</td></tr>
</table>

A molecular model of the structure of DNA or deoxyribonucleic acid. The two lateral chains are wrapped like a double helix and held together by crosswise hydrogen bonds. DNA is the molecule that transmits genetic information.

worked out only for a few proteins because these are very large and complicated molecules. Today, luckily, and thanks above all to the development of analytic techniques such as chromatography, electrophoresis and the ultracentrifuge, it is possible to separate a protein from the thousands of other substances with which it is normally mixed up within the cell, and thus to study its principal properties.

Chemical analysis has shown that in the proteins the amino acids are inter-linked always in the same way, with a single type of bond known as the 'peptide linkage', which is formed between the carboxyl group of an amino acid and the amino-group of the next amino acid, with the elimination of water:

$$NH_2-CH-COOH + NH_2-CH-COOH \xrightarrow{-H_2O} NH_2-CH-CO-NH-CH-COOH$$
$$\quad\ \ |\qquad\qquad\quad\ \ |\qquad\qquad\qquad\qquad\qquad\ |\qquad\qquad\ |$$
$$\quad\ \ R'\qquad\qquad\quad R''\qquad\qquad\qquad\qquad\qquad R'\qquad\quad R''$$

By means of X-ray analysis, it has also been determined further that the geometric resistance of the atoms nearest the peptide linkage ($C-CO-NH-C$) is planar. The most important consequence of this planarity lies in the fact that a protein can be studied as if it were a long chain made of lots of small planar sections, wrapped in various ways around the carbon atoms which are directly linked with the lateral chains of the various constituent amino acids. There are only twenty different amino acids found in the proteins. This is surprising at first glance, given that there are literally millions of different proteins in nature. But in reality these twenty different structures are more than enough to produce so many proteins, given that each polypeptide chain contains a very large number of amino acid residues, usually more than a hundred, for which reason every single amino acid is normally present more than once and can even occupy different positions along the chain.

In the proteins there are various levels of structural organization, of increasing complexity. The first level has to do with the order in which the various amino acids are inter-linked, i.e. the sequence in which they follow one another in the various polypeptide chains. The second level has to do with specific spatial arrangements taken on in some cases by long, in other cases not so long, sections of a chain, resembling a helix or leaves folded like a concertina. The third level of organization is the three-dimensional form that the polypeptide chains may assume, capable as they are of varying between a more or less spherical to a very elongated form. The fourth and final level only applies when two or more polypeptide chains enter into contact and take on a more stable configuration, thus giving rise to a protein formed by two or more chains united by low-energy non-covalent bonds, like, for example, hydrogen bonds or bonds of the electrostatic type.

It is only in fairly narrow pH and temperature ranges, approaching the respective physiological ranges, that the proteins will assume a well-defined spatial form known as the *native form*, and become biologically active. Beyond these ranges they undergo processes of structural disintegration (flaking) accompanied by a loss of functional activity. These processes then lead to less compact and more disorderly spatial forms, known as *denatured forms*. Thus the proteins are fragile molecules and their spatial form is determined by their function. They may be classified in different ways, the most significant of which is that based on their biological function.

In the higher animals the most abundant protein is collagen, situated in the fibrous connective tissues, or in the tendons, bones and cartilages. Another protein with structural functions is α-keratin, which forms wool, hair, and the horns and hooves of animals. Actin and myosin form the filaments of the muscular fibrils and are at the basis of muscular contraction. Among the so-called 'carrier' proteins we find haemoglobin, present in the red blood corpuscles of vertebrates, which conveys oxygen from the lungs to the tissues, and serum albumen, which carries the fatty acids of the adipose tissue into the circulation. The hormones usually have a fairly low molecular weight compared with the proteins, but they are also polypeptide in nature. They are

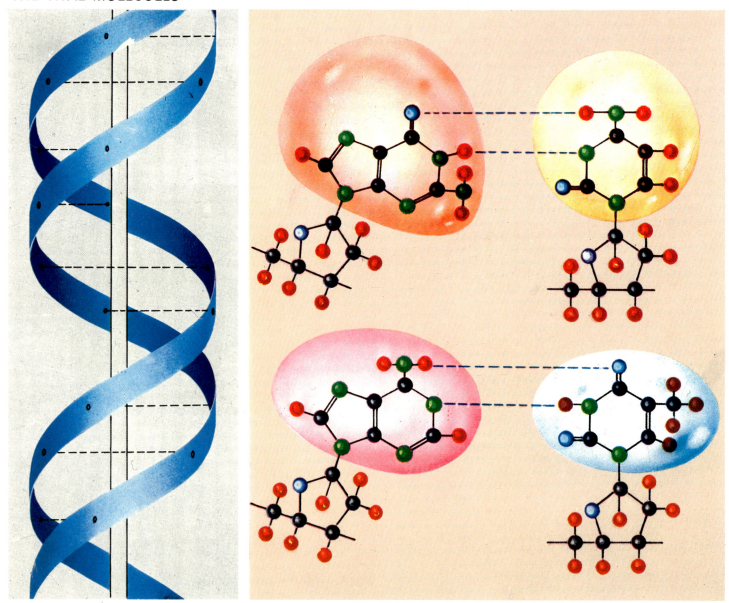

secreted in the endocrine glands, such as the hypophysis, the pancreas or the thyroid gland, and have regulatory functions. For example: the hormone secreted by the hypophysis controls bone growth, while the insulin secreted by the pancreas governs the amount of glucose in the blood; a shortage of insulin gives rise to diabetes mellitus.

Among the protective proteins in vertebrates, a very important class is that of the antibodies or gamma globulin. These molecules can block, by binding together, any alien substance (the antigen) introduced into the blood, thus neutralizing the possibility of disease. The reaction between antigen and antibody is called the 'immunity response'.

Lastly, the enzymes, which are a proteic type of molecule, constitute the cellular machinery by means of which simple chemical compounds are transformed into macro-molecules and vice versa, thus enabling vital chemical reactions to take place – vital, that is, for the normal processes of cellular growth. These have the property of recognizing the substances on which they can act chemically by their form, and greatly accelerating their metabolic reactions.

Let us now turn our attention to the nucleic acids and the way they function

Left: diagram of the double DNA helix wrapped around an ideal axis. Right: four nitrogenous bases present in DNA inter-linked in twos by hydrogen bonds. Top: the guanine–cytosine pair; below: the adenine–thymine pair. Colour key: black – carbon; blue – oxygen; red – hydrogen; green – nitrogen.

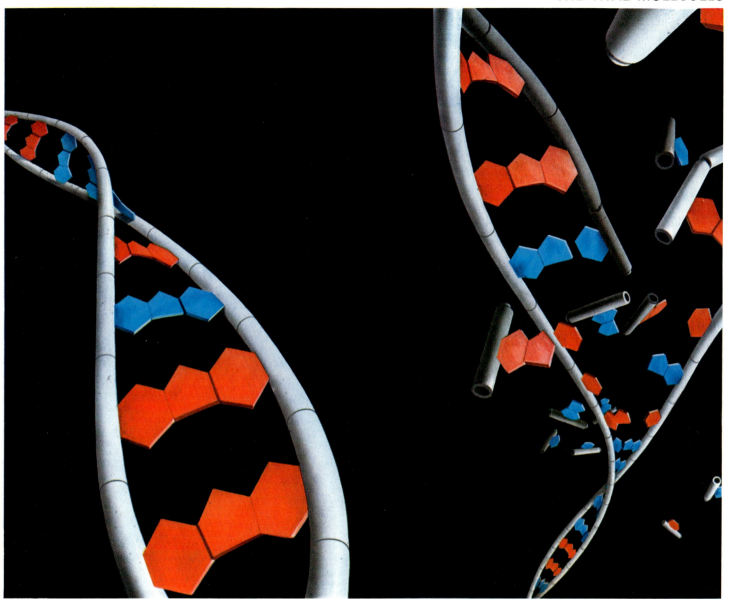

Left: the double DNA helix. Right: the double helix opens, the filaments separate, and each one binds to itself the nitrogenous bases present in the atmosphere, which are already linked to sugars and phosphates, in order to duplicate itself. Thus two double helices are born from just one.

as genetic material. Most of our knowledge about their chemical and biological properties derives from studies carried out on very small organisms such as viruses and bacteria, which are thus much easier to investigate in the laboratory. Not everything that has been discovered about them genetically can be applied to higher organisms, but many of the fundamental processes relating, for example to the inheritability of characteristics, are today taken as universally valid. We shall only discuss these general aspects here. The discovery of the nucleic acids was made more than a century ago by a German chemist, Miescher, who, to begin with, isolated them from pus cells and subsequently, in a very pure form, from salmon sperm. The substance purified by him was white, had acid properties, and contained phosphorus as well as carbon, hydrogen, oxygen and nitrogen. It was called nucleic acid, and more exactly deoxyribonucleic acid (DNA), because it was extracted from the cellular nucleus. Despite the fact that it was then studied in depth chemically, its colossal importance in biological terms was not understood for another 75 years. It was in fact not until 1943 that a handful of researchers showed that DNA was the *transforming factor*, i.e. the factor capable of transforming over generations a strain of harmless pneumococci into virulent forms. This

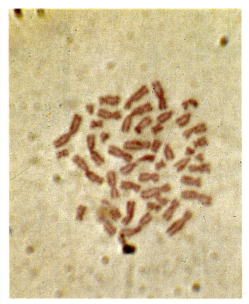

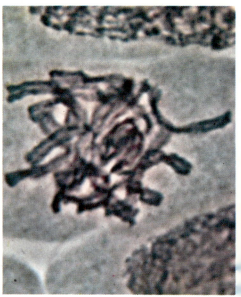

discovery was not instantly welcomed in scientific circles, because too much weight was still attached to the opinion that proteins were more suited as molecules to transmitting hereditary characters.

Nevertheless the alarm bell had been sounded, and it was not long before additional important evidence in favour of DNA as genetic material emerged. Of these, the experiments made by Hershey and Chase in 1952 were the most important. These two scientists used a material that was even simpler and less valued than that previously used: bacterial viruses.

Once these viruses penetrate inside the bacterial cells, they reproduce very swiftly and in large numbers, using to this end the material and machinery of the host-cell, and eventually cause the death of the host. By using the radio-active isotopes of sulphur and phosphorus, the scientists observed that part of the new-formed viruses contained DNA which had already been contained in the original viruses, whereas the protein covering was completely new. As a result they were able to conclude that only the viral DNA had penetrated the bacteria and been used for the reproduction of the viruses.

At that time, and we are talking of just over 25 years ago, it was known that all the cells of a given organism contain the same amount of DNA, except the

Left: a Siemens electronic microscope, by means of which it is possible to observe the shape and dimensions of the very smallest cellular particles. Top right: male cell human chromosomes. Below right: chromosomes of an onion-root cell dividing. The chromosomes present in the cell nucleus contain the genes, which are parts of the DNA molecule, transmitting the hereditary characters of individuals from generation to generation.

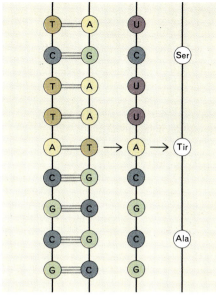

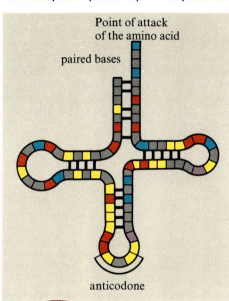

Point of attack
of the amino acid

paired bases

anticodone

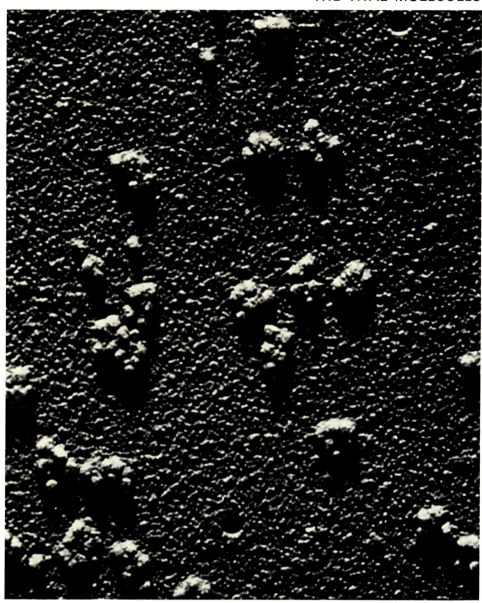

Top left: the flow of genetic information from the DNA to the proteins. The carrier of this information is a molecule of ribonucleic acid or carrier RNA, which forms in the nucleus as a complementary copy of the DNA and then migrates to the cytoplasma where protein synthesis takes place. Below left: an RNA transfer molecule, which is designed to convey amino acids to the ribosomes, where, by means of a complex mechanism, protein synthesis takes place. Right: a microphotograph of polyribosomal complexes during the synthesis of haemoglobin molecules.

gametes (germ-cells) which contain half, but that this did not apply to the proteins. It was also known that the DNA inside each species remained constant, and varied only as the species varied, in terms of chemical quantity and composition.

DNA molecules are very long filaments, formed by a large number of like units called nucleotides. Each nucleotide consists of three chemically distinct parts: a nitrogenous organic base, a sugar and a phosphate group. The sugar present in DNA is deoxyribose (or, more precisely 2-deoxy-D-ribose). There are four nitrogenous bases: adenine, thymine, cytosine and guanine, indicated with the symbols A, T, C, and G respectively. In 1953, on the basis of this and other knowledge, and of the examination of the diffraction of X-rays by a crystalline sample of DNA, Watson and Crick came up with a simple and at the same time brilliant model for the three-dimensional structure of DNA, which caught the attention of scientists. This model made it possible to explain not only the biological significance of a large number of fairly fragmentary data, which had been piling up, but also to tackle the problems of genetics in molecular terms. Watson and Crick proposed a 'double helix' structure for the DNA molecule, in which two DNA filaments are wrapped around each other,

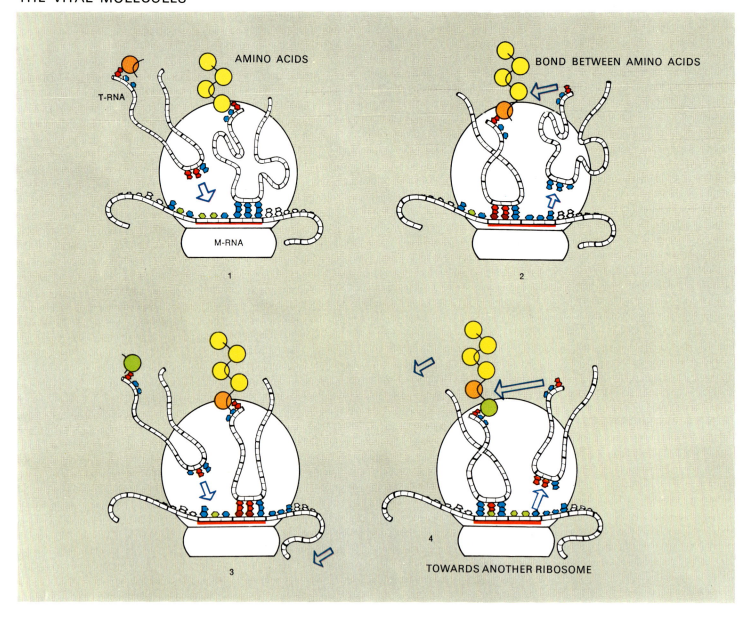

or. the basis of a principle of structural complementarity, whereby A is paired only with T, and G only with C.

A spatial configuration of the bases made in this way agreed satisfactorily with the very important and already known fact that the number of adenine residues found in various samples of various types of DNA equalled the number of thymine residues, with the same applying to the guanine and cytosine residues. Another advantage offered by a model of this type was that it managed to explain, simply, how DNA could be duplicated and then transmit the genetic information contained in it to daughter cells. In fact it was enough to admit that the double helix could open out like a hinge, following the breaking of the hydrogen bonds which keep the bases paired, and that each filament, using the free nucleotides present in the cell and appropriate enzymatic molecules, could form the other complementary filament. In this way there is a passage from one DNA molecule to two. As genetic material, and in addition to supplying an identical copy of itself to the future generation of cells, DNA has the task of supplying to the cell in which it is situated all the instructions necessary for the construction of the proteins. The vital efficiency of a cell, or of a whole organism made up of many cells, in fact depends on the

Protein synthesis occurs in the ribosomes. The RNA transfer molecules, to which the various amino acids are linked, correspond with the carrier RNA already present in the ribosome. Subsequently, appropriate enzymes link up the various amino acids, giving rise to the process of protein synthesis. The specificity of the process is ensured by the complementary pairing between a trio of RNA transfer bases (anticodone) and a trio of RNA messenger bases (codone).

GENETIC CODE*

2 → 1 ↓	U	C	A	G	3 ↓
U	UUU Phe UUC Phe UUA Leu UUG Leu	UCU Ser UCC Ser UCA Ser UCG Ser	UAU Ty UAC Ty UAA Ochre (NA) UAG Amber (NA)	UGU Cys UGC Cys UGA NA UGG Trp	U C A G
C	CUU Leu CUC Leu CUA Leu CUG Leu	CCU Pro CCC Pro CCA Pro CCG Pro	CAU His CAC His CAA GluNH$_2$ CAG GluNH$_2$	CGU Arg CGC Arg CGA Arg CGG Arg	U C A G
A	AUU Ileu AUC Ileu AUA Ileu AUG Met	ACU The ACC The ACA The ACG The	AAU AspNH$_2$ AAC AspNH$_2$ AAA Lys AAG Lys	AGU Ser AGC Ser AGA Arg AGG Arg	U C A G
G	GUU Val GUC Val GUA Val GUG Val	GCU Ala GCC Ala GCA Ala GCG Ala	GAU Asp GAC Asp GAA Glu GAG Glu	GGU Gly GGC Gly GGA Gly GGG Gly	U C A G

Ala	alanine	Gly	glycine	Pro	proline
AspNH$_2$	asparagine	His	histidine	Ser	serine
Asp	aspartic acid	Ileu	isoleucine	The	threonine
Arg	arginine	Leu	leucine	Ty	tyrosine
Cys	cysteine	Lys	lysine	Trp	tryptophan
Glu	glutamic acid	Met	methionine	Val	valine
GluNH$_2$	glutamine	Phe	phenylalanine	NA	non-applicable

*The table is read as follows: the symbols U, C, A, G which appear in the first and last columns and at the top of the table, correspond, respectively, to the bases, uracil, cytosine, adenine, guanine; in each box, beside the amino acid symbol, the trio which codifies it is formed by the symbols of the bases which appear subsequently in the first column, at the top of the table, and in the last column.

The genetic code explains how the four-letter language of the nucleic acids (the four nitrogenous bases) is translated into the twenty-letter language of the proteins (the twenty amino acids). A given amino acid corresponds with every sequence of three nucleic bases; given that the different possible trios are 64 in number, each one of the twenty amino acids is as a rule codified by more than one trio. Three trios do not codify any amino acid and serve only to interrrupt the reading of the genetic message. The genetic code was explained in the 1960s and certainly represents one of the major feats of science

type and number of proteins which it contains. All the cellular processes, from the conveyance of molecules or ions through the membrane, to immunity responses, and from the use of the chemical energy contained in nutritive matter to the biosynthesis of macro-molecules, require proteins, and these, in most cases, must be equipped with catalytic activities, i.e. enzymes. DNA thus represents the language used by nature to conserve life; in fact DNA is life itself in the form of a code; as physiologically active molecules, the proteins execute the message contained in the code.

The genetic information, that is, the messages containing the instructions for the synthesis of the proteins, is enclosed in DNA segments, called genes, each one of which is in its turn formed by a sequence of hundreds, and in some instances thousands, of nucleotide residues. In other words, DNA is the 'book of life', and the genes are the sentences in this book, with each sentence discussing a different protein. The DNA present in the chromosome of a bacterium is formed by a single double helical molecule containing several thousand genes. In the cells of higher organisms each chromosome contains many double helical DNA molecules, each one containing hundreds or thousands of genes. The DNA molecules of a human cell, stretched full-length,

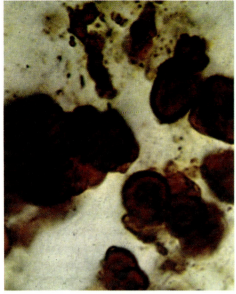

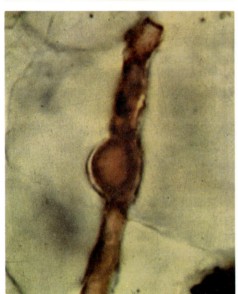

would measure an incredible 1.50 metres (5 feet); the number of nucleic bases contained in it is approximately 10,000 million and the quantity of genetic information contained in these is equivalent to a library of 2000 300-page books.

In conclusion, we shall mention a very topical problem: the origin of life on Earth. According to current opinions, the first forms of life made their appearance on our planet about 3500 million years ago. At the time, the Earth, which was cooling down, already had an atmosphere, albeit rarefied, consisting mainly of simple gases such as methane and ammonia, as well as steam. It is interesting to note that the elements present in these substances – i.e. C, H, N, O – are those which, alone, account for 98 per cent of living matter. The energy coming from the sun in the form of rays and the effect of other physical agents, such as electric discharges during storms and volcanic eruptions, brought about the fission of these simple molecules and the formation of more complicated structures, such as, for example, the amino acids and the nucleic bases. These new substances were then carried by rain into the sea, where they are thought to have organized themselves inside lipidic droplets. These structures then gradually evolved until they formed the first

Left: the remains of the simplest known forms of life to have appeared on Earth; (top) the fossil structures discovered in the Fig Tree in Africa; (below) the fossil of an organism which lived more than 2000 million years ago. Right: the fossil imprint of a jellyfish, with a diameter of 12 cm. (5 in.) found in Australia, which dates back 600 or 700 million years.

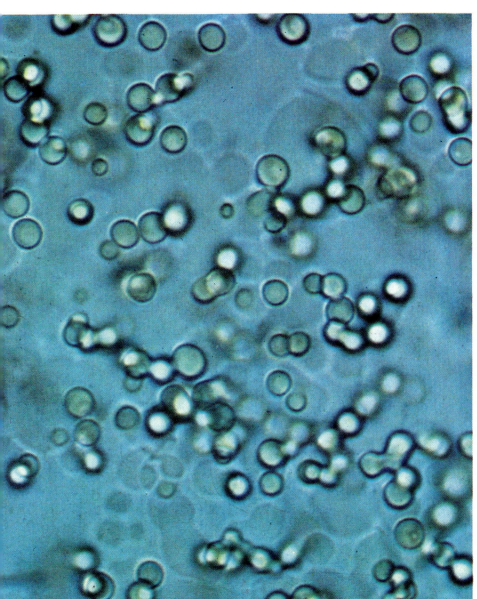

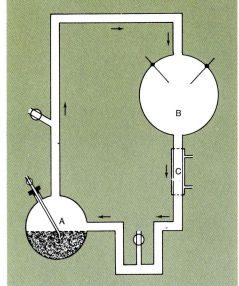

The origin of life on Earth. The first scientist to tackle this problem in scientic terms was the Russian biochemist Oparin (below right) in about 1930. Top right: later experiments, like those carried out by Miller, involve creating a contact between the vapours coming from the bowl A and H_2, CH_4, NH_3 or H_2O present in B and subjecting them to strong electric charges. The organic substances so formed are condensed in C. Left: micro-particles of proteinoids, i.e. of poly-amino acids obtained by heating up a mixture of amino acids.

rudimentary and primaeval cells. These cells grew, using the various organic compounds dissolved in the marine environment, and deriving chemical energy and raw material from them. But because of the slow but sure reduction of the organic materials scattered across the oceans, only the most developed cells, i.e. those equipped with more efficient enzymatic systems to exploit the energy, were able to survive. The most successful organisms were those which managed directly to transform solar energy into chemical energy. They managed, in other words, to synthesize organic compounds such as glucose, starting from the carbon dioxide and water present in large quantities in the atmosphere, by means of the process of photosynthesis. The importance of this process is such that most forms of life currently existing on Earth rely on it.

SPACE

ASTRONOMY

Astronomy has always been one of the most intriguing branches of science and also, there can be no doubt, one of the most popular. In fact it is within everyone's reach to observe the sky and even without instruments it is possible to discover interesting laws for oneself, such as the variation of the height of the Sun, the length of the day in the various seasons, the phases of the Moon, and the movement of the planets in relation to the fixed stars. For these reasons the science of the sky has always been a leading science, and there is not a single culture or civilization in either the New or the Old World which has not given it a great deal of attention. It is also the science that numbers among its practitioners the greatest number of amateurs (a better term might be 'astrophiles', or 'stargazers'), whose consistent contributions to knowledge have been far from negligible (one could say that the sky is so large that there is room for everyone there). With no more than the naked eye it is possible to single out some 2000 stars and see the Milky Way; and with just a modest telescope the number of visible stars increases enormously.

We can divide the history of astronomy into three major periods: the first period, which we shall call the *astronomy of observation*, extended from antiquity to Newton, and its dominant characteristic was the observation and description of the movements of the stars and the planets. Using language drawn from physics, this can be defined as *cinematic astronomy*, and its culmination was the work of Copernicus and Kepler. The second period, which we shall again define with the terminology of physics as *dynamic astronomy*, runs from Newton to the birth of astrophysics and had as its primary aim the study of the application of the universal law of gravity to the heavenly bodies. The third and last period runs from the birth of astrophysics, to the present day. It began with the study of the physical characteristics of the heavenly bodies through the use of the spectroscope by Father Secchi in about the middle of the last century. Each one of these three aspects of astronomy is still very much alive (even though astrophysics is now obviously the crucial discipline).

Observations of the sky continue with increasingly powerful instruments of the most varied types: not only by optical, reflecting or refracting telescopes, but also radio-telescopes, special infra-red telescopes, X-ray and gamma ray telescopes, instruments mounted on meteorological balloons to reduce the screening action of the atmosphere, and instruments even mounted on man-made satellites.

As far as *dynamic astronomy* is concerned, this is consistently applied not only to calculate the movement of the 'natural' heavenly bodies, but now, and above all, for analyzing the often highly complex orbits of the various objects launched by man into space. For such calculations the electronic computer has now become a vital tool, making it possible to solve with a very high degree of accuracy problems which hitherto could not have been tackled at all.

Nowadays, the calculation of the orbits of comets or artificial satellites has become almost a routine task. It is hard to realize, now that the laws of physics are applied to both atoms and galaxies without second thought, how startling it must have been for the astronomers of two centuries ago, to realize that the three laws of dynamics which were valid for Earth were equally valid for the skies, for so long the inscrutable domain of the gods.

The crucial discipline nowadays is nevertheless astrophysics, that is, the physics of the heavenly bodies. This is so not only because of the intrinsic importance of the subject but also because physics itself has, for several years, turned to astronomy as a source of information which cannot be had on Earth. Many years before particle accelerators with thousands of millions of volts were built, the cosmos was already the ideal natural laboratory for high-energy physics, and it is precisely from cosmic rays that some of the major discoveries in the physics of elementary particles have been made, such as the discovery of the *mu* and *pi* mesons, the positive electron, and the annihilation between matter and anti-matter. Perhaps the greatest feat of astrophysics, however, was the discovery by Bethe and von Weizsäcker of plausible thermonuclear reactions which explain the enormous and apparently inexhaustible energy of the stars.

Even though the Bethe-Weizsäcker model, which relates to a reaction in which four hydrogen atoms merge to give a nucleus of helium, has by now undergone various modifications, and other reactions have been proposed, we can without doubt take this discovery as one of the most far-reaching conquests of science; because this discovery, as it were, brought physics back from the stars and down to earth, and linked phenomena in the two spheres, just as the spectroscopic discoveries, which had shown that the elements which form the stars are exactly the same as those already known on earth, had earlier made a similar link between earth and the stars.

All three types of astronomy – *observational, dynamic* and *astrophysical* – contribute to the attainment of what has for all time been the ultimate objective, to have knowledge of the universe as a whole (cosmology) and to explain the origins of it (cosmogony). In effect, as the science historian H. Dingle has observed, astronomy has been nothing more than cosmology for most of its history. For the ancients, the heavenly bodies were not seen as important in themselves but were studied rather as means to an end, that end being the knowledge of the whole universe, and recent interest in cosmological and cosmogonical problems can thus be seen as a return to the problems that occupied the early astronomers.

The mainspring for this current interest of science in cosmology is the same as spurred on the ancients, but there is one significant difference: the acquisition of certain well-proven experimental data, and first and foremost the acknowledgement that those vast and faintly glowing objects, described for the first time by De Maupertius in 1742, were in effect vast clusters of stars like our own Galaxy. Thus modern science knows that there are many galaxies, while the ancients believed there was but one. This hypothesis, which Kant had already hinted at, was later made more specific by Humboldt, who gave it the intriguing name: *The Theory of the Island-Universes,* which has remained with us up to the present day. Once the elements of the Universe had been singled out into island-universes, i.e. into galaxies, the next step was the discovery of their reciprocal movement of separation and hence of the expansion of the Universe itself.

The experimental data has been constantly increased in recent years by more and more specific and accurate observations, and it can indeed be said that the ancient dream of a consistent cosmogonic theory is beginning to have a solid experimental base. It is clear that the more distant in space (and thus in time, given the finite speed at which electromagnetic waves spread) the objects observed, the more valid the experimental analysis, aimed at the theoretical reconstruction of the birth and evolution of the Universe. To this end, investigations made with gigantic radio-telescopes capable of picking up signals coming from very remote objects – and not only from stars or galaxies, but also from more mysterious bodies such as *Quasars* and *Pulsars* – have been essential for the formation of twentieth-century cosmology.

The study of the structure of the Universe, stimulated by new experimental discoveries, is, one might say, the study of the structure of space, time and matter, that is, of the fundamentals of scientific knowledge.

THE STAR-STUDDED SKY

Studying the heavens, in ancient times and right up to the arrival of modern astronomy, has always involved two aspects which are distinct but not easily separated: the religious, and the specifically scientific. On the one hand, men recognized immutable laws in the fact that day follows day and season follows season – and these laws undoubtedly represented the earliest core of scientific knowledge, on the other hand the first attempts at explaining these laws had recourse to the divine.

The Chinese civilization produced many astronomers, of whom, unfortunately, we know very little. Of the Western World, the greatest astronomers of ancient times were the Babylonians, the race that inhabited the region of Asia bounded by those two mighty rivers, the Tigris and the Euphrates, the region otherwise called Mesopotamia (*mesopotamia,* in Greek, means 'between rivers') which corresponds roughly to the modern state of Iraq. We tend to call these peoples by the generic name of Babylonians, but in effect there were many different races and houses; there were indeed Babylonians, but there were also Assyrians, Chaldeans and Sumerians (the latter being probably the most ancient, of pre-Semitic origin). The gods of the Sumerians were astral: the main god was An, god of the Sun, together with Nin-Anna, goddess of the star Venus, who was also the progenitrix of the human race.

It is known for certain that the peoples of Mesopotamia, some 3000 years before the birth of Christ, were acquainted with many of the more important constellations, and documents have come down to us which reveal that the priests in the temples, who were also astronomers (the biblical Tower of Babel was quite simply a grandiose observatory), knew about the movements of the seven stars roaming the skies, or planets (*planetes* in Greek means a wanderer), namely, the Sun, the Moon, Mercury, Venus, Mars, Jupiter and Saturn; we also know that they had realized that the planets themselves moved in a specific belt of the 'vault of heaven', not unlike a circle. This belt, otherwise known as the Zodiac, was divided by the priests of Mesopotamia into twelve parts, with each part taking the name of the corresponding constellation. The names of these divisions, or signs, have come down to us via the culture of ancient Greece. The essentially religious nature of Mesopotamian astronomy can be seen from their picture of the cosmos, according to which the Earth was a great disc surrounded by the ocean, beyond which a soaring mountain-range held up a solid hemisphere in which the stars were situated. Moreover, for every great city on Earth there was a corresponding, and finer, city in the sky.

Perhaps the major contribution of the Mesopotamians to astronomy lies in the accurate logging of the eclipses of the Sun, which were used, 1000 years later, by Ptolemy and Hipparchus. The Egyptian civilization, which was more or less contemporary with the Babylonian civilization, did not – at least as far as the evidence goes – develop a specifically astronomical science, nor were its cosmological ideas very sophisticated: the sky was a goddess (Nut) who stood on her hands and feet, and the Sun was a navigator sailing on a celestial river. But the Egyptian civilization was responsible for the most accurate observations, in ancient times, concerning the length of the year, and this is easy to understand if one thinks of the importance of the alternation of the seasons for an essentially agricultural society, the economy of which relied

Opposite: the nebulae Trifid and Laguna photographed from Mount Wilson Observatory in California.

substantially on the periodic floods of the Nile; the Egyptian year lasted 365 days and was governed by the appearance, at dawn and before sunrise, of the star Sothis (Sirius), in the constellation which they called the Great Dog, because it was dedicated to Anubis, the god often depicted with the head of a dog (or jackal). But they also knew that the 365-day year was shorter, in length, than the solar year, and that every four years the appearance of Sirius (or the Dog Star) on the horizon was considered to be one day late: the Egyptian year was thus 365 days and a quarter. But it should not be thought that this astronomical knowledge was accessible to all Egyptians; the job of keeping the calendar fell exclusively to the priests, and the store of astronomical knowledge, so vital to a land-based society, was jealously guarded and kept secret, on pain of death, by a caste which used it as an effective means of wielding political power.

Astronomy was certainly held in high esteem in the great civilizations of the East as well, such as the Indian and Chinese cultures. It is worth remembering also that what has survived as evidence of other civilizations, which did not reach any high degree of development in other spheres, almost invariably contains some evidence of astronomical observations. The greatest megalithic

Left: the astronomical observatory at Jaipur, India. Top right: sunrise at Stonehenge (England), site of an ancient sun-cult, and (below) general view of the huge megalithic complex.

Versions of the Zodiac. Top left: a detail of an Egyptian Zodiac (first century B.C.). Below: signs of the Zodiac in an Arab treatise. Right: the twelve signs of the Zodiac in a mediaeval miniature. Today it is thought that the Zodiac was Babylonian in origin, on the basis of evidence provided by inscriptions dating back to the fourteenth century B.C.

monument in the world – Stonehenge, in England – was built in about 2000 B.C. (the date has been fixed by carbon dating – modern methods based on measuring radioactive carbon) and is in fact an astronomical observatory, and a highly developed one at that, in which the alignment of the colossal stones produces a very accurate calendar, based on the observation of the rising and setting of the Sun, and of other visible stars. The astronomical ideas best known to us, however, are undoubtedly those of ancient Greece, the principal source being Aristotle's *De Coelo*. The father of Greek science was Thales (624–565 B.C.) who, on a business journey to Mesopotamia, became acquainted with the bases of Babylonian astronomy, and returned home predicting the eclipse of 585 B.C. This prediction in turn aroused the interest of the Greeks in the systematic study of the skies.

It would appear that Aristarchus of Samos (310–230 B.C.), the Copernicus of Antiquity as he is called, declared that the Earth made an annual revolution round the Sun, as well as a daily rotation on its own axis (this had also been the view of Herclides Ponticus (388–315 B.C.). He also asserted that the light of the Moon was nothing more than light reflected from Earth.

At about the same time Eratosthenes (276–194 B.C.) had accurately

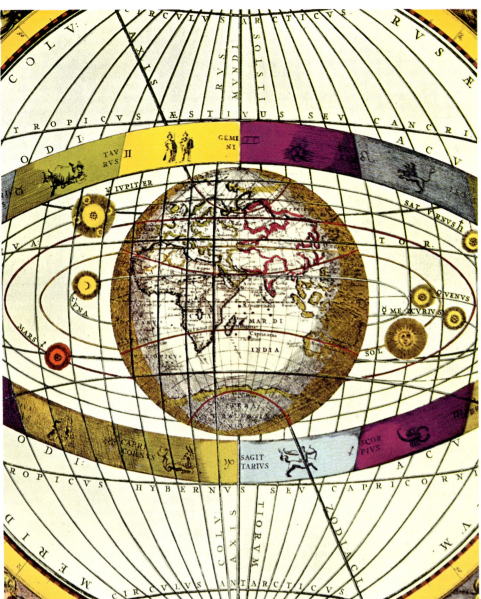

calculated the circumference of the Earth. There were numerous other Greek astronomers who were ahead of their time, but here we shall just briefly discuss the concepts of Aristotle, because it was these which influenced the whole of the subseuqent development of astronomy. Aristotle adopted the ideas of Eudoxus, later modified by Callippus, according to which the stars and the planets moved with a circular, uniform motion along concentric crystalline spheres; the centre common to all these spheres was the Earth, which was thus placed at the centre of the Universe.

The idea of concentric spheres, crystalline and incorruptible, was clearly derived from Pythagoras and was to dominate the scientific world for 2000 years, until 1600 A.D., particularly because of the work of Ptolemy (Claudius Ptolemaeus) of Alexandria in Egypt, who lived in about the second century B.C. Ptolemy's most important work is the *Almagest*, one of the scientific writings which, in the whole of history, has most influenced subsequent thought. According to Ptolemy, the Earth is at the centre of the Universe, surrounded by the seven planets and by the sky with its fixed stars, but its position is not exactly central. By complicating the movement of the planets a little more (by means of epicycles), Ptolemy was able to accommodate the Greek

The Danish astronomer Tycho Brahe (top left) carried out many precise astronomical observations which enabled his disciple and pupil Kepler to deduce the laws of the movements of the planets. Below: one of the telescopes used by him. Right: his system, still essentially geocentric, in which the Sun and the Moon revolve around the Earth, while the planets revolve around the Sun.

Nicolaus Copernicus was the true founder of modern astronomy (right: his death, in a painting by N. Lesser). Once and for all he moved the Earth from the centre of the Universe, and proposed the heliocentric theory, according to which the Earth, like all the other planets, moves around the Sun. Left: two diagrammatic drawings of the Copernican system.

experimental observations of the planet's movements which were made with a degree of accuracy comparable to that achieved 1500 years later by Kepler. That both Ptolemy and Kepler should have worked on the same problems, within the same conceptual framework, shows just how long-lived was the geocentric (Earth-centred) view of the Universe. It should also be mentioned that, from a purely mathematical point of view, it is, of course, irrelevant whether the Sun moves round the Earth, or the Earth round the Sun. In the *Almagest* there is also a list of visible stars and numerous astronomical instruments are described in great detail; among these are the armillary astrolobe and the *Triquetrum*, which made it possible to measure, directly, the angles of the greatest importance for astronomical calculations.

Ptolemy's cosmological views were, to remain unchanged for many centuries; it was not in fact until the sixteenth century, with Copernicus, that there was a real revolution in the science of the skies. Nicolaus Copernicus (1473–1543) was born in Poland and undertook most hif work in Italy; he was also a physician (and not very popular because he did not ask his patients to pay him). In his scheme of the Universe, he moved the Earth from the centre and reduced it to the status of a simple planet, revolving, with all the other

153

planets, around the Sun, which was fixed and motionless. This *heliocentric* theory (i.e. placing the Sun in the centre) was published in 1543, the same year that he died, in *De Revolutionibus orbium coelestium,* but even before its publication, many friends and disciples of Copernicus were aware of it. In fact, in 1539, Joachimus Rhacticus, a professor at Wittenberg, had already published a succinct exposition of the new theory. Copernicus wrote '. . . if we transform the annual revolution of the Sun into a revolution of the Earth, then the rising and setting of the constellations and stars will appear, morning and evening, in the same manner', and '. . . the standstills, and backward and forward movements of the planets are not peculiar to them, but only, clearly, apparent movements, caused by the movement of the Earth. Eventually, the Sun will be seen to be the centre of the Universe . . . .'

Although giving a much simpler and clearer explanation of the movement of the planets than Ptolemy's scheme of the Universe, the new system met with hostility from the Church, which saw man removed from his place at the centre of the world, and thus, it seemed, demoted to a secondary role. Luther said that Copernicus was a madman who took no account of the fact that the Holy Scripture said that Joshua had stopped the Sun, not the Earth. Despite this

The star-studded sky can easily be illustrated with the use of celestial globes, which were common up to the last century. Left: a seventeenth-century globe, maker unknown. Top right: a globe from Valencia made by the Arab Ibn as Sahil. Below: a globe designed by Tycho Brahe.

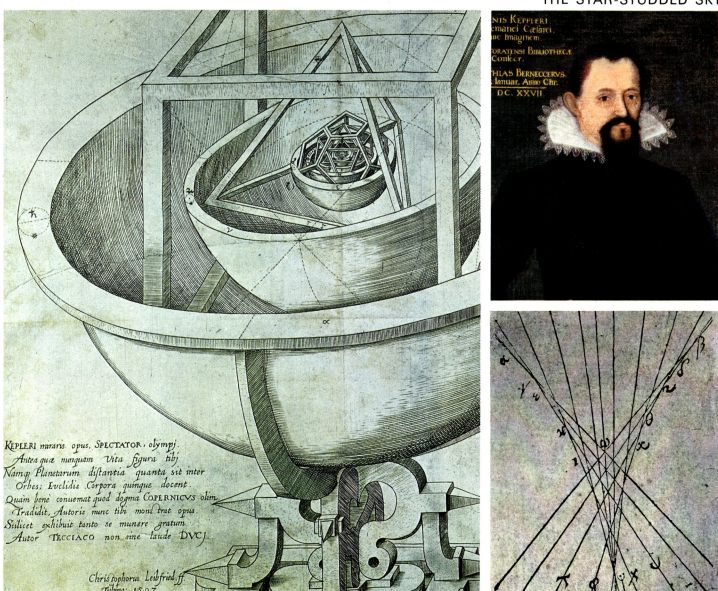

Johann Kepler established the laws of planetary movement, from which Newton deduced the law of universal attraction. Left: a plate from his work Mysterium Cosmographicum, *in which the five regular polyhedra are placed in relation to the five already known planets. Top right: a portrait of the astronomer. Below: a page from* Ad Vitellionem paralipomena.

opposition, the Copernican system gradually gained ground, not least because of the highly accurate observations of a convinced anti-Copernican, but a sincere scientist, the Danish astronomer Tycho Brahe (1546–1601); his observations reached the limit of what it was possible to see with the naked eye, or, at best, with instruments which did not magnify. The man who was the real link between classical and contemporary astronomy was Johann Kepler, a collaborator of Tycho Brahe and recipient of the famous Danish astronomer's last and most advanced observations. With his mystical and speculative spirit, Kepler adopted the Copernican system as a young man, and had tried to explain the harmony of the Universe by comparing the dimensions of the celestial sphere to those of five Platonic solids: this, more or less, was the synthesis in his first cosmological work, the *Mysterium Cosmographicum* (1596). Once acquainted with Brahe's accurate observations, he realized that a circular movement by the planets was very difficult to maintain. By introducing elipses in the place of circles, Kepler (still adopting the heliocentric theory) showed that the apparent movement of the heavenly bodies could be explained with just eight elipses instead of the 34 circles used by Copernicus. Copernicus' research led to a further revolution: the movement of the planets was in fact

neither uniform nor circular. The Sun occupies one of the foci of the elipse which constitutes the trajectory of each planet, and the planet itself moves more quickly, the nearer it is to the Sun, in such a way, that is, that it covers equal areas in equal times. This description of planetary movement, which gave his first two famous laws, was published in 1609. In 1618 Kepler put forward his third law, that given two different planets, the square of the time taken to travel their orbits is proportional to the cube of the greater axis of the elipse which constitutes their orbit around the Sun.

With Kepler it can be said that the path to the dynamic interpretation of the Universe was opened: it was in fact on the basis of his laws that the brilliant Newton deduced the law of universal attraction, which then gave a simple and clear explanation of the movement of the planets in the context of laws of physics verifiable on Earth, and removed astronomy, once and for all, from the divine role which it had enjoyed until then in respect of the other sciences.

It is not hard to understand Kepler's enthusiasm about his own discovery; as he himself said: '... my book will be read by posterity, and now, it matters little. It may well await a hundred years for a reader, because God has awaited 6000 years for someone to contemplate His own work.'

Concentrations of stars in the Galaxy in three photographs that have been joined up. We can see a conspicuous increase in the density of the stars in the vicinity of the galactic equator. The zone illustrated lies between the constellations Cygnus and Cassiopeia.

Some of the earliest telescopes, optical instruments conceived in the early seventeenth century by Dutch technicians. Their use led to considerable progress being made with astronomical observations. Top left: Galileo's telescope of 1609. Below: the telescope with which Galileo discovered Jupiter's satellites. Right: Isaac Newton's telescope.

It is quite surprising that these descriptions of the movements of heavenly bodies which are so accurate and still valid were deduced from observations made, to all intents and purposes, with the naked eye. A huge leap forward came with the introduction of instruments designed to increase the range of the human eye, that is, the introduction of telescopes. It can be said that all astronomical progress that followed was conditioned by progress made with these instruments. It seems that the first telescopes were invented in Holland in the early seventeenth century, and they were certainly known about in 1610 by Galileo who presented such an instrument to the Doge of Venice, Leonardo Donato (an act which earned him the doubling of his salary). Galileo's telescope consisted of two lenses, a convex lens and a concave eyepiece, and it was Kepler himself, in his *Dioptrica*, who first explained the rudiments of the theory. It is impossible, here, to follow the historical development of the improvements made to this basic tool of astronomy. Suffice it to say that while Galileo's first telescopes gave an enlargement of about 30 diameters, which already made it possible to discover the mountains on the Moon and the satellites around Jupiter, by 1655 the first of Saturn's satellites (Titan) had been discovered and by 1659 the nature of Saturn's rings had been revealed by

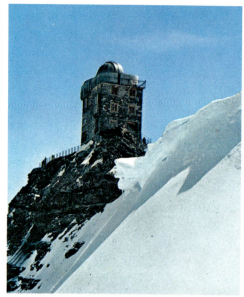

observations. All the early telescopes were of the refracting type, that is, they used transparent lenses which necessarily produced deformed images through the phenomena of chromatic and spherical aberration. Despite these limitations, until about the mid-sixteenth century refractors were built with huge focal lengths (up to 200 metres [665 feet]) which did not improve performances appreciably. It should be remembered, however, that even with the early telescopes it was possible to see that stars remained no more than points of light and must for that reason be very remote, whereas some zones of diffused light could be broken down into fairly dense clusters of single stars. Chromatic aberration was in the end eliminated, thanks, particularly, to Newton: lenses as such were no longer used, being replaced by mirrors which could concentrate light in the same way, and were easier to make. The use of mirrors still did not eliminate spherical aberration, but this was eventually removed by constructing parabolic mirrors. In fact the history of the telescope is much more complex, and even nowadays both reflecting (using mirrors) and refracting (using lenses) telescopes are built, but the most recent instruments, which reach extremely advanced technical levels, are all based on reflection. Major advances in astronomy have been achieved in recent years by the

Observation of the stars is carried out in specially designed buildings known as Observatories, usually situated at high altitudes. Top right: the Observatory on the Jungfrau in Switzerland. Top left: the Observatory at Chichèn Itzá, used by the Mayas. Below: the Uraniborg Observatory where Tycho Brahe made many of his precise observations. Centre: the solar tower of the Rome Observatory. Below right: the Parkes radio-observatory in Australia.

A picture of the mirror of what is currently the world's largest optical reflector, at Zelenciuskaja in the Caucasus, in the Soviet Union. Note the closing device.

colossal reflectors at Mount Wilson and Mount Palomar, USA, built respectively in 1917 and 1948, and great things are now being expected of the very recent Soviet Zelenciuskaja instrument. But as we shall see, the greatest limitation of the optical telescopes, i.e. those which simply magnify the power of the human eye, lies in the fact that the range of wavelengths which they can use is extremely limited as compared with the range of the rays which the stars themselves send down to us. The crucial step forward will occur when the radio-telescope and other instruments capable of picking up non-visible rays are fully exploited.

With the introduction of optical instruments it was possible to start thinking of solving some of the major problems of astronomy. How far away were the stars? What was the origin of their light which never seemed to dim? The most immediate and certainly the most ancient method of determining the distance of heavenly bodies, based on the same principle as is used for measuring large distances on the earth's surface, was the method of triangulation. As we know, if we know the length of the base of a triangle and the measure of the two base angles, we can, by simple trigonometry, calculate the length of the other sides, and also the distance of the apex of the triangle from the base. To measure the

distance of a star, one takes as the base of the triangle the distance between the Sun and the Earth. This is known to be about 150 million km (93 million miles). The accurate calculation of the distance between Sun and Earth, which, because of its importance has been called an Astronomic Unit, deserves a chapter all to itself. Sufficient to say here that in recent times this measurement has been calculated with great precision by the use of radar. In using triangulation, it is only possible to obtain the measurements of distances for relatively close stars. In effect, with modern methods, and a ten per cent degree of accuracy, the distances of some 700 stars have been calculated, with the nearest star, Alpha Centauri, being the incredible distance of 40,000,000 million km. away (25,000,000 million miles)! It is evident that for distances of this order the kilometre or mile are rather inconvenient units and astronomers prefer to use the unit of the *light year,* the distance which light travels in the space in one year. With this new unit of measurement (the equivalent of about 10,000,000 million km. or 6,250,000 million miles) Alpha Centauri is about four light years away. To give an idea of the distances of the other stars, it is worth remembering that the brightest visible star in the northern hemisphere, Sirius, is about 8.5 light years away, and the North Star is about 400. This latter

From earliest times man tried to gather the brightest stars into constellations. Left: the Pleiades cluster. Right: two old pictures of constellations in the northern sky. Opposite: Hertzprung-Russell diagram of stellar evolution. By following the arrows on the various lines, we can see how three different types of stars vary in luminosity and temperature as they age. (The luminosity increases from top to bottom, and the temperature from left to right.)

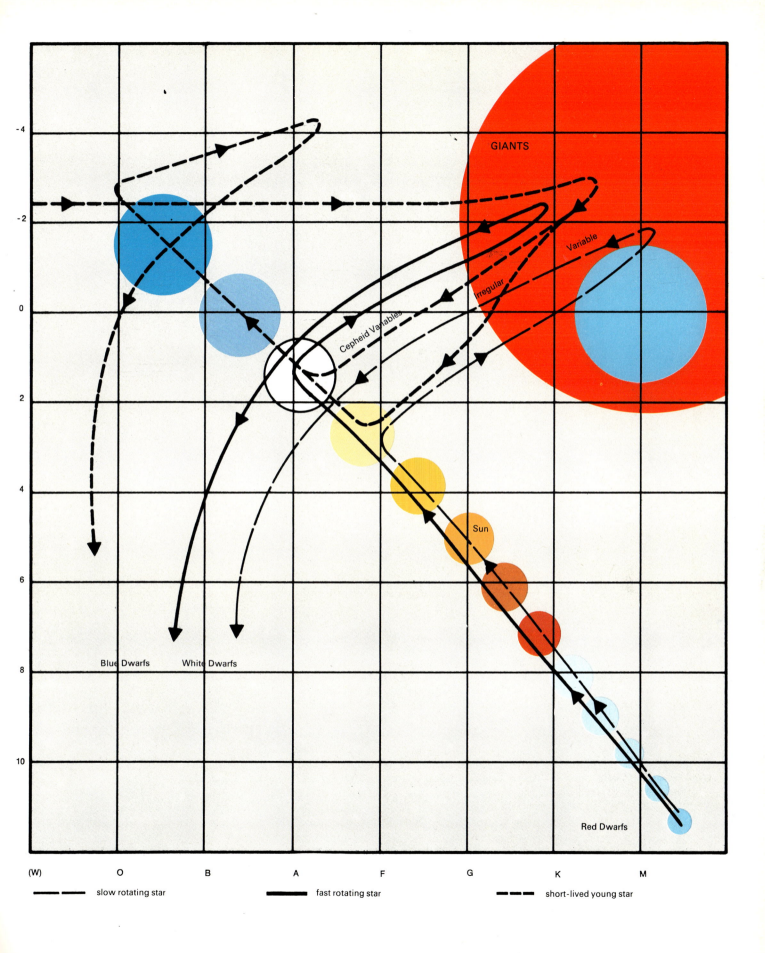

GIANTS

Variable

Irregular

Cepheid Variables

Sun

Blue Dwarfs   White Dwarfs

Red Dwarfs

(W)   O   B   A   F   G   K   M

slow rotating star   fast rotating star   short-lived young star

distance, however, is about the furthest that can be calculated by triangulation; beyond that other methods have to be used. If all the stars were of equal strength it would not be difficult, once one knew the distance of one star, to determine the distance of all the others. In fact, it is only human instinct that suggests that the fainter stars are more remote from, and the brighter stars nearer to, the Earth. Unfortunately stars differ greatly in the strength of the light they give out, hence the link between a star's apparent brightness and its distance from the Earth is complicated. The calculation of a star's distance away from the Earth involves not only its brightness, but also other factors. A very important one, for example, is its colour, or more precisely, the distribution of the various frequencies which constitute the light emitted by the star: with appropriate instruments known as *spectrometers,* linked to telescopes, it is possible to trace, accurately, the distribution of these frequencies. The laws of physics then allow the colour of a star to be related to its temperature. Stars come in a variety of forms, from very hot white stars, down to relatively cold red stars. It has been established that there is a rough correlation between the colour of a star and its luminosity for those stars whose distance from the Earth have been determined by trigonometry. Thus if, in the

Stars are also alive, because they are born, they evolve, and then they die. But not all of them undergo the same evolution. Top: the evolution of a 'normal' star, like our Sun, which ends with the White Dwarf stage. Below: the evolution of a large star which finally explodes as Supernova.

Luminescent gases in the Sagittarius constellation. It is thought that a star is born from the concentration of dust and gas clouds possibly like those seen in this photograph.

case of a star whose distance we cannot determine by trigonometry, we know the spectral distribution, that is, the colour, then it is possible to establish its actual luminosity, and then calculate its distance from the Earth on the basis of its *apparent* luminosity. Another method employed relies on the fact that the luminosity of certain stars, known as Cepheid Variables, undergoes regular variations (for example, Delta Cephei pulsates regularly about every six days). It has been observed that the pulsation period is associated, approximately, with the intrinsic luminosity: by a process of similar reasoning, once a Cepheid has been isolated in the sky, it is possible to deduce its distance and the distances of the other stars, which for some reason, we believe to be near it. This method has made possible the measurement of distances of up to twenty million light years, but as we shall see, even this distance is fairly small, in the cosmological scale. We have seen that to the question: how far away are the stars? it has been possible to give a feasible and reliable answer. We are now left with the other basic question, namely: whence do stars derive that enormous fund of energy which they continue to shine down on us for billions of years? Or, the even more basic question: have the stars always been the way we see them, or have they changed in the course of time? Modern astronomy has

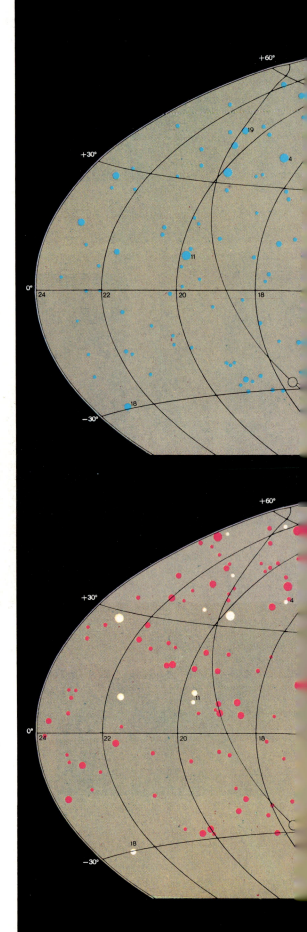

1. Sirius
2. Canopus*
3. Alpha Centauri*
4. Vega
5. Capella
6. Arcturus
7. Rigel
8. Procyon
9. Achernar*
10. Beta Centauri*
11. Altair
12. Betelgeuse
13. Southern Cross*
14. Aldebaran
15. Pollux
16. Spica
17. Antares
18. Fomalhaut
19. Deneb
20. Regulus

The heavenly bodies send electromagnetic rays of every conceivable wavelength, from the very shortest (gamma rays) to the very longest (radio-waves), but the human eye is only capable of seeing a very narrow band, i.e. the visible band. If our eyes were sensitive to another band of wavelengths, the sky would undoubtedly look very different to us. Fortunately, the eye is not the only means of seeing available to us, and with other instruments we can 'see' the sky in different bands, for example in ultra-violet or infra-red.

The diagram on the right shows the sky as seen from the Mount Palomar Observatory in the visible band (top) and in the infra-red band (below).

The difference between the two maps is immediately evident.

In the list above are the twenty brightest stars in the visible band: those marked with an asterisk are not visible from the latitude of Mount Palomar, and are thus not shown in the map.

One can see that in the lower map, the infra-red band, only twelve of the fifteen brightest stars are marked, which means that three of the brightest stars in the sky do not give off infra-red rays, or only emit them in very small quantities.

It is thought, in general, that only one quarter of the visible stars emit infra-red rays.

Observations of the sky derived from infra-red rays (likewise ultra-violet rays or radio-waves) are extremely important sources of information about physical processes taking place in stars and which are revealed through the electromagnetic rays to which these processes give rise. Infra-red astronomy, which requires special instruments both for observation and for recording, seems likely to make significant contributions to our knowledge about the processes of formation and evolution not only of heavenly bodies such as stars and galaxies, but also of the Universe as a whole.

The numbers on the maps refer to the corresponding stars listed above.

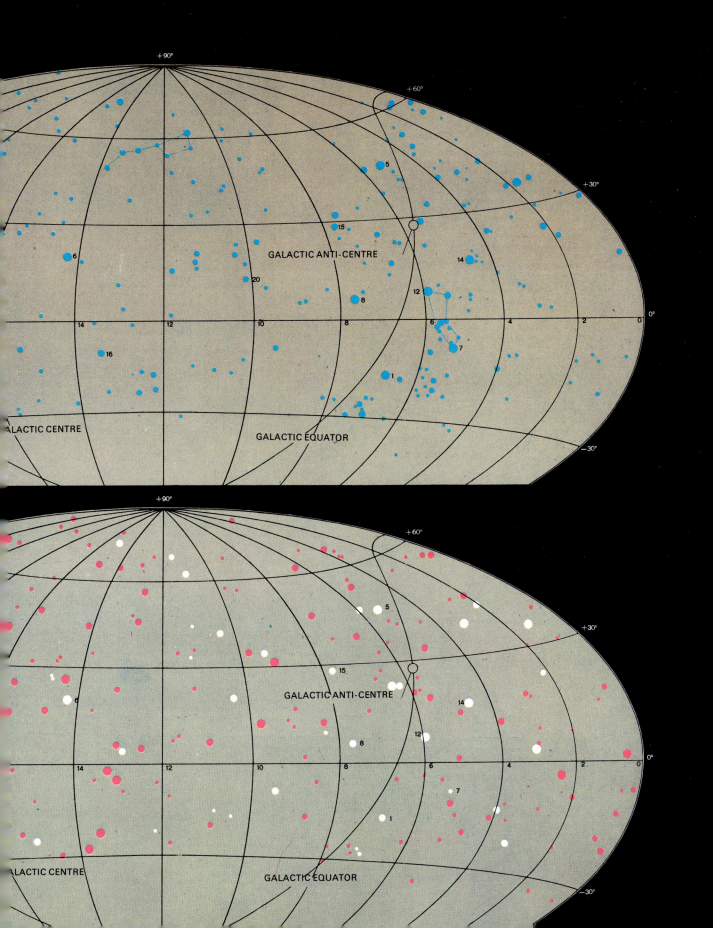

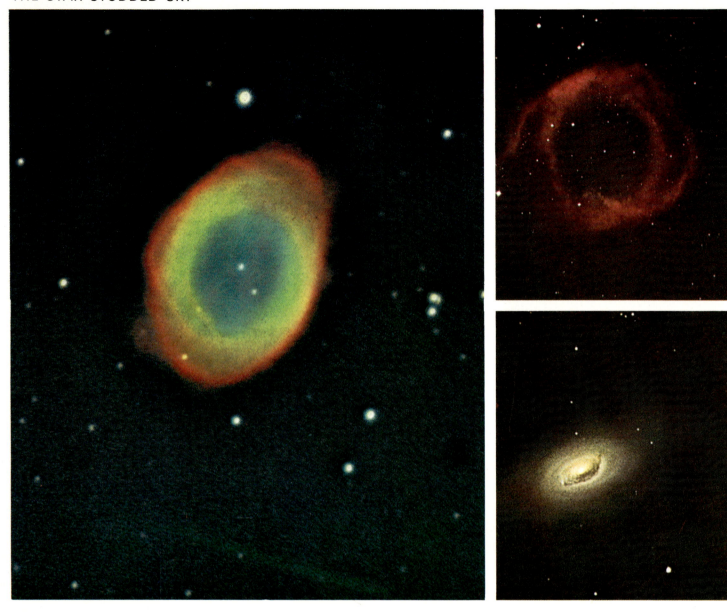

offered answers to this question. The stars, like living organisms, are born, live, and die, even if the time-span for this chain of events is so long that we cannot directly observe the stages of evolution in any one star. Fortunately, however, the panorama of the skies is so vast and so diversified that we can observe newly-born stars, young stars, old stars and dying stars and reasonably link together a pattern. The mystery of the incredible store of energy released by the stars has only been unravelled recently, after science understood thermonuclear reactions.

Spectroscopic analysis has made it possible to determine the material of which the stars are composed: most stars are formed of hydrogen (60–90 per cent), helium (10–40 per cent), with the other elements representing not more than three per cent of the total mass. In interstellar space, hydrogen and helium are the atoms which are essential and they represent the material from which the stars are formed. It may happen that in some region or other this 'raw material' starts to grow denser, in a process which becomes gradually more effective because it increases gravitational attraction. By means of a secondary process, clouds of dust and gas are then formed which in turn, under gravitational attraction, contract, thus giving the initial form to what will

Left: the annular Nebula of Lyra M.57. This is one of the clearest examples of a planetary nebula, i.e. of a faint star surrounded by a 'wrapper' of expanding gases. Top right: another planetary nebula in the constellation Aquarius. Below: the famous Coma Berenices, known as early as the third century B.C. near the galactic North Pole.

Left: part of the nebula ring of the Velo in the constellation Cygnus, caused by the explosion of a Nova. Right: the M.51, spiral galaxy in the constellation Canes Venatici. Seen from without, our Galaxy would look very similar to this.

eventually become a new star. Because of this contraction, the temperature of the new heavenly body increases and may become so high that it will trigger off violent thermonuclear reactions.

Depending on the composition of the original gas cloud, the stars can be divided into two major classes: those which have condensed from clouds of hydrogen, helium and other material expelled from other stars, these constitute the so-called Population I, and those which are rich in hydrogen and relatively poor in helium, with very small quantities of other different atoms, these constitute Population II.

But what happens when the nuclear fuel on a star is exhausted? Astrophysics suggests that such a star may have one of two very different ends. It may explode, releasing a colossal amount of energy (it is said to explode in the form of Supernova); or it may die more quietly, leaving behind the inert and dark residue of the nuclear reactions. Which type of evolution occurs depends essentially on the mass of the star. If, on a temperature-luminosity diagram, we represent the temporal evolution of a star, we can follow its destiny in time (which obviously runs into thousands of millions of years). For a quite normal star such as the Sun, after about 10,000 million years a new thermonuclear

reaction is triggered off which will make it become bigger and redder. Having remained in the Red Giant stage for a further 10,000 million years, all the nuclear fuel will be used up, and the Sun will contract, at the same time expelling illuminated gaseous material from the central focus (in other words we have a *planetary nebula* which envelops the nucleus of the star). The evolutionary process then continues, the star becomes smaller and denser (i.e. a White Dwarf), and lastly cools down completely, turning from white to red, to become an inert Black Dwarf, i.e. the residue of what had once been a star. This is the destiny of our Sun.

The nebula of Orion can be seen with the naked eye in the beautiful winter constellation of that name: it is the central star of the three which form the sword of the mythical hunter loved by Aurora. Orion is more than 1500 light years away from us, but it spreads over a vast area, equivalent to about 100 light years.

THE SUN AND THE PLANETS

The Sun is a star, and seems the largest and brightest star of them all; in fact, however, it is just an ordinary star which is not even particularly bright, but it appears so because it is a great deal nearer to us than any other. While the light from the nearest star (Alpha Centauri) takes more than four years to reach the Earth, the light from the Sun reaches us in about eight minutes. The Sun is the primary source of every form of life on our planet, and even the sources of energy most widely used by us (i.e. petroleum and its derivatives) are nothing other than the fossil remains of living forms which grew by means of the Sun's light. It is only very recently that man has learnt to build his own energy-generating 'private sun' in the form of nuclear power-stations. In primitive societies with no scientific knowledge about the complex nuclear reactions which give rise to the solar light and heat which keep us alive, the Sun is deified and worshipped, for its function as a source of life is appreciated. The attributes which various diverse cultures and civilizations have bestowed upon it are surprisingly similar. As well as exalting its power as a source of light and heat, the most varied races have attributed to the Sun the power of healing sickness, and purifying sinners. Its daily disappearance and re-appearance gave rise in many peoples (such as the Egyptians) to the myths of death and resurrection, and at the same time to numerous magic rites designed to delay its disappearance at sunset, and invoke its appearance at dawn. If we take a deeper look at many myths and religions which are current today, we can see that they derive, to a considerable degree, from ancient sun cults. The pagan rite of celebrating the winter solstice as the birth of the Sun has become the Christmas of Christianity. But what do we know about the Sun as a star? Unfortunately it must be admitted that, despite the enormous and important progress made especially since the Second World War, the Sun is still somewhat shrouded in mystery. In fact, as almost invariably happens in science, the very fact that we have improved our capacities for observation has solved some old problems, but revealed other even more complex ones.

Let us start by taking a very brief look at the features of the Sun which we know about with the most certainty.

First and foremost, it is fairly easy to calculate the Sun's apparent dimensions, that is, the angle from which it is visible from Earth: this angle is slightly more than half a degree (in fact it lies somewhere between 31'28" and 32'32", this variation being due to the different distances between Sun and Earth depending on the time of year). If, in addition, we know the Astronomical Unit, we can then determine its effective diameter, which turns out to be about 1,400,000 km (875,000 miles), i.e. more than 100 times the diameter of the Earth. This is a huge figure; to make it easier to appreciate imagine that the Earth is a ball 5 cm (2 in) in diameter. On this scale the Sun would be a ball with a diameter of more than 5 metres (approx. 17 feet), and stand about half a kilometre (1700 feet) away. By applying the law of universal gravitation, it is possible to calculate the mass of the Sun and its density, which is about one and a half times the density of water. The Sun is therefore much less dense than the Earth, and this is easy to appreciate if we think of it in terms of a gaseous star.

Another interesting fact which can be deduced from the laws of dynamics is

the value of the gravitational pull on the Sun's surface; that is 28 times greater than that on the Earth's surface. A man weighing 70 kilogrammes (157 lbs) on the Earth would weigh about 2000 kilogrammes (4500 lbs) on the Sun! In so far as they are deduced from visual observations and celestial mechanics, all these facts were well known soon after Newton's time, and with the arrival of the first optical instruments, certain other features of the Sun, which is anything but unchanging, were before very long recorded. First and foremost, the Sun has a thin outer atmosphere, called the corona, which is clearly visible during eclipses and which, for this reason, must have been known about from earliest times. Although it gives off only a relatively faint light, more than 500,000 times less than that of the Sun's surface, the corona has posed one of the major problems in astrophysics because it was discovered in 1933 that its temperature is something approaching $1,000,000°$, whereas the temperature of the Sun's surface is only about $5000°$; thus the corona cannot be heated by the surface of the Sun itself, unless one of the fundamental principles of physics were to be overthrown, i.e. the second law of thermodynamics. The explanation of this very high temperature was given by the great Swedish astrophysicist Hannes Alfvén: he suggested that the heat of the corona comes directly from the

Worship of the Sun always played an important part in ancient civilizations. Top left: an Egyptian picture of the Sun god. Below: a detail of the Sun Stone *from the stone Aztec calendar of the* History of the World. *Right: an illustration of the Sun from the* De Sphaera *codex. Opposite: a poetic photograph of a sunset (top left); and the midnight sun on Ross Sea (top right); an eclipse of the Sun (bottom left); ice crystals mirror the dawn (below right).*

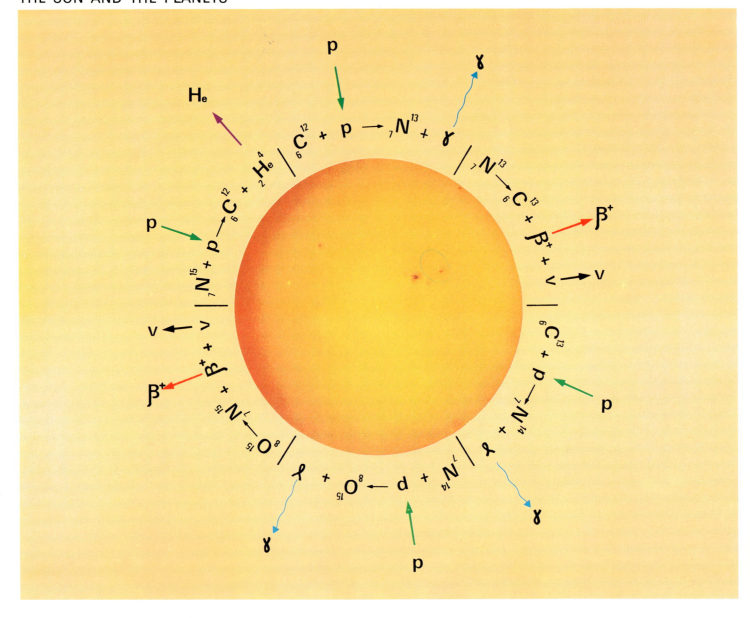

nucleus and not from the Sun's surface. In addition it was discovered in about 1960 that the corona expands, and generates a kind of very strong 'solar wind' (up to 600 km [375 miles] per second) which is the cause of those spectacular phenomena visible in the far latitudes and called aurora borealis or Northern lights.

The temperature of the Sun's surface is about 5000° (to be precise 5600° Kelvin); the surface itself is usually called the photosphere (i.e. sphere of light) but it should not be thought that the photosphere is a bright, uniform surface; in fact even with an ordinary telescope numerous oddities can be observed (although the curious reader should not test the truth of this statement without first equipping himself with a pair of very dark glasses). It is in fact possible to see lighter areas called *faculae* and a large number of small spots (small, that is, from where we are, although their actual diameter is about 1000 km [625 miles]) called *granulations*, which are always present. The most distinctive feature however is the spots which are darker areas, fairly irregular in shape, with a temperature of about 4000° (they thus appear darker only by contrast with the brighter background). These spots are one of the most interesting features of the Sun, and are still the object of much research, even though they

The Carbon-Nitrogen Cycle put forward by Bethe and Weizsäcker. Four Hydrogen nuclei (protons indicated by 'p') merge to form the nucleus of Helium (He): the Carbon and the Nitrogen (C and N) assist in the nuclear reaction as catalysts. Positive electrons and neutrinos are also produced (β^+ and V)

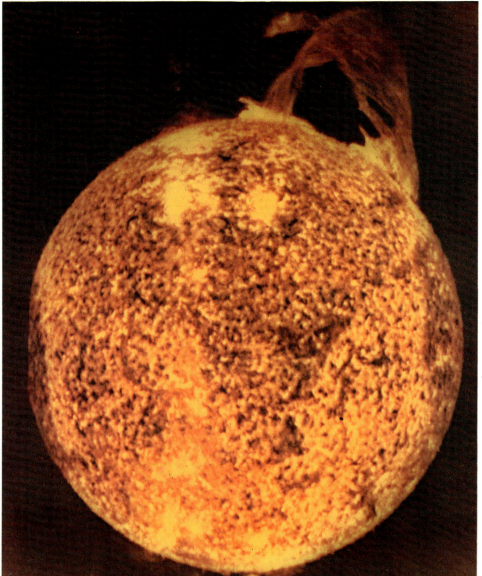

Top left: an annular eclipse, in which the Moon does not entirely cover the sun. Below: various phases of a total eclipse in a multiple-exposure photograph. Right: a spectacular eruption on the Sun's surface, photographed by Skylab on 19 December, 1973.

were discovered many many years ago: Galileo had observed them in 1611, but it seems certain that Chinese astronomers had described them some 1000 years prior to that. The spots are directly linked to all the various phenomena of solar activity, some of which are extremely spectacular, such as, for example, the huge tongues of fire (the *prominences*) which sometimes soar up from the Sun's surface for hundreds of thousands of kilometres (on 21 August, 1973, one such prominence was observed which rose up 400,000 km (250,000 miles) above the surface). As well as taking part in the rotating movement of the Sun around its own axis (a movement already observed by Galileo which takes about 25 days), the spots also have specific movements from the poles towards the equator. This movement is also periodic and lasts about eleven years, although there have been numerous irregularities in more than 300 years of observation. For example, from 1645 to 1715 there was an apparent absence of sun-spots and it was precisely during this period that the solar corona was not observed during the eclipses which occurred, nor were there any Northern lights, at least within inhabited latitudes.

From what has been said, it is clear that the whole of the Sun's surface is anything but static: in fact it is the scene of phenomena of inconceivable

violence in comparison with anything that happens on the Earth. For example, in the prominence observed in 1973, the Earth would have been like a speck of dust in a flame in a fireplace.

From the value of the 'solar constant,' that is, the amount of energy which falls in one second on to an area one centimetre square, it is easy to calculate the total energy given off by the Sun in one second: this, in physical terms, is its power. The value obtained is quite literally astronomical: the Sun has a power of 100,000,000 million horsepower (or more exactly 3.7×1023 kilowatts). No chemical process, such as combustion, can explain such a degree of constant power, thinking of the conversion of matter into energy (the mathematical formula being $E - mc^2$, perhaps the most famous formula in physics), the only possible answer is a nuclear reaction. The patterns of nuclear reactions have led scientists to consider various possible models, some more complex than others. At the present time it is thought that the most plausible solution is a fusion reaction in which the hydrogen is converted into helium within the Sun's nucleus. This basic step forward was made in 1939 by the German physicist Hans Albrecht Bethe, who suggested two main cycles: the so-called *nitro-carbon cycle* and the *proton-proton chain*. In this way, and taking into account

A special telescope aimed at the Sun to observe an eclipse in North-equatorial Africa. The study of this phenomenon, which is of great interest for observations of the activity of the Sun's surface, entails frequent changes of position for the observers to obtain the best observation point. Opposite: an overall view of the planets in the solar system.

the total mass of the Sun, the total amount of energy released and Einstein's equation, ($E=mc^2$), it is possible to predict that the Sun may have a life of 10,000 million years. But it should be said that in recent years Bethe's models have been questioned, and that the problem of how the Sun generates its own energy is still essentially open.

Nevertheless, the star which gives us our light and heat is interesting for another reason: around it, and in addition to the Earth, revolve numerous other heavenly bodies which do not have their own light – the planets. Those known from earliest times are Mercury, Venus, Mars, Jupiter and Saturn.

In 1781 William Herschel discovered Uranus, while the discovery of Neptune, which was observed for the first time by Galle in Berlin on 23 September 1846, constituted a real triumph for Newtonian astronomy.

This discovery had been predicted by the French mathematician Le Verrier who, by applying Newton's laws of motion, had observed that the upheavals in the orbit of Uranus could only be explained by the presence of an unknown body upsetting things. Then Galle found the new planet less than one degree from the place allotted to it by the complicated calculations made by Le Verrier. In 1930 Percival Lowell discovered an even more remote planet, which

Top: a photograph taken at the Munich planetarium, which shows the orbits of the various planets. Below: the two largest planets in the solar system: Jupiter (left) and Saturn (right) with its distinctive rings formed by countless particles of dust and ice.

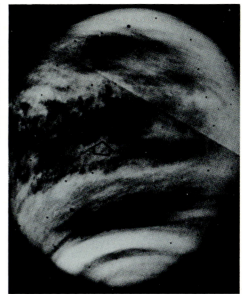

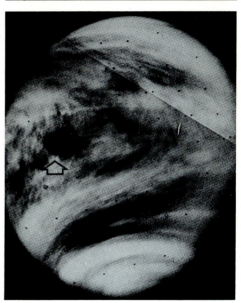

A series of photos of Venus taken by the American space probe Mariner 10, when, on 5 February 1974 it passed within about 700,000 km. (440,000 miles) of this planet on its journey to Mercury. The colours are the result of computer processing: in fact all these photos were taken by ultra-violet light.

was named Pluto. As well as these planets we should remember that we are in the company of numerous other bodies revolving around the Sun, namely the satellites of the planets themselves, the planetoids, the asteroids, the comets (even though these latter may often only be fleeting companions). Let us now briefly describe some of the main properties of the various planets and try to compare them where possible with our own planet. First of all it should be borne in mind that despite the recent spate of space explorations we still know little about the planets lying beyond Saturn.

The planet nearest to the Sun is Mercury (its average distance is about 60 million km [37.5 million miles], as opposed to the Earth's 150 million km ([93 million miles]); it is a very small planet, its diameter is less than 5000 km (3125 miles), with an average density about equal to that of our planet, which means the surface gravity is very low. A man weighing 70 kilogrammes (157 lbs) would weigh just 26 kilogrammes (58 lbs) on Mercury. The Mercurian year lasts 80 days, and its day 59 earth-days; this is a recent discovery, because until not so long ago it was thought that the Mercurian year was the same as the Mercurian day (or, more precisely, that the period of revolution and the period of rotation coincided) so that Mercury was always facing the Sun with the same

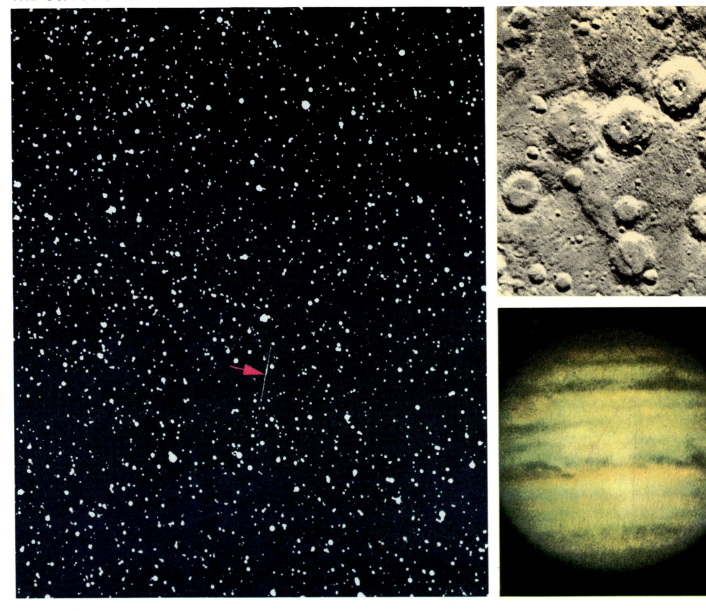

surface. It is a planet which, according to our yardstick, is more or less uninhabitable. It has no atmosphere, the daytime temperature rises to 350° C (662° F), and at night drops to −150° C (−238° F).

The second planet is Venus which is about 110 million km (70 million miles) from the Sun, has a year of 225 earth-days, and a day of 243, with the curious feature of backward rotation (which it shares with Uranus). Its dimensions are roughly the same as the Earth's, as is its mass (both are slightly smaller) and its gravity is consequently 88 per cent that of Earth. Unlike Mercury, Venus is covered by very thick clouds, containing a lot of carbon dioxide, which produce a kind of 'glass-house effect', which is sadly familiar to us on Earth too, especially in highly polluted zones; the result is a ground temperature approaching 500° C (932° F).

The third planet is the Earth and, obviously, its importance is such, for us at least, that a whole chapter has been devoted to it. Situated 250 million km (150 million miles) from the Sun, we find the fourth planet, Mars, which has stirred the imaginations of scientists and science-fiction writers alike, ever since Schiaparelli believed he had seen on it kinds of canals which seemed to have been made by intelligent creatures. Mars has a year of 687 days and a day

The asteroid belt is possibly the remains of a planet which exploded ages ago. Left: part of the orbit of the asteroid (or planetoid) Icarus. Top right: Mercury seen from Mariner 10 at a distance of 7500 km. (4700 miles). Below: Mars, the next likely step for astronauts.

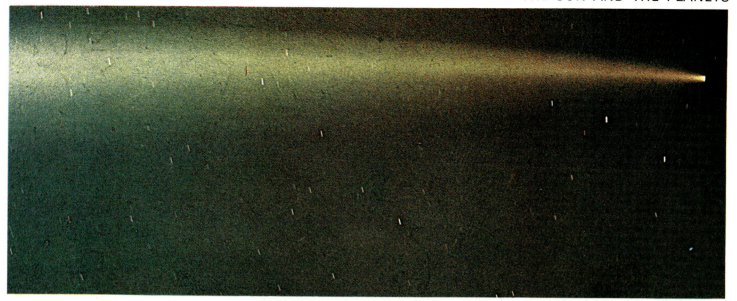

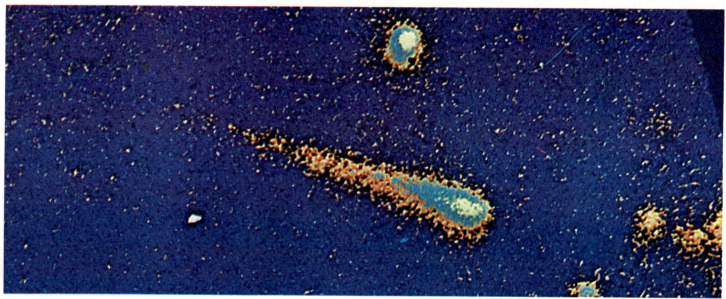

The comets are also part of the solar system. These are nebular masses with a tail sometimes millions of kilometres long, which travel in very elongated elliptical orbits. Top: the comet Ikeya-Seki, in a photograph taken on 29 October 1956, eight days after the disintegration of its nucleus, caused by solar action. Below: an ultra-violet photograph of the comet Kohutek, taken by Skylab on Christmas Day 1973. The colour picture has been processed from the original.

which differs from the earth-day by slightly less than half an hour; it is smaller and lighter than the Earth, and less dense too (gravity here is 38 per cent that of Earth); it has a thin atmosphere containing a lot of carbon dioxide and argon and is fairly cold, $-23°$ C ($-9°$ F). It has two satellites Phobos and Deimos.

Far away from these four planets (800 million km [500 million miles] from the Sun) is the gigantic planet Jupiter, an enormous body with a mass 320 times that of Earth and greater than that of all the other planets put together. It is essentially liquid, with perhaps a small solid nucleus and a dense atmosphere. Its day is very short, less than ten hours, which makes it very flat in shape at the poles. It has an array of thirteen satellites, four of which were discovered by Galileo in 1610, and two of which are larger than the planet Mercury. Another remarkable feature of Jupiter is a strange red spot discovered in 1878 and still somewhat mysterious, but which is probably a kind of huge storm. We still know very little about the other planets, apart from their distance from the Sun and their period of revolution. After Jupiter there is another giant, Saturn, familiar to all with its distinctive rings, with ten satellites; Saturn must be cold and forbidding, and rotating very quickly on its own axis (turning once in just over ten hours). We know even less about the last three planets, Uranus,

179

Neptune and Pluto. The first two have a mass about fifteen times that of Earth, a temperature below $-200°$ C ($-328°$ F) and five and two satellites respectively. If estimates are correct, Neptune's largest satellite, Triton, is the largest satellite in the solar system, with a diameter of 6000 km (3750 miles). Together with the actual planets, many other bodies revolve around the Sun including the comets, with their very elongated orbits in the form of a very flattened elipse (it is as well to mention that the orbits of the planets, with the exception of Pluto, are elipses with the Sun in one of the foci, but so slightly flattened that they are only barely distinguishable from perfect circles).

Among the other bodies keeping the Earth company in the solar system we should not forget the asteroids, which occupy a belt lying between the orbits of Mars and Jupiter; there are more than 1000 of them, with orbits generally quite unlike those of the planets, and some asteroids are quite large.

Earth is being constantly bombarded by a hail of meteorites which, luckily for us, usually disintegrate in the atmosphere. On the 8 March 1976 the explosion of one of these objects, at an altitude of about 20,000 metres, caused an actual meteoric shower over an area of northern China. Left: the famous crater at Devil's Canyon, Arizona, caused by a gigantic meteorite which is mentioned in Navajo legends. Top right: a meteorite made of iron and nickel which fell to earth in Siberia in 1947. Below: meteoric iron found in Arizona.

GALAXIES—QUASARS—PULSARS

If one observes the sky with the naked eye, especially on a clear winter's night, it is easy to detect a faintly glimmering strip running across it. According to Greek mythology, this is the remains, scattered throughout the firmament, of a drop of milk spilled by Juno as she was nursing Hercules; for pre-Colombian peoples it was a huge brother of the rainbow; and for the ancient Germans in the days of Julius Caesar it was the path of winter hoar-frosts. ... In almost every mythology there is a reference to this band of light spanning the night-sky, which we nowadays call the Milky Way or Galaxy, although, as we shall see, it would be more accurate to call it 'our' Galaxy.

With the invention of the first telecopes it was easy to identify the true nature of this majestic formation: in fact the Milky Way is formed (as Democritus had supposed in ancient times) by a vast number of individual stars, as well as more compact groups of stars and clouds of dust and gas.

Today, in practical terms, we know that our solar system is part of the Milky Way (so we are not in a particularly good position from which to study it!), and it is thought that the Milky Way itself has the form of a flattened disc with a diameter of about 100,000 light-years and a breadth of some 20,000 light-years. Seen from above it would resemble one of those toy windmills children play with. Our Sun is simply a not particularly bright star which is situated towards the outer edge of the arms of the windmill. In addition to the Sun, the Milky Way is made up of hundreds of billions of other stars; we are thus some way from the anthropocentric and geocentric view that dominated man's concept of the Universe up to the time of Copernicus.

In fact in terms of the cosmos and its 'economy' we are even less important, because our Galaxy is not alone, and it is not even special: with the Mount Palomar telescope with its diameter of 5 metres (17 feet) billions of galaxies like ours have been observed; we have, therefore, to acknowledge, very modestly, that we are simply the product of a chemico-physical process (life) which is probably quite common, and which has occurred in our case on a planet belonging to a normal star belonging to a similarly normal galaxy.

All this knowledge, which we have sqeezed into just a few lines here, is in fact the fruit of a vast amount of dedicated research, with many a thrilling moment, and many a crisis, all fairly recent. It was in fact not until 1924 that the great American astronomer Edwin Hubble managed, on the basis of observations made with the Mount Wilson telescope, to determine the approximate distance of certain Cepheid Variables belonging to diffused objects (i.e. not points of light, like stars, but extensive luminous clouds, formed by many stars, with measurable angular dimensions). Hubble's measurements showed quite clearly that the objects to which the Cepheids belonged (we should remember that the period of this type of star is one of the classic methods of determining the distance) were very far away from us indeed, and could not be part of the Milky Way. Thus there were other groupings of stars or, as they were called, other Island-Universes, some of which were similar to our Galaxy. From then on it was necessary to adopt a larger scale to represent the dimensions of the Universe: in fact the nearest galaxy to our own is more than two million light-years away.

It has been proved among other things that the galaxies nearest to us (let us

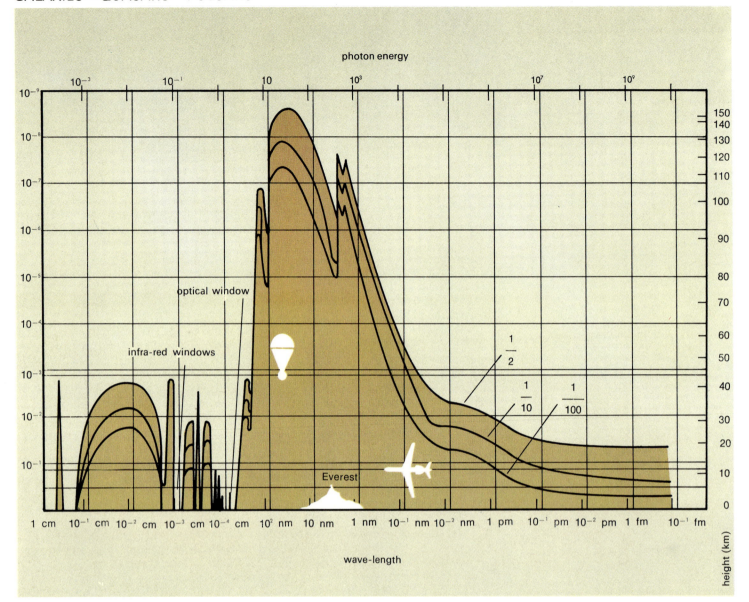

This diagram shows the absorption of electromagnetic radiation by the atmosphere on the basis of wavelength. The higher the peaks, the greater the absorption. The 'holes' between one peak and the next correspond to zones of transparency or 'windows'. The largest window corresponds to the radio-wave zone. Opposite: the huge radio-telescope at Arecibo, in Puerto Rico.

say those not further than three million light-years from Earth) form, together with our Galaxy, a group which is quite separate from the others: this is the so-called Local Group. This gathering of enormous objects into groups seems to be a fairly common occurrence in the Universe given that several galaxy groupings have been identified, from the smallest, such as ours which contains seventeen, up to the gigantic groups which run into thousands. In our quite normal Milky Way there are 100,000 million stars.

But let us now try to give an indication of how it has been possible to make these discoveries which, like discoveries made about our own tiny world, stir our imagination so strongly. A real leap forward was made when new techniques of non-optical observation were introduced. The principle tool involved in these techniques is the radio-telescope, which is capable of 'star-gazing' with 'eyes' which are sensitive to wavelengths both the human eye and the photographic plate cannot 'see'. The human eye of course sees only what lies in the band between about 0.0004 and 0.00008 millimetres of a wavelength, the first wavelength corresponding to violet light, and the second to red light but the skies send us information in the form of electromagnetic waves of every frequency, from radio-waves (measurable in kilometres, metres, centimetres

and millimetres, or miles, yards, feet and inches) to gamma-rays measuring some ten-billionths of a centimetre. It is thus clear that our eyes, and likewise the largest optical telescopes, can receive only an infinitesimal part of the information sent to us from the cosmos. It is as if we had a radio in which the tuning-knob was virtually stuck.

Modern radio-telescopes, now to be found throughout the world, work on the same principle as a normal optical reflecting telescope, that is, they concentrate energy issuing from the skies into a confined space. The basic nature of the rays the two types of instrument concentrate is also the same (electromagnetic) but the difference between wavelengths involved entails significant differences in construction. A radio-telescope can tolerate greater errors than its optical counterpart (it can thus have a much larger surface), but it pays for this advantage by having a lower level of definition, which means, broadly speaking, that it sees heavenly bodies as much more blurred. For example, in West Germany, at Effelsberg, there is a colossal radio-telescope some 100 metres (350 feet) in diameter which operates at a minimum wavelength of three centimetres.

The spiral structure of our Galaxy has been determined by radio-

The galaxies, immense heavenly bodies containing billions of stars, reveal their structure when observed with powerful telescopes. Left: a typical spiral galaxy – the Sculptor. Top right: a detail of the Milky Way, the galaxy to which the Earth belongs. Below: the Triangulum spiral galaxy.

Stars often gather in much less ordered shapes than actual galaxies; these are called star-clusters. Here we see the Messier 3 cluster.

astronomical observations at a wavelength of 21 centimetres, a particularly interesting wavelength because it can cross through the Earth's atmosphere and because it corresponds to a precise 'quantum jump' in the hydrogen atom. From examination of the characteristics of this line it has also been possible to determine the presence of magnetic fields in our Galaxy, another valuable source of information about its structure and movements.

On the basis of numerous observations, we know today that our Galaxy has a spiral structure which rotates on its own axis and makes a complete revolution about every 200 million years. We also know that it has an overall mass equal to 200 million times the mass of the Sun.

The diffused nebulae which we now know to be Island-Universes similar to our own Milky Way come in numerous and diverse shapes and sizes which suggest there is some common pattern of evolution for such universes: when analyzed optically and with radio-telescopes, we find elliptical galaxies, spiral galaxies, barred spiral galaxies and other totally irregular galaxies; our Galaxy belongs to the spiral category, and to a distant extra-galactic observer would look like the spectacular Andromeda Nebula (known scientifically as M 31), which lies slightly more than two million light-years away from us.

It should not be forgotten that the further we manage to see into space, the further back we go in time. The Sun we see is in fact the Sun as it was eight minutes ago because this is the time taken by its light to reach us; so when we see Andromeda we are in fact seeing it how it was two million light-years ago. It is thus evident that the concepts of observation in space and evolution in time are, in astronomy, inextricably linked. But are we sure of the distances that have been assessed? We have already seen that the methods in which we in some sense put our greatest trust, because they apply to the Earth's surface (the triangulation methods, essentially) have only a very short range, somewhere in the order of a hundred light-years or so, at most. All other methods used are indirect. They are based on certain phenomena relating to heavenly bodies with a known distance which have then been extrapolated and applied to what are presumed to be more distant bodies. A new and promising way of obtaining measurements for very distant stars (though subject to all the limitations of indirect measurements) has been supplied by the determination of the *red-shift* of the spectral lines of the light emitted from distant galaxies. The discovery of the *red-shift*, made by Edwin Hubble in the early 1930s, indicated that the further a heavenly body was from Earth, the more the electromagnetic rays

The various types of galaxies may be the result of a dynamic evolution which causes a progress from the cluster to the spiral. This idea, put forward by Edwin Hubble, has been questioned recently.

The Jodrell Bank radio-astronomical centre in Great Britain. Left: the 38-metre (125-foot) diameter radio-telescope at different angles. Right: the lower section of the huge 76-metre (250-foot) reflector. For a long time this was the world's largest adjustable radio-telescope.

that we receive from it were shifted towards the greater wavelengths (in the optic region towards red). This effect is well known in physics, and is called the Doppler effect, and more or less all of us have tangible experience of it (or at least those of us who travel by train). If a train approaches, whistling, and, still whistling, draws level and then away, we can easily pick up a variation of the 'acuteness' or 'shrillness' (the correct terms where frequency is concerned) of the whistle of the train passing by us. To be more specific, before the train draws level with us, we hear a shriller sound (that is, one shifted towards shorter wavelengths compared with those we would hear if the train were stationary), whereas after passing us we hear a lower-pitched sound (that is, on a longer wavelength). We now know, on the basis of the laws of quantum mechanics, the 'stationary' frequency of many spectral lines of the elements which form the heavenly bodies. Hubble in fact observed that the more distant the body, the lower pitched it is, in other words, the more its wavelength increases. In optical terms this means that the spectral lines characteristic of each element are shifted towards the red part of the spectrum, in proportion to the distance between the heavenly bodies and the Earth. This gives rise to a very simple mathematical relation: $V = Hr$, where V is the moving-away speed

(directly measurable from the *red-shift* of the spectral lines), *r* is the distance of the heavenly body from the Earth and *H* is a constant which is, not coincidentally, Hubble's initial. The value of *H* is clearly the crucial value for establishing distances and has recently been the object of fairly drastic criticism. Nevertheless, the $V = Hr$ law, which we now call Hubble's Law, seems to be standing up to all types of criticism. With the currently accepted values of *H* it has been possible to calculate distances in the region of 5000 million light-years.

The fact that the further away the objects observed by us lie, the greater their speed moving away from the Earth, might at first sight suggest that once again the Earth is being taken as the centre of the Universe, so that the old 'geocentric' view is gaining ground again.

This is not so. Let us imagine that we have a huge elastic sheet on which countless dots (the galaxies) are marked; the sheet is being pulled at all four corners by four invisible people; let us imagine that we are on one of these dots. We shall see all the other dots moving away from us and the same reasoning would be made by every inhabitant of every dot/galaxy in the sheet/universe. Thus this modern view is in no way a re-assertion of a geocentric universe. The

The huge Andromeda spiral galaxy is the best-known nebula and can be seen with the naked eye in the constellation of the same name. Here we see it in a colour photograph. On the outside it is blue because here we find the greatest number of thickly-clustered young, hot stars. Towards the centre it tends to look red, because here the colder stars predominate. This is also the nearest galaxy to Earth.

M.31 in Andromeda (left) is some two million light-years away. Top right: the M.51 galaxy in the constellation Canis Major, which is about 9,400,000 light years away, is formed by about 100,000 million stars. Below: the so-called Trifid (three-pointed) nebula, situated in the southern hemisphere.

same kind of analogy can be put forward for three-dimensional space. If we take our Galaxy to be one currant in a loaf cooking and rising in the oven, it will be seen that every currant (galaxy) will see every other currant moving away from it, as the loaf rises.

Other explanations have been put forward to account for the *red-shift* of the spectral lines of remote heavenly bodies (for example, an 'ageing' of the photons emitted), but to date the idea that it is caused by the expansion of the Universe seems the most probable one. The Universe as a whole is thus ever expanding. The main problem remains: is the Universe finite, or infinite? Will the expansion ever stop? There are basically two theories about the origin and structure of the Universe. One is the so-called *steady state* theory which suggests that the Universe has always been the way we see it today, and always will be: it will always be expanding but will at the same time always maintain a uniform density of matter. It is evident that in order to make provision for the void left by this expansion, it is necessary for new matter within the Universe itself to be created by some unknown mechanism. This theory, which was set forth by the great astronomer and author Fred Hoyle, is undoubtedly at once acute but simple. On the basis of observations made in recent years it would

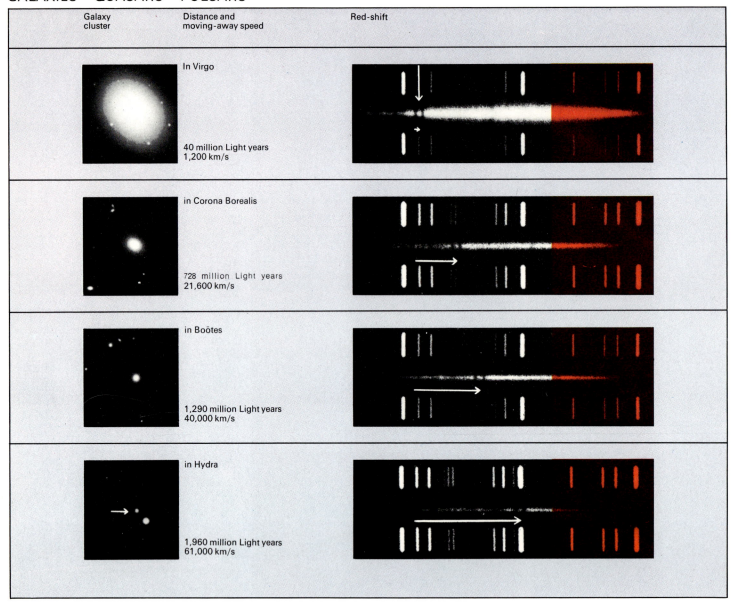

Galaxy cluster	Distance and moving-away speed	Red-shift
In Virgo	40 million Light years 1,200 km/s	
in Corona Borealis	728 million Light years 21,600 km/s	
in Boötes	1,290 million Light years 40,000 km/s	
in Hydra	1,960 million Light years 61,000 km/s	

seem, however, that it would have to yield to the alternative theory, the *Big Bang* theory. According to the *Big Bang* theory, which was proposed in 1948 by the physicist George Gamow, the Universe had an actual beginning, about 12,000 million years ago: then all matter was concentrated in a single place in the form of a primordial ball-of-fire. The explosion of the ball-of-fire gave birth to the Universe and the galaxies are thus the fragments of the explosion, hurled into outer space. The density and temperature of the primordial ball-of-fire were colossal, and respectively 10,000,000,000,000 million (ten billion billion) tons per cubic centimetre and 100 million °. According to the Gamow theory, all the chemical elements were formed from nuclear reactions occurring at the moment of the *Big Bang*; today, this original theory has been modified to some extent, because we know that the synthesis of chemical elements, by means of nuclear reactions, is a process which occurs continually in stars, and that the stars can explode and thus as it were put heavy elements back into circulation in the Universe; and these heavy elements are then incorporated by newly-forming stars. The *Big Bang* theory seems to have survived experimental tests carried out so far, but there is still undoubtedly a need for further confirmation of it. As we shall see, however, the most convincing results in its favour have

Various galaxies, with their distance from Earth and their moving-away speed in km. per second. We can see how the speed increases with the distance (Hubble's Law). The speed measurement is made by observing the red-shift of the spectral lines (Doppler effect).

We should not think that the Universe is static and immutable: the galaxies which form it are constantly moving further and further away from each other, and the further they are from Earth the greater their escape velocity appears to astronomers. In the drawing we see the galaxies 'in flight' from a central point, where our own galaxy is situated. Every shape represents a galaxy and the trail indicates the distance covered.

emerged from experiments made with non-optical instruments – usually with radio-wave detectors.

When we talked of the probable way in which our Sun might die, we saw that, once through the Red Giant and White Dwarf stages, it would go out once and for all and remain as an inert black mass in the sky. But this fate is not common to all the stars: in fact those with a mass equal to at least one and half times that of the Sun will die in a much more sensational manner. Instead of passing gradually through various stages, the large stars will, at a certain point in their life, literally explode, and the explosion will be so violent that the light thrown off in this phase will be 1000 million times brighter than that originally emitted: this is called the explosion of a Supernova.

It is a fairly rare occurrence; since 1054 just three Supernovae have been observed with the naked eye. In that year Chinese astronomers recorded an unforseen increase in luminosity in the constellation Taurus. Today we can see the remains of this explosion in the form of a luminous, expanding cloud, the famous Crab Nebula. The other two explosions of Supernovae recorded in our Galaxy date back to 1572 (observed by Tycho Brahe) and 1604 (observed by Kepler).

Perhaps of even greater interest is what remains after the explosion of a Supernova. According to certain theories, originating with the Russian physicist Lev Landau, it is possible, after the explosion, to obtain conditions in which the central nucleus of the exploded star is composed solely of neutrons. This would imply an extremely small but also extremely dense *neutron star*. Landau's hypothesis, dating back to the 1930s, remained a hypothesis pure and simple until July 1967, when a group of radio-astronomers from Cambridge, led by Anthony Hewish, discovered to their surprise that a heavenly body situated outside the solar system was sending extremely regular radio-signals at intervals of about one second. Landau's theory then came back into favour, once the science-fiction possibility that these were radio-transmissions from extra-terrestrial intelligences had been abandoned. This radio-source, which had been christened a *pulsar* (short for *pulsating star*), could be generated from the enormous magnetic field of a quickly rotating neutron star. Today several *pulsars* have been found and the analysis of the signals suggests that they are only a matter of some kilometres in diameter. A more sensational discovery was finding a pulsar actually near the centre of the Crab Nebula. It is currently thought that the origin of the cosmic rays present

The Crab Nebula, the visible remains of the explosion of a Supernova described by Chinese astronomers in 1054. This is one of the most studied of all heavenly bodies: the arrow indicates the position of a faint star which sends intermittent but very regular radio-signals. This is the pulsar NP 0532, formed by a neutron star rotating quickly on its own axis.

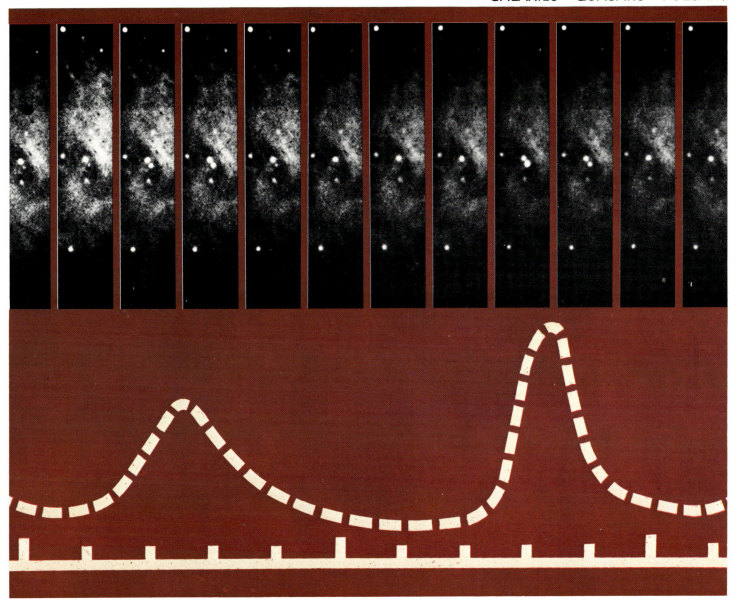

A graph of the electromagnetic 'flashes' emitted from the pulsar in the Crab Nebula. In the graduated scale the interval between 0 and 1 corresponds to 33/1000ths of a second.

in the Galaxy might reside in this strange type of star.

There are yet other stellar objects in the cosmos: of those the most interesting are the *quasars* (an abbreviated form of Quasi Stellar Radio Source), discovered in 1963 by Maarten Schmidt, who aimed the Mount Palomar telescope, then the largest in the world, in the direction of a radio-source detected by the Jodrell Bank radio-telescope. In the position of the radio-source Schmidt saw a somewhat insignificantly small star, but with an enormous *red-shift*; according to Hubble's Law, it must have been about 1500 million light-years from the Earth and moving away at the incredible speed of about 50,000 km (30,000 miles) per second. Although emitting an energy 1,000,000 million times greater than that emitted by the Sun, quasars cannot be very large, and are thus more akin to stars than to galaxies. The source of their huge amount of energy is still a mystery. Other objects even more mysterious than the *quasars*, whose existence had also been predicted in theory, are the so-called Black Holes. When a star with a very large mass has reached the phase of 'collapse', and in some sense caves in on itself because of the force of gravity, it can acquire such a density that it will be unable to 'flee' from anything, including light. We know that for every heavenly body there is a given 'escape

velocity'; if a body is launched at a speed lower than this given speed, it inevitably falls behind (for example on Earth the escape velocity is about 11 km [7 miles] per second). The escape velocity depends both on the dimensions and on the mass of the heavenly body: if, for example, our Earth contracted and kept its own mass unchanged, the escape velocity would continue to increase. When the radius of the Earth was reduced to the point where the escape velocity reached a value equal to that of the speed of light (which according to the theory of relativity is an impassable limit), nothing would be able to escape any longer, not even the photons, and the Earth would have then turned into a Black Hole. The fact that light, too, is affected by the action of gravity is again a consequence of Einstein's theory according to which every electromagnetic ray has an 'equivalent mass'.

We have already remarked that the Black Holes had been the object of hypothesis for some time, by physicists. Now they are back in fashion because it would seem that certain experimental observations made in recent years have confirmed their real existence. But how is it possible to detect such a weird object which swallows up everything like a bottomless pit and never gives anything back? One way is this: we know that every time an object deviates

The hypothesis that the Universe came into being as a result of a Big Bang *has been recently confirmed, experimentally. With the trumper receiver (left) it is possible to measure the basic noise of the Universe, the features of which tally quite well with those which the echo of the great primordial explosion must have had. Top right: J. Weber's gravitational radiation detector. Below: the radio-telescope at the Max Planck Institute, showing the parabolic dish and the swivel-system.*

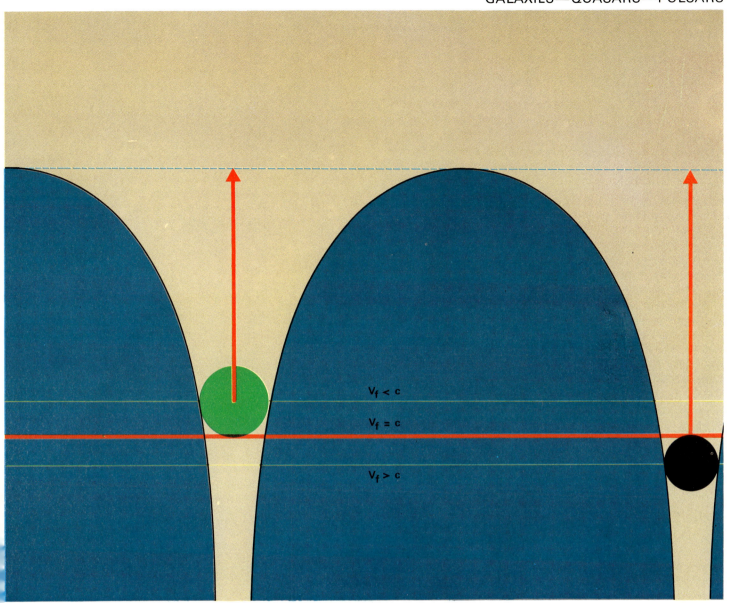

$V_f < c$

$V_f = c$

$V_f > c$

When the escape velocity (vf) of a heavenly body exceeds the speed of light (c), nothing, not even photons, can escape from its surface, and the body becomes a Black Hole, which is completely invisible – a sort of bottomless pit which swallows up everything, and lets nothing go: recent observations have, it would appear, confirmed the existence of such phenomena, which have been suggested theoretically for several decades.

from movement along a straight line, there is a force coming from somewhere which causes this deviation. In the cosmos the dominant force is gravitational. If, therefore, one could observe a star turning around 'nothing', that nothing would have to be the cause of the force, i.e. something with a mass, and a possible candidate would certainly be a Black Hole. In the constellation Cygnus a star of this type has been discovered which emits a considerable quantity of X-rays. The most likely hypothesis is that the star is part of a binary system, along with a Black Hole. The X-ray observed would be a consequence of the entry of stellar matter into the Black Hole.

MAN VERSUS SPACE

The possibility of overcoming the Earth's gravitational attraction and going off in person to explore the heavens has been one of mankind's most ancient dreams.

In practice, man only fulfilled this dream a few years ago; it is only a few years since the science and technics of space-flight – astronautics – achieved success. The date of the first successful flight can be put at 4 October 1957, when the Soviet Union put Sputnik I into orbit. This was the Earth's first artificial satellite, which travelled on an elliptical orbit at a maximum distance of about 9400 km (5850 miles). The complete realization of the age-old dream – the first manned space-flight – was achieved again by Soviet technology on 12 April 1961, when the capsule Vostok I, with Yuri Gagarin on board, completed an entire orbit of the Earth, and then landed successfully. The impact of this conquest of space was tremendous, not least because it brought to light the unsuspected supremacy of Soviet technology.

After a few years, during which space-flights followed one another in close succession, for a variety of purposes, the world's enthusiasm cooled considerably, and news of them was demoted from head-lined front-page articles to brief bulletins hidden in the depths of daily newspapers. The subject hit the headlines once more with the first landing on the Moon by American astronauts in the Apollo programme.

The dwindling of popular interest in space-flights does not mean that their importance has diminished from the scientific point of view – quite the contrary. We are much better acquainted with our solar system, precisely because of the vast harvest of data which has been made available by ventures which may have been hardly spectacular, but which have been painstakingly programmed.

Before we discuss the space conquests of the last twenty years, we should briefly trace the history of man's attempts at extra-terrestrial exploration; this is essentially the same as the history of 'vectors', that is, of rockets capable of putting space-capsules into orbit.

The invention of the rocket probably cannot be exactly dated, but it is the fruit of a long, slow process of development. There is, however, no doubt that in the thirteenth century the Chinese were already acquainted with the rocket and used it as a weapon, whereas in Europe its appearance dates back only to the seventeenth century, and thereafter interest in it was only spasmodic. The true founder of modern astronautics was a young Russian mathematician, Konstantin Eduardovich Tsiolkowski, who, in 1895, published an article about journeys into space in which he proposed, well ahead of his time, the use of liquid fuels instead of solid fuels – i.e. liquid oxygen and liquid hydrogen. Tsiolkowski subsequently made extremely important contributions to the theory of space-flight, but his natural modesty and the language-barrier meant that his work was paid very little general attention for many years.

In practical terms, the fathers of modern astronautics are generally considered to be the German Hermann Oberth (another mathematics teacher) and the American Robert Goddard, after whom one of the major space-research centres in the United States is named.

Oberth was principally concerned with inter-plantery flights and manned

The Earth seen from an altitude of 44,000 km (27,500 miles) in a photo taken by Apollo 10.

space-stations, which, in his view, could be used for meterological observations (he thought of an iceberg warning-system, with the tragedy of the *Titanic* in mind). Oberth's first work – *The Rocket in Interplanetary Flight* – was published in 1923. Three years later Goddard launched a rocket which used as fuel liquid oxygen and gas oil, but it only reached an altitude of 60 metres (200 feet). The importance of the work done by Tsiolkowski, Goddard and Oberth, however, lies in the fact that their hypotheses and proposals were not the product of wild imagination but of a truly scientific rationale backed up by reliable, and still valid, calculations. The first major developments occurred shortly before the Second World War, when German scientists opened the way to modern space technology.

It can be said without any doubt that the parent of all modern carrier (or mother) rockets was the notorious *V2*. The then youthful Wernher von Braun had had a part in its construction, and when the war was over he moved to the United States. The *V2* was a rocket which used, as fuel, liquid oxygen and a mixture of water and ethyl alcohol. It could reach a top speed of 6000 kph (3750 mph) and was widely used during the last war to bombard London, showing itself to be a lethal weapon, not least because its speed, which was considerably

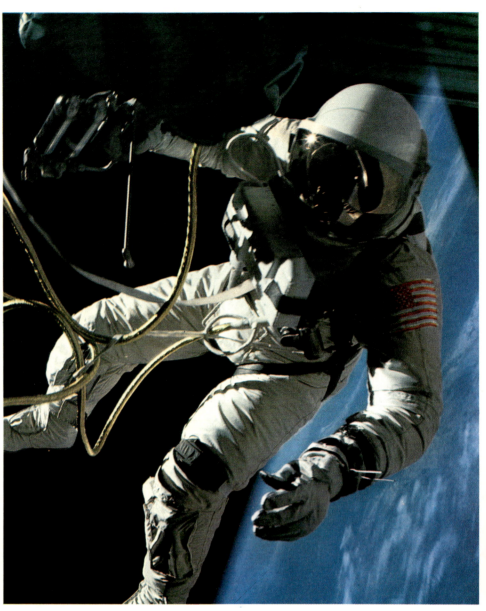

faster than the speed of sound, made it difficult for any warnings to be given to the population. As well as these rockets, the Germans developed other notable types of rocket, among them the *Rheinbote*, a four-stage rocket which was very advanced for its time.

At the end of the war both the Soviet Union and the United States profited from the store of German knowledge about rocketry, by importing the most qualified rocket experts and studying the components of captured *V2s* (the Russians even constructed modified, improved *V2s*).

It would take too much space to give the whole history of rockets here; sufficient to recall that as early as 1949 the United States had reached an altitude of 400 km (250 miles) with their WAC rocket, mounted as a second stage on a *V2*, while the Soviet Union, about whose research in this area the West knew very little, stunned the world with the successful launching of the first artificial satellite, Sputnik I, in 1957.

It should be said that for many years the Russians showed considerable superiority over the United States, putting into orbit satellites which were much heavier than American models; the first American satellite was in fact launched only on 31 January 1958 (after a second Sputnik weighing more than

The first orbital flight was made on 12 April 1961 by the Russian Yuri Gagarin (top left). Subsequent improvements enabled the Russians to carry out the first 'space-walk' on 18 March 1975. Below left: the astronaut Leonov walking in space. Three months later the Americans achieved a similar feat when White, floating in space, emerged from the Gemini capsule (right). Opposite page; right: the launching ramp of the Saturn V carrier rocket. Top left: an astronaut in the Mercury capsule and (below) two phases in the link-up between Gemini 10 and the Agena missile.

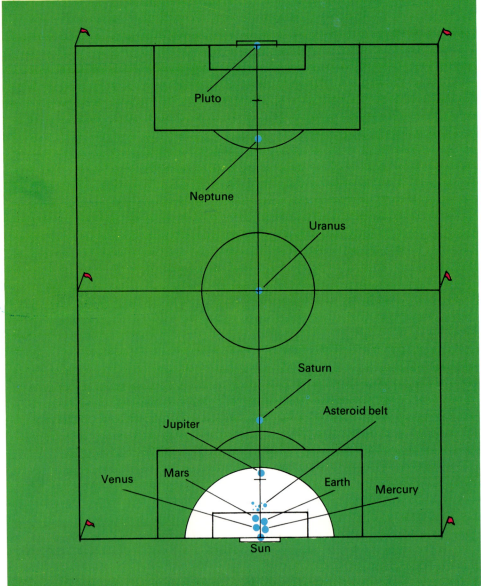

half a ton and carrying a dog on board had already been put into orbit). The first American satellite, Explorer I, weighed just 15 kilos (34 lbs). After these first launchings there was a race between the world's two major powers which was closely watched by the rest of the world. In this race the Soviet Union was always ahead, and this lead was consolidated further when the first man, Major Yuri Gagarin, was sent into space on 12 April 1961. By this date some twenty space-missions had already been completed, some of which were of great scientific interest: for example, the Van Allen belts had been discovered (1958), the first transmissions of messages from space had been made, and the Moon had been encircled, with television recordings of its hidden side.

It was not until 20 February 1962 that the United States first managed to achieve a manned orbital space-flight, piloted by John Glenn Jr. who, on board the capsule Friendship-7, orbited the Earth three times before landing safely in the sea (another aspect of Russian technical superiority lay in the fact that while American capsules had to end up in the sea, thus requiring a sometimes hazardous recovery operation, the Russian capsules came down on land). There would be little point in making a list, here, of the countless space-missions which followed and which are still being made, not only by the two

Left: the relative distance of the planets in the solar system compared with the size of a football pitch; despite the huge progress made in recent years, exploration of the planets with automatic space-probes has not yet gone beyond the penalty area. Right: the Lunar Excursion Module (LEM) in flight and hooking-up.

200

The exploration of the Moon has benefited from specially designed mechanical aids in both the Russian and the American missions. Top left: the Lunakhod, an automatic Soviet vehicle. Right: the military salute given by moon-explorer Irwin, to the American flag. Below: sample collecting from the Moon's surface by the Apollo 17 mission. The vehicle on the right is the battery-powered Lunar Rover.

major powers, but also by other technologically advanced nations.

But one general note is necessary: the emphasis on propaganda gradually subsided and as from the 1960s there was an intensification of launchings made for scientific research. By now the Earth's gravitational field is crisscrossed by a vast number of satellites which carry out tasks involving meteorological reconnaissance and act as television and communications relays in general. It has become commonplace for us to watch television programmes 'via satellite' about events of international interest happening thousands of miles from home. It is precisely the scientific aspect which we want to deal with in the few pages to come.

Mankind's age-old dream of exploring, directly, the planets in the solar system was taking shape, and it is natural that the object of the first research should be the heavenly body nearest to Earth – the Moon.

As early as 1959 the Russians had sent an automatic space-station to the Moon, which had discovered that our satellite was surrounded by a layer of ionized gas, and the following month, as we have mentioned, the space-probe Lunik IIII had sent back to Earth photographs of its far side.

Here too the Americans were lagging some way behind. In January 1962 the

attempt to land the automatic space-probe Ranger-3 on the Moon ended in disaster, because the capsule missed the target by 40,000 kilometres (25,000 miles). But they caught up handsomely when the first man, Neil Armstrong, set foot on the Moon in July 1969.

The fact that the first human landing on the Moon was transmitted direct to every television set in the world contributed considerably to boosting the prestige of the United States in the field of space-research. The exploration of the Moon, started more for reasons of national prestige than out of real scientific interest, was part of an ambitious project (the Apollo programme) which, as a result of numerous missions has succeeded in giving an extremely detailed picture of our own satellite. We now have access to highly detailed photographs taken either directly on the surface or at various altitudes, and an enormous amount of data gleaned by the most varied and sophisticated instruments imaginable, such as seismographs, spectrometers and magneto-meters. In addition, laboratories in many universities have had access to samples of lunar rocks for direct analysis and investigation.

But man could certainly not sit back when the nearest heavenly body had been visited; before long his gaze turned to the planets. This involved a large

The now numerous artificial satellites are in a wide variety of shapes and sizes, depending on their function. Left: a satellite for cosmic ray research. Top right: Echo, a satellite acting as a passive relay reflecting radio-waves. Below: Early Bird, acting as an active relay, is used for telecommunications.

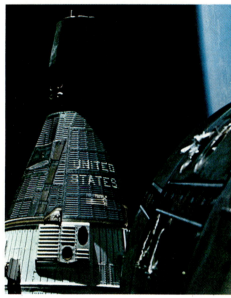

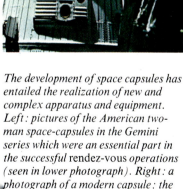

The development of space capsules has entailed the realization of new and complex apparatus and equipment. Left: pictures of the American two-man space-capsules in the Gemini series which were an essential part in the successful rendez-vous operations (seen in lower photograph). Right: a photograph of a modern capsule: the fuel tanks are visible.

jump in terms of distance. While the Moon is about 38,000 km (23,000 miles) from us, the nearest planet, Venus, is about 50 million km (30 million miles), under the most favourable conditions.

The first attempt at interplanetary exploration dates back, in fact, to February 1961, when the Russians launched the space-probe Venera I in the direction of Venus with the aim of measuring the interplanetary magnetic field. This was the first space-station to complete a hyperbolic orbit, i.e. an open orbit (all the previous probes had made elliptical, i.e. closed, orbits). Venera I went within about 100,000 km (60,000 miles) of Venus, but unfortunately radio contact was lost when the probe was 7 million km (4.3 million miles) away from Earth. The American interplanetary satellite, Mariner II, was more successful, and managed to fly within 35,000 km (22,000 miles) of Venus and also to record some interesting scientific data relating principally to the planet's temperature.

The exploration of Venus continued in the years that followed, with both Russians and Americans taking part, but it was the former who obtained the most important data. In March 1966, four months after being launched, Venera III made the first landing on the planet. Although it failed to send back any data, success in this respect came in the following year with Venera IV

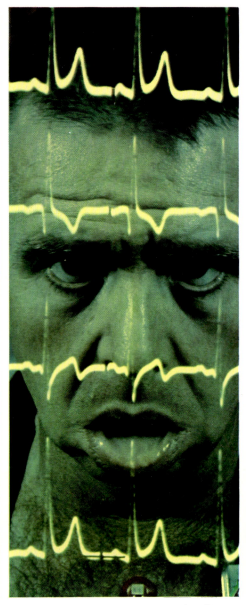

which transmitted findings of great interest for more than 90 minutes during its parachute-assisted descent into the extremely dense atmosphere of Venus. To crown these Russian successes, the landing module of Venera IX came gently to rest on the surface of the planet on 22 October 1975, sending back photographs of the planet's terrain, which is invisible, because of the thick cloud-cover, not only to the most powerful of optical telescopes on the Earth but also to orbiting space-capsules. Three days later, and about 2000 km (1250 miles) from the spot where Venera IX had landed, the module of a second probe, Venera X, also landed on Venus, and sent back a second photograph which differed little from the first.

If the exploration of Venus is almost completely in the hands of Russian space-research teams, the exploration of Mars is due, in the main, to the Americans.

In July 1965, having been launched in November of the previous year, the space-probe Mariner 4 flew over the planet at a maximum distance of 10,000 km (6000 miles) and sent back 22 television pictures of the surface of Mars. The Mariner project was followed up in later years and the subsequent missions produced more precise pictures of a planet which has always aroused man's

Left: an experiment simulating conditions of weightlessness to which astronauts are subjected for long periods of time. The human body seems to handle these anomalous and irregular conditions fairly well, but problems resulting from very rapid acceleration are more taxing. Right: the effect of acceleration and the recording of physiological data.

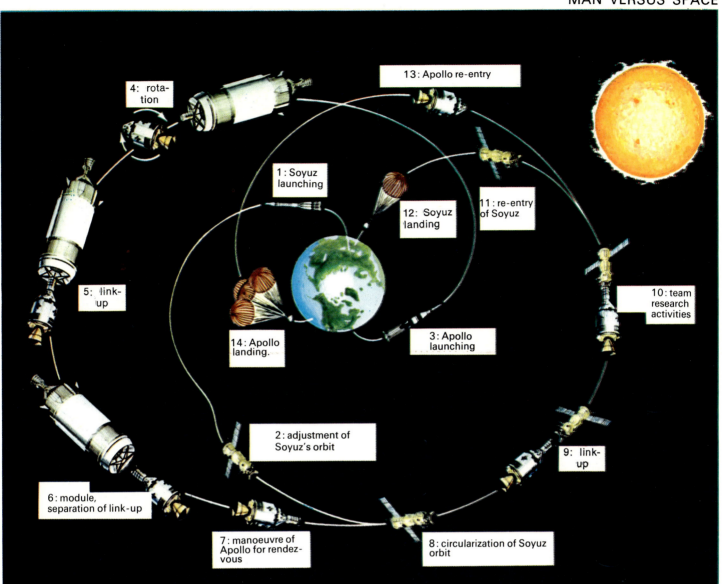

4: rota-tion

13: Apollo re-entry

1: Soyuz launching

12: Soyuz landing

11: re-entry of Soyuz

5: link-up

14: Apollo landing.

3: Apollo launching

10: team research activities

2: adjustment of Soyuz's orbit

9: link-up

6: module, separation of link-up

7: manoeuvre of Apollo for rendez-vous

8: circularization of Soyuz orbit

The sequence of the various launching and flight-manoeuvre phases of the two space-craft Apollo and Soyuz during their mission. In the phase of linked-up flight the teams had at most two days in which to undertake technical and scientific experiments and research activities.

curiosity.

In August 1969 Mariner 6 and Mariner 7 came within 3500 km (2200 miles) of Mars, and sent back in all 202 pictures, while in November 1971 Mariner 9 was the first space-craft to enter into orbit round another planet. In July 1976 Viking I landed on Mars, sending back the first photos showing a stony desert beneath a drab sky.

The last relatively close planet in the solar system is Mercury. Only one exploratory missions has been made to this planet, by the probe Mariner 10, which flew over it in March 1974 at an altitude of about 700 km (440 miles). This was a combined mission because as well as sending back some 8000 pictures of Mercury, the probe also passed quite close to Venus and transmitted data about that planet as well.

Between Mars and Jupiter there is a considerable 'gap': for more than 150 million km (95 million miles) there are no planets, just planetoids or asteroids which are to be found in a vast belt. The exploration of Jupiter with space-probes thus involved the unknown risks of crossing through this belt. The first launching towards Jupiter occurred in March 1972, in the form of Pioneer 10 which successfully negotiated the asteroid belt and came within 130,000 km

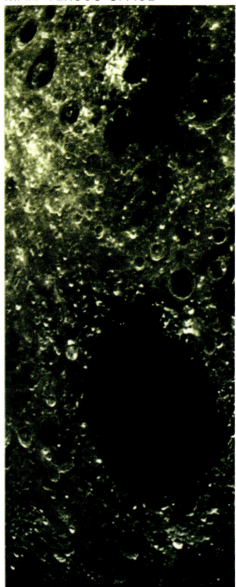

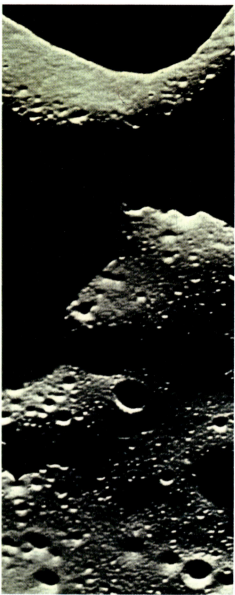

(80,000 miles) of the gigantic planet on 4 December 1973, sending back useful data about its magnetic field. Pioneer 10 was the first space-craft to make use of the field of gravitational attraction of another planet. A month later it was followed by Pioneer 11, which came even closer and is, at the present time, on its way to Saturn.

After this summary of the various space ventures, we might well ask ourselves what their value is, at a time when the Earth itself is in economic and social crisis, for the cost of these programmes is, literally, 'astronomical'. There are of course ideological and propaganda reasons, there is the satisfaction of scientific curiosity, there is a significant spin-off in technological advance of use in everyday life; there is also the advance that has been made, in space-medicine, in understanding the human body.

One of the conditions to which man accustoms himself most readily is that of the Earth's gravity, which has a value of about 9.8 metres per second (32.7 feet); this is indicated, usually, by the symbol g, and is the scientific cause of our weight. During space-flights, be it on take-off, landing or in orbit, we find gravity conditions which differ sharply from those on Earth. There is a transition from a complete absence of gravity while in orbit ($g = 0$) to values of

Aspects of the Moon's surface. Left: the Sea of Crisis as seen by the Apollo astronauts. Centre. a geological map of the region near Kepler's crater. Right: the Langrenus crater in which the central cone is clearly visible.

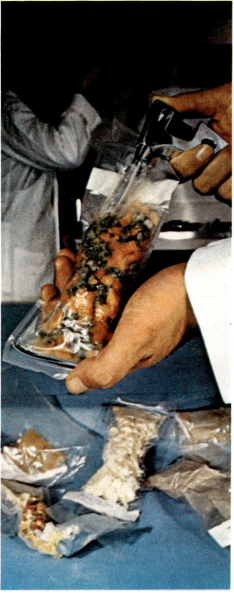

For long manned interplanetary flights, space-craft must be a tiny, self-sufficient world. Top left: a marsh algae capable of producing a considerable amount of oxygen, which is vital for the human metabolism. Below and right: the use of dehydrated foods. After the addition of water the food must be sucked from appropriate containers, because of the absence of gravity.

about $7 g$ during the phases of take-off and landing, that is acceleration and deceleration. The astronaut experiences a sensation similar to, but much more marked than, the feeling we have when we are pressed against the floor of a high-speed lift as it starts to move upwards. An acceleration of about $7 g$ means that the weight of the individual increases seven times, and this applies to all the tissues and to the internal organs as well. For example, the specific weight of the blood in these conditions is almost the equivalent of the specific weight of liquid iron. Fears for the lives of astronauts was, logically enough, considerable.

In effect the human body has shown itself to be much more hard wearing than expected and (for brief periods) will tolerate accelerations up to *20 g*.

We thus find the almost paradoxical phenomenon that the limits imposed on acceleration are determined more by the structures of the rocket than by the structure of the human body. The conditions of total weightlessness pose, for space-medicine, questions which are even more far-reaching than those posed by very high acceleration. First of all, while the periods of acceleration are fairly short and can be simulated on Earth, weightlessness is the condition in which the astronaut lives for days on end, and it cannot be simulated on Earth,

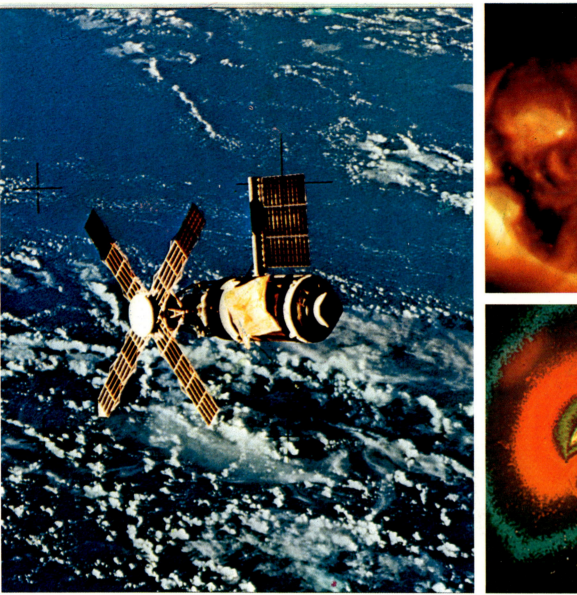

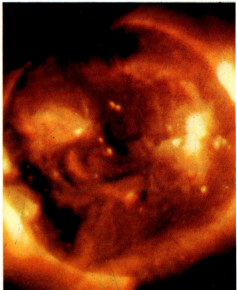

except for brief periods. At the same time it should not be forgotten that in an unnatural situation of this type there may well be fairly considerable psychological problems: in fact, the adaptation to the condition of weightlessness involves continual stress, which leaves little room for adaptation to the other variations in environmental conditions, and induces a state of genuine over-tiredness. For these reasons, and even though, to date, experience has shown that the human body will stand up quite well to a prolonged state of weightlessness, it is necessary, for future projects, for space-craft to be equipped with some system for creating an artificial gravity. The only possible system is to have a vehicle which is radially symmetrical and rotates on its own axis. The astronauts inside such a vehicle will be subjected to a centrifugal force (proportional to their distance from the axis of the vehicle and to the square of the frequency of rotation) which will push them outwards, giving them the impression of being in a gravitational field, i.e. of having weight. With a circular space-craft with a radius of, say, ten metres (33 feet), it would suffice for it to make one complete revolution on its own axis in slightly more than six seconds, for there to be an artificial gravity, inside it, equivalent to the gravity of Earth. One of the best models of a space-craft of this type can

Left: Skylab photographed by the astronauts themselves: the solar panels arranged in the form of a cross are clearly visible. Top right: an X-ray picture of the solar corona taken with the Skylab telescope. Below: an ultra-violet photo of the light emitted by the Earth, taken with the photo-telescope installed on the Moon by the Apollo 16 astronauts.

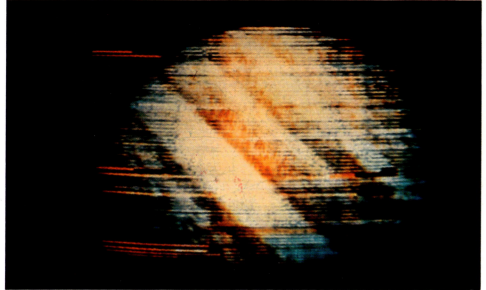

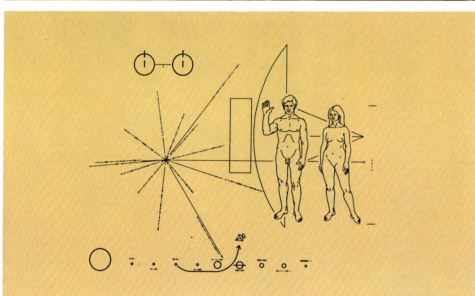

The Jupiter mission of the space-probe Pioneer 10. Top left: photograph of the planet taken from the space-probe. Below: a drawing carved on an aluminium gilded plaque attached to Pioneer 10. The radial lines represent the position of fourteen pulsars which would enable intelligent beings to detect the position of the solar system. Right: the take-off of a carrier rocket, a three-stage Atlas-Centaur.

be seen, and admired, in Stanley Kubrick's fine film *Space Odyssey*, which is meticulously accurate scientifically. There is a scene in this film which arouses curiosity and also much laughter. We see an astronaut walking around inside the circular space-craft, which the audience sees from the front. At one point he walks on the 'ceiling'; in reality, because, inside the space-craft, the artificial gravity is always aimed outwards and hence the bottom is always the actual outside point, the attitude of the audience is like that of the person amazed that people living in the southern hemisphere do not 'fall off' the Earth.

All the major problems to do with man's survival in space will certainly be multiplied when the time comes to tackle interplanetary flights, because even the shortest flight (to Venus) will inevitably last for months. For such lengthy periods it will be necessary to include in the crew a space-doctor and an engineer, because the problems of optimizing the weights and volumes transported will come into play. It is in fact possible to calculate that each astronaut needs, every day, about 5 kilos (11 lbs) of oxygen, water and food, and that almost half this weight is accounted for by water. And this figure does not account for the water needed for ordinary and vital hygiene. It is thus evident that for fairly long flights it will be crucial to recycle various products

209

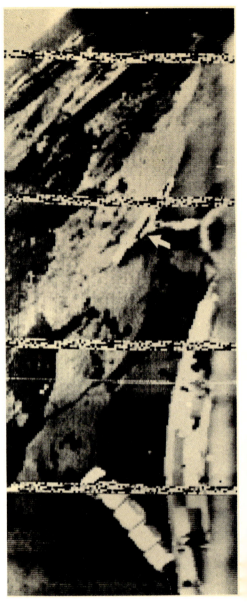

The planet Venus has been of the greatest interest to Soviet space technology. Left: assembly of the automatic probe: Venera; right: the surface of the planet photographed from Venera 10.

(by this term we mean the partial regeneration of a product so that it can be re-used). For water the problems of regeneration seem to be fairly well resolved, but the greatest problem is that of oxygen, which is vital for the metabolism of the body. An interesting prospect seems to be a system which will use the photosynthesis of certain algae which, using carbon dioxide and water, produce oxygen by taking energy from sunlight, something that is certainly not lacking on board space-craft.

Another method might be that of thermal decomposition for separating oxygen from carbon: in this case fairly high temperatures would be necessary which, in principle, could be obtained by focusing (with parabolic mirrors) the Sun's rays.

However, there are other dangers in store for astronauts, first and foremost of which is the danger of radiation, be it electromagnetic (X- and gamma-rays) or corpuscular (above all protons and electrons). We know that protons and electrons are continually emitted by the Sun and that, therefore, an adequate screening is necessary, but the danger still remains, because the structures of the space-craft can absorb the particles, but the impact of the electrons against metal structures can produce penetrating X-rays which could seiously damage

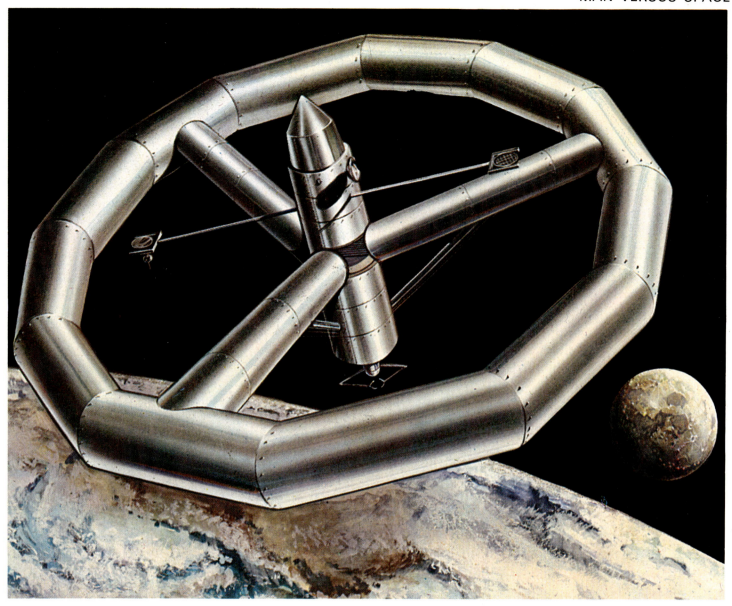

The model of a large rotating space-station. The rotating movement creates inside the station a field of apparent forces which simulate gravity. Future astronauts will not have to cope with the problems of weightlessness.

human cells. On the other hand, damage from primary cosmic rays seems less likely.

However, all these dangers, and possibly even greater unknown hazards, have certainly not discouraged man from his desire to venture towards the depths of the cosmos, just as explorers and conquerors in history have not been discouraged from venturing into unknown and dangerous lands.

In the years to come we shall certainly witness new and major astronautical developments; man will set foot on other planets as he has already done on the Moon, but let us hope that he will not be forced to colonize them by having made uninhabitable the planet on which he was born and has evolved, the Earth.

THE EARTH

GEOLOGY

Geology is the science of the study, composition and structure of the Earth's crust; it provides a chronological reconstruction of the processes which, in the course of hundreds of millions of years, have modified the crust's original appearance and nature.

The Earth's crust now has complicated corrugations or folds which are seen as the main mountain ranges, and vast areas which are flattish, where the original surface has been relentlessly whittled away to varying degrees by the erosive action of water, ice and wind.

In the geological history of the Earth we find four major mountain-forming cycles: evidence of the most ancient, which has been largely obliterated by erosion, is to be found in Canada and northern Scotland. The most recent is the Alpine cycle, responsible for the formation of the principal present-day 'folds' on the Earth: the Alps, the Himalayas, the Rocky Mountains and the Andes. The continual and gradual evolution of the Earth's surface produced the great variety of unusual forms which give it, as a whole, its at once picturesque and somewhat tortured aspect.

The geological history of the Earth is documented by the fossil remains of organisms which existed in past epochs. The first single-celled beings which inhabited the original oceans and seas gradually and progressively evolved towards more and more complex and varied forms of life, which only emerged from the water some 450 million years ago to conquer both dry land and the air. The first traces of life are therefore to be found in the sedimentary rocks now situated or originally deposited under the sea. In the rock formation of Swaziland, in southern Africa, dating back 3400 million years, curious microstructures have been found with a roundish form; these are considered by some scientists to be single-celled blue algae. Laminar structures certainly associated with the activities of algae and bacteria have been observed in rocks some 3000 million years old in Rhodesia.

Evidence of subsequent stages of the evolutionary process is found in more recently formed layers of rock wherever the original layering has been conserved. On the basis of the principle of superimposition, according to which the more recent strata are superimposed on the more ancient, in a normal layering of rocks, it is possible to establish a *relative chronology*. Quite often the chronology is made possible by the presence of the same fossil remains found, even on a continental scale, in layers of rocks of different origin. Then more specific analyses of the rocks enable us to single out their probable area of origin and the environment in which they were deposited, thus building up a palaeogeographical picture of the conditions in which sedimentation took place. We then know which came first, which was the second layer, which the third, and so on, but this does not allow any layer to be dated. That is why this research only gives us a sequence of events, a chronology relating one layer to another. The problem of working out an *absolute chronological scale* giving dates has exercised the minds of many eminent scholars from the last century onwards. Studying the evolution of molluscs from the start of the last Ice Age and comparing it with that of previous eras, Lyell gave the Myocene Epoch a duration of twenty million years. On the basis of related studies he put the beginning of the Palaeozoic Era at 240 million years ago, and the start of the

Cainozoic Era at 80 million years ago. The last few decades have seen, for the dating of the absolute age of rocks, the introduction of methods based on the slow and natural radioactive decay of certain elements. Because we know the speed at which this phenomenon occurs, it is possible to calculate the absolute age of rocks. Depending on the case, the method used is chosen on the basis of the mineralogical composition and the likely age of the rock itself. On the results obtained from relative and absolute chronology, it has been possible to divide the Earth's history up into *eras*, *periods*, *epochs* and *ages*.

At this point we should introduce a new concept, that of *geological time*. The normal perception of time means that we are capable of assessing, at the most, the duration of historical periods equivalent to not more than four or five millenia. Vice versa, when dealing with the geological processes, we have to refer to fairly long periods, in the region of millions or even billions of years: we thus have to enter a new chronological hierarchy.

Another concept, which is part and parcel of the study of geology is the *principle of actualism*, set forth by Lyell. This principle states that the same causes currently at work with, for the most part, modest intensity, if carried on for very long periods of time will bring about major phenomena such as the accumulation of sedimentary nappes, the formation (folding) of entire chains of mountains, and the formation of huge lava plateaus or tablelands. For example, if today we put some sea shells in a bowl of water, they would sink to the bottom. Lyell's principle means that over millions of years, the same simple operation, of empty sea shells falling to the bottom of the sea, and then lumping together, led to the formation of certain rocks.

This principal contrasts sharply with a previous theory elaborated by the palaeontologist G. Cuvier who, in order to explain the wide variety of fossil remains, suggested that from time to time apocalyptic cataclysms had occurred which, he maintained, had caused the overnight extinction of various living species, followed by the emergence of new species.

The theory of *continental drift*, set forth by A. Wegener in 1912, appears to fit in perfectly with the principle of actualism. Wegener maintained that the continents were floating on the ocean-bed and explained their current position by suggesting a slow movement of drift which started some 200,000,000 years ago, when, presumably, they were all joined in one continental block.

Recently this theory has been developed more specifically, thanks to certain sensational discoveries made as a result of deep-sea exploration. Recent bathymetric findings and reliefs have brought to light the curious confor- mation of the ocean-bed, revealing a complex system of submarine ridges and trenches which are associated with areas of intense seismic activity. Where the ridges are concerned we find large lava flows (effusions) which cause the constant expansion of the ocean-bed; in the vicinity of the trenches the bed sinks into the plastic substratum, called the *mantle*. The relatively recent age of the thin ocean crust confirms the hypothesis of an incessant process of regeneration and consumption.

Decisive proof comes from the magnetometric findings which have detected zones, symmetrically positioned in relation to the axis of the trenches, in which the direction of the Earth's magnetic field, alternatively normal or inverted, has remained 'frozen' in the magmatic rocks when they solidified. This periodic inversion is the key for the reconstruction, in chronological terms, of the evolution of the ocean-bed. The system of ridges and trenches subdivides the Earth's crust into six principal blocks in which the continents are the area above water. Not unlike spherical vaults (or segments) running across the mantle, these blocks grow around the ridges and disintegrate by withdrawing near the edges of another advancing block. The corrugations of the Earth's crust can be attributed to the interaction of the various blocks.

Geology has recently become important as a science which aids the location and systematic exploitation of natural resources. Considerable importance attaches to oil-prospecting, for example with the use of sophisticated equipment which makes it possible to detect an ever-growing number of oil- fields.

THE EARTH: FROM ITS BIRTH TO THE PRESENT DAY

On the basis of current knowledge, it is not yet possible to give a definite answer to questions about the birth of our planet. From various standpoints by different methods, geologists are investigating the history of the Earth to see if chronologically it will be possible to pinpoint the beginning of the existence of our planet. It has not yet of course been ascertained when exactly the galaxies formed in the Universe, although their birth is thought to date back to between twelve and sixteen billion years ago. According to the latest scientific views, the Earth originated from the original solar nebula as the result of fragments being hurled into space and then gathered together to form our planet. Subsequently, and because considerable overheating occurred, the whole formed by these fragments developed the properties of a spindle, or half-spindle, producing, within the Earth, a concentric and orderly arrangement of matter. As a result of the force of gravity and the rotating movement, there was an accumulation of the heavier elements at the nucleus (or core) and a gradual distribution of the lighter elements towards the outer areas. The Earth must have looked something like a slowly cooling ball of fire. Once a relatively low temperature had been reached, portions of the crust then started to solidify and form; these portions were like islands which subsequently sprang up and characterized the whole Earth's surface. The geological history of the Earth starts from this point.

To the major questions such as the origin of the first continental masses – using the accepted meaning of the term – or of the great oceanic basins, there have to date been only tentative replies. According to the most recent theories, the Earth reached its present shape about 4,600,000,000 years ago. If we took as valid the latest theories about global tectonics, which are the only theories capable of coherently explaining the eruptive, sedimentary and metamorphic phenomena of the last hundreds of millions of years, the problem would be partly solved, although it would then be necessary to check to what point the principles on which this hypothesis is based could be taken back in time. Results obtained to date suggest that one can consider that the mobility of the Earth's crust, as early as in the Precambrian period, was similar to its recent mobility, although the phenomenon, then, was of a different intensity.

According to modern authors, the great rocky masses which we, today, find in all the continental areas, already existed 3,000,000,000 years ago. In that period, in addition to their primary rocks (from the solidifying of the crust) there were also changes taking place because of particularly high temperatures and pressures. However, it is not possible to know the extent or position of the continents and oceans at that period; nor is it possible to establish if the thickness of the crust masses could be compared with their present-day thickness. On the basis of current knowledge it can be presumed that the continental crust was formed at the very beginning of the history of the Earth and that subsequently it underwent modifications because of pressures and movements. The evolution and development of the crust have happened at a regular pace, but have followed step by step the patterns caused by pressures.

If we assume that the Earth's crust was affected in its early days by the same sort of activity as we now see in volcanoes, in a time-span of about 3000 million years a continental crust would have been formed, but of an unstable nature; it

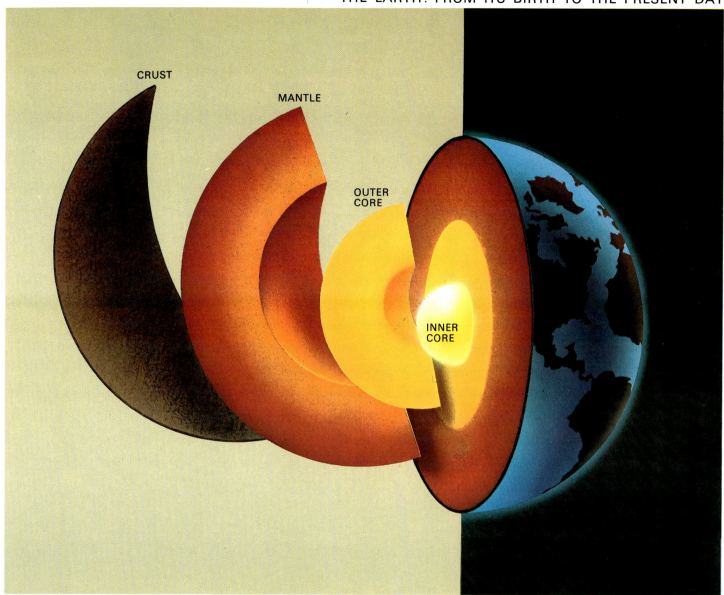

CRUST

MANTLE

OUTER
CORE

INNER
CORE

A diagrammatic picture of the Earth, with half of it in cross-section to show the inner structure. The nucleus *or* core *at the centre is divided into two parts: the inner core, which is probably solid, and the outer core, which is fluid. Then around the core we have the* mantle, *and lastly the Earth's crust, the thickness of which varies from region to region. Scientists consider that enormous and very slow* convector currents *are moving in the mantle.*

would have been made up largely of basic rocks, although these were later subject to change. The first stable rocks in the crust must have been sialic, that is of the granite type, and must have been formed during a particular phase of continental growth. Because of the lower density, these were able to float above the lower basaltic zones, and the beginnings of the hydrosphere would have coincided with their formation.

As time passed, the Earth tended to assume the form which we, today, can reconstruct reasonably exactly. Harking back to Wegener's theory, in the Palaeozoic era the continental masses formed one enormous block, the *Pangea*, which then gradually split up, first into two large masses, the *Gondwana* in the south and the *Laurasia* in the north. These then broke up into the continents as we know them today. The mass of water, which started off as a single vast ocean, the original Pacific Ocean, also tended to separate, and created new ocean basins, which gradually spread outwards.

The formation of the Atlantic and Indian Oceans started in the Secondary (or Mesozoic) Era with the appearance of the fractures which split Gondwana and Laurasia into various parts. The expansion of these fractures is connected with the future mid-oceanic ridges which currently constitute a continuous

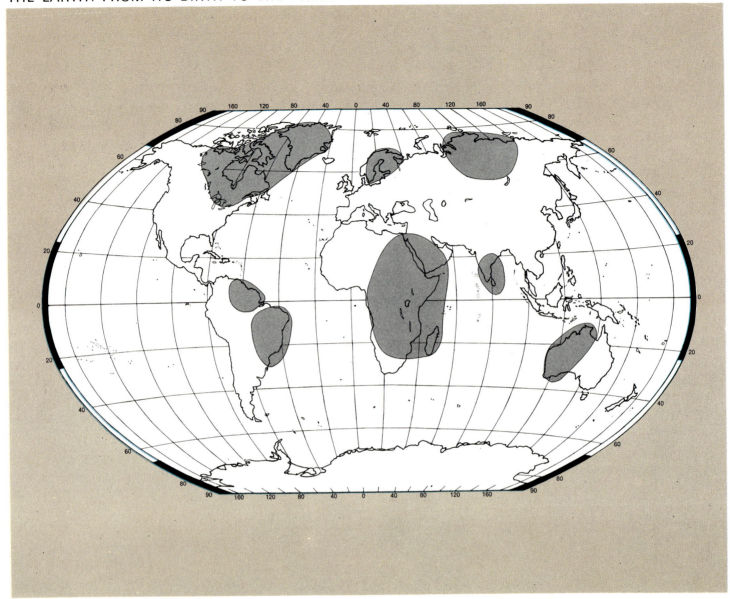

system 40,000 km (25,000 miles) in length, stretching right across the oceans. While the Atlantic and Indian Oceans spread until they reached their present size, the Pacific Ocean on the contrary became smaller and smaller because of the encroachment of the western shores of the Americas towards the Philippine archipelago. The expansion of the Atlantic and Indian Oceans was helped considerably by volcanic activity in the vicinity of the ridges, with the creation of a new ocean crust. The shrinking size of the Pacific Ocean, on the other hand, is due to the swallowing up of the old crust in the oceanic trenches in the vicinity of the zones of withdrawal. As far as the future is concerned, a new ocean will theoretically come into being between the Red Sea and the Gulf of Aden as a result of Arabia drawing away from Africa. This drift, which started some twenty million years ago, will involve eastern Africa along the line of the deep rift valley in which the great African lakes are situated.

By this very gradual process, the Earth has thus reached its present state. Internally, it consists of a series of almost concentric layers of different rock, the density of each layer being greater the nearer it is to the centre. By resorting to indirect methods such as the analysis of the pattern of propagation of the seismic waves, it has been possible to ascertain the existence of these layers

A schematic map of the dislocation of the Archeozoic shields. Formed for the most part by igneous and metamorphic rocks, the shields are the oldest and most stable areas of the world, and form the nucleus of the continental masses. They are gently rolling lowlands and have been whittled down by meteoric erosion.

Top left: sedimentary rock in polished cross-section clearly showing a number of fossils. Here we have Eocene nummulites. Below: a thin, magnified section of crystalline limestone with muscovite. Top centre: tormaline-bearing pegmatite. Intrusive rock deriving, by differentiation, from a granite and characterized by large crystals with quartz and feldspar associations. Below: thin magnified section of a porphyritic rock. Top right: gneissic-type schistose metamorphic rock. Below: thin magnified section of muscovitic gneiss with augen texture.

(involucra) which are formed by physically and chemically different substances and separated by areas of discontinuity. The first discontinuity we find going deep down is the Mohorovicic, also known as the Moho, which separates the surface crust, which is formed mainly by basaltic and granitic rocks, from the *mantle* beneath it.

The thickness of the crust varies between 5 km (2 miles), at the ocean-bed, and 50 km (30 miles) beneath the continents. This difference in thickness is explained by the principle of *isostasy*, according to which the lighter sialic continental masses must have a greater thickness to balance the heavier simatic masses, which are present immediately below the oceans. The *mantle*, which is shown up by the increasing speed with which the seismic waves pass through it extends to a depth of 2900 km (1800 miles). After this we have the nucleus or core. The limit between these two layers is shown by a clear discontinuity, known as Wiechert-Gutenberg's discontinuity.

The *mantle*, with its thickness of 2900 km (1800 miles), accounts for more than 80 per cent of the Earth's volume and it seems to be composed mainly of a wide variety of types of *dunite* with incorporated remnants of peridotite and eclogite.

The *nucleus* (or core), which, like the mantle, we only know about indirectly, is reckoned to be in the state of cast iron, with small quantities of nickel and cobalt. This composition is explained by the high density necessary in the nucleus to harmonize with the average density of the Earth. In fact, at the presumed pressures and temperatures inside the Earth the iron would reach a density of about 9.5 per cent but still remain in the liquid state. It would, however, be very tacky. As a result of the most recent seismographic studies, there is a theory about an intermediate discontinuity, about 1300 km (800 miles) from the centre of the Earth, which would suggest an *outer nucleus*, with a thickness of about 2200 km (1400 miles), and an *inner nucleus*. The difference between the two nuclei might, some maintain, be purely chemical in nature, and, in the view of others, purely physical, i.e. with the inner nucleus solid rather than fluid.

Returning to the crust, its composition is the result of rocks belonging to three large major groups: *eruptive* or *igneous*, *sedimentary* and *metamorphic*.

The igneous rocks originate from the magma, that molten gaseous mixture which, from inside the Earth, rose from time to time towards the surface, cooling all the time. The different rate of cooling and the different pressures to

Left: wind-erosion in Bryce Canyon National Park (Utah, USA). Top right: Grand Canyon National Park, Arizona, USA. The Colorado River has, over a period of some ten million years, carved out a regular series of sediments to a depth of about 1500 metres (one mile). The picturesque steps are due to the different resistance put up by the rocks to physical alteration. Below: steppe near Shahreza, in Iran.

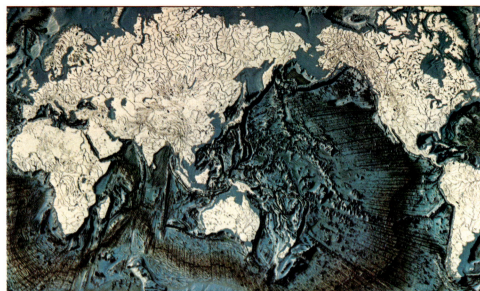

Top left: silicified fossil tree trunks in the Algerian Sahara. This phenomenon is due to the replacement of the original vegetable tissues with mineral particles, which keeps the structure of the wood unchanged. Top right: the morphology of the ocean-bed; we can see the long-submerged reliefs in mid-ocean. Below left: dry mud from the Devonian period, in a lake. Centre: a present-day desertscape: the Tassili (Sahara). Right: sediments from the Carboniferous period clearly showing the sedimentary cycles.

which the molten material was subject gave rise to different rocks according to the size and nature of the crystals forming them. We can thus distinguish *intrusive* from *effusive* rocks. The former solidified deep down, whereas the latter spilled out on to the surface before solidifying. Depending on how exactly the magma surged to the surface, it formed large, bulky bodies, called *plutons*, or smaller bodies, called *seams* or *veins*. These are the outer layers of the plutons and were produced through the magmata filling up cracks in the rock, which is why they may be irregular and of no fixed pattern. The *seams* were formed while the plutons were becoming solid and can be considered as *undifferentiated* when their composition is the same as the magma, and *differentiated* when they are of a different composition, having been changed when they solidified: this is the case with *haplitic seams*. In some cases, during the cooling phase, there was a contraction on the edge of the magmatic mass, which led to the formation of fissures in which the magma infiltrated, giving rise to more seams.

In the final phase of the consolidation of a pluton, when large quantities of steam containing various chemical elements were released, either *pneumatolytic* or *hydrothermal seams* were formed, according to whether steam and gas

at a high temperature or water at a low temperature were involved. This type of seam is particularly important, for in them are found most metal deposits.

On the basis of their silica content, we can distinguish between *acid rocks*, where there is more silica, and *basic rocks*. Among the more typical acid rocks are the *granites*, and among the basic rocks the *gabbros*. These two groups are easily identifiable: in the acid rocks there is a predominance of white or colourless crystals, whereas in the basic rocks we find many coloured crystals.

Because of the rapid cooling undergone by them, the *igneous* rocks have a micro-crystalline or vitreous structure which often makes it impossible to identify the components and, hence, their acid or basic character. We find the *trachytes* among the igneous acid rocks, and among the commoner basic rocks the *basalts*, with a distinctive blackish coloration and a prismatic-columnar crack formation which sometimes makes them quite striking in appearance: particularly striking are the basalts of Scotland, and those found on the mighty plateaus of Africa and the Deccan, or Indian peninsula.

Sedimentary rocks originate from deposits formed from other rocks which disintegrated or underwent some chemical change. The final form of these sediments depended on many factors: the nature of the pre-existing rock, the

Left: a salt pool in the Dallol in Danakil. With a temperature of about 80°C (176°F), the water gives off into the air a strong sulphurous odour. Top right: the same phenomenon, but one in which the entire landscape is completely changed. The ground temperature makes it possible to fry an egg. Below: this is how stalactites and stalagmites are formed: drop by drop they gradually piece together the unworldly landscapes found in karstic caves. Opposite: the marvellous 'salt-falls' of the Dallol. This is a saline phenomenon connected with intense hydro-thermal activity.

way in which the sediments were laid and moved, the modifications undergone by the deposits during and after their positioning, the physiographical and climatic environment, and so on.

The sedimentary rocks can be roughly subdivided as follows: *clastic rocks*, *rocks of chemical origin* and *rocks of organogenic origin*.

The clastic rocks derive from the simple deposit of material broken off or worn away from rock by the erosion of water or some similar natural phenomenon. Such sediments are usually carried by rivers, ice or wind. Where a river is involved the sediment may be carried *in suspension*, *in solution*, or *by being rolled* along the river-bed. These rocks, depending on the size of the particles, are called *conglomerates*, *sandstones*, *silts* and *clays*. The conglomerates are formed by sometimes quite large pebbles or stones cemented together by calcium carbonate or silica. They are divided into *pudding-stones*, when the elements are rounded, or *breccias*, when the elements are angular and irregular. In addition there may be *monogenic* and *polygenic* conglomerates depending on whether the pebbles are similar to each other or different. Polygenic pudding-rocks can be found at the bottom of transgressive series, indicating a coastal type of deposit, or in piedmont alluvial cones.

Soaring mountains are not just barren geographical points, useless prominences on the Earth's surface, or inhuman barriers of rock and ice, sheer faces and pinnacles to be climbed, or snow-covered ski-slopes. They are also something less practical than this, something very valuable, in that they hold vast symbolic meaning for man. The photograph shows the pinnacles of Chamonix, with a squall whipping up huge columns of powdery snow, looking like a huge fire.

Left: typical erosion furrows which, given the clayey and sandy nature of the terrain in which they form, are not easy to drain off. Erosion furrows are groups of small, extremely steep valleys, situated very close together, usually carved out in clayey and sandy slopes and hillsides by rain and streams. Right: erosion in Turkey. Erosion affects the surface section of the Earth's crust and is closely allied to natural agents which move, i.e. water, wind and ice.

The breccias, however, are found mainly at the foot of steep mountain-sides, in the form of scree, or in areas affected by glaciers, where they are found as morainic deposits. The sandstones are formed principally by granules of quartz, feldspar, mica and other minerals and are classified according to the material that binds them together; this is usually *calcareous* or *siliceous*. Characterized by regular stratification, they are present throughout geological history in most parts of the world.

The silts and clays are the finest sediments and differ from each other in the average diameter of the particles of which they are made up, which, in the case of the silts exceeds 0.002 mm, and in the case of the clays is smaller than that.

The rocks of chemical origin derive from the precipitation of substances in solution in water, as a result of evaporation or by chemical reaction. In basins where there is considerable evaporation, evaporites are formed, represented by salts such as rock-salt (halite), gypsum and anhydrite. Important rock-salt deposits exist at Stassfurt and Halstatt, where salt has been mined since the Iron Age. Gypsum is common in Italy, where it is mined industrially.

When we find concentrated solutions of calcium carbonate, as in the vicinity of mineral sources, for example, encrustant calcareous deposits are formed

known as *travertines* (*Lapis tiburtinum*), used by the Romans as building material for such important monuments as the Colosseum and the temples at Paestum, in Campania. Step-like deposits of variously coloured travertine occur at the Mammouth Terraces in Yellowstone National Park. In grottoes in karstic zones we find the formation of concretionary deposits which give rise to the well-known *stalactites* and *stalagmites*. These are formed when water containing calcium bicarbonate trickles through crevices and fractures of rock. Partly because of evaporation and partly because of the release of the carbon dioxide contained in the water, a certain amount of this bicarbonate turns into insoluble carbonate which is then deposited. Stalactites and stalagmites, which form from the roof and floor of caves respectively, and sometimes even join together, produce fantastic unique forms known to us all.

The most beautiful *organogenic* rocks are to be found in coral-reefs and atolls, which are simply circular reefs, elliptical reefs or reefs in the form of a horseshoe with an inner lagoon communicating with the open sea by narrow, deep channels.

They are widespread in the central and western Pacific and in the Indian Ocean, where, because of the pleasant inter-tropical climate, they are also

Left: Ordovician sandstone in the central Sahara, seen from the air. The strange morphology seems due to erosion caused by the melted water from the retreating Ordovician glaciers. Top right: ripples (the imprint of currents) in late Ordovician deposits. These forms result from strong currents, also probably caused by the glaciers' melting. Below: beautiful tower-like eroded shapes, formed by Permian red sandstone, in Monument Valley between Arizona and Utah, USA.

Top left: the Domus de Janas *'fungus' at Tafani in Sardinia, moulded by corrosion. This phenomenon is due to the grinding action of sand carried by the wind. Below: erosion at work in the Bay of Ixanumera, Pine Island, New Caledonia. The formation of the water-line overhang is caused not only by the mechanical action of the waves, but also by the nature and degree of crack-formation in the rock which is subject to abrasion. Right: earth pillars in incoherent morainic deposits in which, incorporated in the clay mould, we may find quite large stones which have stopped the rock beneath from being eroded.*

tourist resorts. As far as their genesis is concerned, Charles Darwin (1809–1882) explained that the atolls developed around islands which were constantly sinking, but at a sufficiently slow rate to allow for the compensatory growth of the coral. Later, R. A. Daly advanced the hypothesis that the formation of coral islands could be associated with the rising level of the sea because of the thawing of the quaternary ice-caps.

The metamorphic rocks are nothing more than rocks, already formed but then subjected to processes which have modified their original mineralogical and petrographic composition. As the magma rises in the earth it finds its way between chimney rocks, and so rocks with a lower temperature, and magmatic rocks with a higher temperature, come into contact.

The heat experienced by these rocks, often accompanied by gaseous emissions from the magma as it cools, causes the disappearance of some minerals and the appearance of others, and thus the creation of a different rock dozens or even hundreds of yards thick. This process of alteration is known as *metamorphism*.

As well as *thermal metamorphism* there are cases of *regional metamorphism* which may be *kinetic*, due to directed and localized pressures, *dynamo-thermal*,

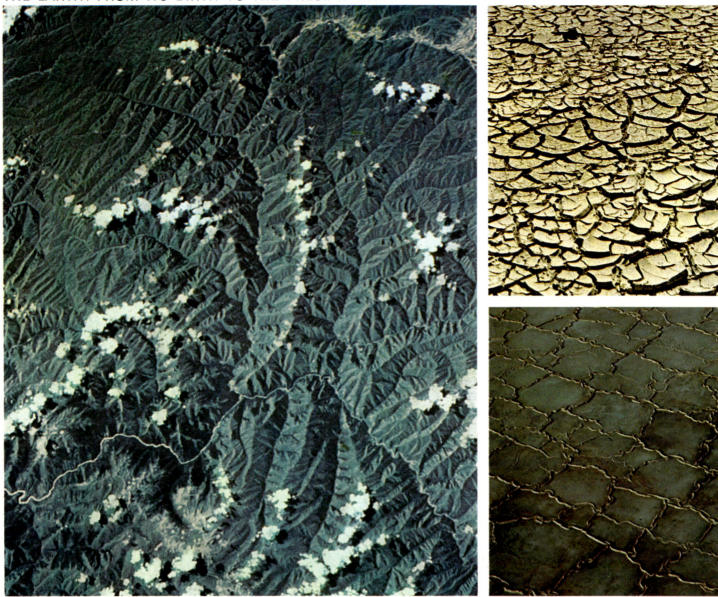

when pressure with high temperatures are involved, or *plutonic*, when pressures and temperatures are both fairly high. The first is responsible for the formation of breccias and mylonites, the second for slates, phyllites, mica-schists, gneisses and saccharoid marbles, and the third for granulites.

The corrugations (folds) to which the Earth's crust has been subjected have caused various morphological structures, all of which have been subjected to a moulding evolution which gradually causes them to disappear. The principal moulding agent may certainly be considered to be water, which erodes continually, conveys and deposits, in such a way that it progressively evens out reliefs and depressions. A humid climate, however, where the vegetation is most dense, protects the soil from this erosion, which is sharper in arid regions.

Accelerated erosion first triggers off a widespread trickling process which carries off the finest particles; subsequently, the water concentrates in rivulets giving rise to increasing erosion. This type of trickling in rocks liable to erode, which are impermeable and without vegetation, is responsible for the formation of *erosion furrows* and *earth pillars*.

Depending particularly on latitude, the different regions of the world enjoy different climatic conditions which in turn determine the landscape and

Left: a photograph taken by an artificial satellite over the Himalayan chain. Top right: a polygonal crack-formation in clay. Below: a galla, an association of salt forms in Lake Assaleh (Danakil). Opposite: an example of marine erosion (on the Dutch coast). The low tide enables us to single out deep furrows (the dark areas) carved out of the mud and sand down the centuries by the slow but steady action of the sea (aerial photograph).

Left: a detailed view of the sandy Algerian desert. The crescent-shaped dunes, also known as barchans, *are convex on the windward side, where the gentler slope is. Right: the beach at San Sebastian in Spain. The nearly-parallel wrinkles, lying vertical to the wind-direction, echo the movement of waves and indicate a similar dynamic origin.*

environment: some are unique.

In regions with an arid climate, where the low rainfall discourages animal and vegetable life, deserts have formed which, in all, occupy an area of more than 21 million sq km (14 million sq miles). Most of this area is made up of the hot deserts scattered along the subtropical belt, between Lat. 15° and Lat. 35°. The world's major deserts are the Sahara, the Arabian desert, the Indian and north American deserts in the northern hemisphere, and the Kalahari, Australian and Atacama deserts in the southern hemisphere. The Sahara, with its area of 8.5 million sq km (5.3 million sq miles) is undoubtedly the world's largest desert and accounts for about one third of the whole of the African continent.

In more extreme latitudes, on the other hand, we find the so-called *cold deserts* represented mainly by the Gobi and Turkestan deserts in Central Asia and the Patagonian desert in South America.

The origins of the world's deserts are still often hard to determine. Contrary to what has been thought in the past, the deserts are now thought to be a relatively recent development. The discovery of fossils and even of archaeological remains has shown, for example, that in the Sahara until about

*Left: the Arizona desert (USA).
Right: a large dune with sandy
undulations in the arid Asian desert in
Turkmenistan (USSR). Because of
the sharp seasonal range of
temperatures, this region is
particularly hostile, even for nomads.*

4000 years ago there were contained lake-basins and dense forests. What is more, and again contrary to what one usually imagines, the desert often has an uneven and broken structure which may vary even within a radius of a few kilometres. We can find, here too, mountain ranges deeply eroded, to form valleys in which there may be lakes or more likely dry water-courses. Wherever the development of the desert is more advanced we see stony or duny deserts, landscapes more generally associated with the word 'desert'. Stony deserts are caused by denudation, by the action of the wind which constantly carries off the fine particles which themselves develop from the mechanical break-up of the rocks, usually quite considerable because of the wide temperature range experienced in such parts.

Duny deserts, on the other hand, are those areas where the wind deposits material, i.e. the place where the sand and dust carried by the wind (often from stony deserts) pile up and form *dunes*. These are found most often in flattish areas, in the form of crescent-shaped accumulations called *barchans*, which form vast expanses sometimes stretching for thousands of kilometres. Their presence denotes a relatively thin covering of sand and strong prevailing winds. If the wind blows in different directions *star-shaped* dunes are formed.

Contrary to what one might expect, the desert environments are not lifeless; they house many thriving vegetable and animal organisms, which are highly specialized and organized so that they can survive in what is really a very hostile habitat. The animals, for example, confine their activity to the cooler times of day, emerging from their lairs in the late afternoon or evening, and returning to them next morning. In future economic plans the deserts are seen as vast agricultural reserves from which man will draw in abundance. It is clear, however, that to restore fertility to these areas enormous quantities of water will be needed, and this will only be possible with the help of highly specialized techniques.

Radial dunes form a backcloth to a Libyan oasis. The oases are situated in depressed areas in deserts where the topographical surface intersects the water-table. As well as allowing the development of lush palm trees, water also summons desert nomads to use oases as staging-posts. In the larger oases there may also be stable settled populations. At the present time many oases are turning into tourist resorts.

VOLCANOES AND EARTHQUAKES

The great fractures in the Earth's crust, usually occurring at the edge of the various oceanic and continental shelves, and along the mid-oceanic ridges, are the areas where volcanic activity is most frequent and widespread. The presence of volcanic and seismic areas at the edge of these shelves does not, however, exclude the existence of extinct or active volcanic areas in inland continental areas, as in the case of eastern Africa, or mid-oceanic areas, such as the Hawaiian Islands.

Volcanic activity and earthquakes are not casual and independent episodes, but rather reflect the geotectonic conditions of the zones in which they occur and on which they depend.

Volcanism, according to the most up-to-date geological concepts, is an indication of the geodynamic phenomena which are constantly modifying the Earth's crust. The distribution of the volcanic areas and the nature of the magma are not haphazard, but allied to structural aspects of the Earth's crust.

The composition, in chemical terms, of the magma also influences the type of activity and the actual morphology of volcanic structures.

At the present time there are more than 700 active volcanoes; but there are also numerous examples of volcanic activity throughout geological history which suggest that on the Earth there has always been continuous and intense volcanic activity. Vulcanites – volcanic materials – cover vast areas of the ocean-bed: the Hawaiian Islands, the islands of Melanesia, the Sunda Isles, the Philippines, the Japanese islands, and the Aleutians are all tangible evidence of this impressive activity. Huge lava flows, in the past, have also covered vast continental areas of the Deccan in southern India, and of eastern Africa.

The *magmatic basin* is the site of the fluid materials destined to erupt; it lies at an average depth of 30–40 km (20–25 miles). The *volcanic chimney* is the conduit through which the magma, gases, vapours and other materials rise to the surface and it ends in the crater. The *external structure* is almost invariably formed by the accumulation of the materials thrown up by the volcano in the past. Often the structure has a conical shape, but it may also take on various other shapes, depending on the nature of the matter erupted. We find tapering or depressed conical forms, depending on whether the magma is, respectively, acid or basic. In fact the more fluid and gas-rich the magma, the swifter the flow of lava, which, before solidifying, may extend over considerable areas, and create rather flattened volcanic cones which are called *shield volcanoes* or *lava shields*, or *Hawaiian-type volcanoes*, because of their frequency in the Hawaiian Islands. When the magma is acid, and thus more sticky (viscous), pinnacles (aiguilles) may form, of the type seen in Mt. Pelée in Martinique: these are solid protrusions of magma already solidified inside the conduit and then pushed outwards by the pressure of the gases. These volcanic forms should not be confused with similar forms found in other regions: the Devil's Tower in Wyoming, USA, the pinnacle of St Michel D'Aguille at Le Puy in Auvergne and the Pic Laperrine in the Hoggar (Sahara) are in fact examples of lava-filled ancient chimneys, the outer structure of which has been destroyed by erosion. In the same volcanic areas we can also find dome-shaped lava forms due to the viscous magma-flow ceasing.

There is also another type of volcanic structure, of which Vesuvius is a

typical example: the *strato-volcano*. Characterized by explosive phases alternating with phases of slow, fluid eruption, the structure of these volcanoes is formed by pyroclastic materials such as lapilli (cinders), ash, scorie and bombs, interspersed with lava flows.

Etna (3300 metres – 10,902 feet), Fujiyama (3778 metres – 12,388 feet), Mt. Hood (3730 metres – 12,200 feet) in Colombia, and the Pico de Teide, Tenerife, Canary Islands (3716 metres – 12,162 feet), are some of the Earth's most impressive strato-volcanic reliefs. Unlike other features on the Earth destined to evolve gradually, the volcanoes can be formed and transformed in fairly short periods of time, because of the matter thrown up from them. In such cases the volcanic structures can be formed very quickly indeed, as happened with Mt. Paricutin, in Mexico, which was born on 20 February 1943, and summoned vulcanologists from all over the world to witness the event.

In a matter of days, a cone was formed of more than 150 metres (500 feet) which then progressively grew until it reached its present height of 450 metres (1500 feet).

And just as they can develop in a very short time, so volcanic structures can also vanish, just as quickly. This may be the result either of rapid erosion of

Left: a volcanic eruption at Heimaey, in Iceland. This island, situated at the northern tip of the mid-Atlantic ridge, is affected by active volcanism of a widespread and frequent kind. Top right: a fossil sand volcano, of tiny dimensions, near Gara Zaharzack in the Tassili desert (Algeria). The structure can be attributed to the outflow of subterranean water under pressure, in a submarine site. Below: the crater of Stromboli, in the Lipari Islands, erupting. This volcano, 925 metres (3038 feet) high, has a predominantly explosive activity.

The birth of Surtsey Island, south of Iceland, on 14 November 1963. This land-mass emerged from the ocean in a single night, and over the course of the following twelve-month period reached a height of 170 metres (565 feet). This is a typical example of rapid formation by the accumulation of volcanic matter.

loose material or the destructive consequences of the explosive phases which can demolish the summit of the volcanic structure. The island of Ferdinandea, which formed in July 1831 between Sicily and Pantelleria, disappeared within a few months. Often the volcanic structures found in oceanic areas gradually disappear because of the slow subsidence of the ocean-bed.

These slow sinking movements, known generically as *positive bradyseisms*, are countered by similar slow lifting processes, known as *negative bradyseisms*. These two movements, which are almost always associated with volcanic or post-volcanic activities or phenomena, may alternate in one and the same area. A typical example of these combined effects is given by the Pozzuoli region in Italy, which sank several metres after the days of the Roman Empire to the point where the monuments existing at that time were under the sea (as evidenced by the traces left by marine date mussels) and then re-emerged during the Middle Ages.

Lava surfaces basically fall into two types. The terms for these come from Hawaii: 'Pahoehoe' lava which is undulating but with a continuous skin; and 'a-a' lava which is rough surfaced like cinders. Within the lava masses cooling may last several months. The cooling process is followed by a contraction of

the mass which sometimes gives rise to prismatic-columnar crack-formations, typical in the basaltic rocks and well exemplified in the Hebrides, Scotland and Ireland.

If one takes into account the distribution of earthquakes, we can see that, across the globe, there is a correspondence between the seismic areas and the volcanic areas which demonstrates the close connection between the two phenomena.

By correlating earthquakes with volcanic activities it has been established that the former are generally linked with the initial phases of eruptions in relation to the shift of the magmatic masses and to the very considerable stresses acting on the eruptive conduit. But this does not exclude the possibility of seismic phenomena associated with successive or deeper magmatic activities. The above type of earthquake, known as *volcanic*, may be followed by the *tectonic* type due to the stresses to which the Earth's crust is subject as a result of the shifting or settling of crust masses which have in time become unbalanced. The earthquake caused by the *San Andreas Fault* in California is sadly famous; on 6 April 1906 it destroyed the whole city of San Francisco. It was a truly disastrous earthquake, because an enormous number of people

Top left: a smallish volcanic crater, filled with bubbling mud, at Reykiahlid in Iceland. Below: the volcano Nyragongo in equatorial Africa. The lake of boiling lava is surrounded by walls and terraces from which there are continual emissions of gas and vapour. Top right: a jet of lava during the eruption of Kilauea, in Hawaii. Below: Etna in explosive mood, hurling cinders and ashes, mixed with vapours, from the north-east crater, at the end of the summer of 1968.

Top left: ropy lava near Reykjahlid, Iceland. This phenomenon is caused by the superficial cooling of a lava flow which has little gas, but is still in motion. Below: the front of a glowing lava-flow on the slopes of Etna. Top right: lava-flow from Etna's north-east crater. The liquid, incandescent magma, forms a long narrow tongue. Below: a mass of cooling basaltic lava. One can see how closely it resembles clinker. On the pages following: the mighty chasm of the volcano Nyiragongo in eastern Zaire.

died and because of the material damage caused above all by fires which raged uninterruptedly for fully three days and nights. Short circuits and broken gas-pipes caused these fires, which engulfed the city, and against which the inhabitants could do nothing, because there was no water. Other catastrophic earthquakes have occurred at Messina (1908), Avezzano in the Abruzzi (1915), Tokyo (1923), Agadir (1960), Alaska (1964), and, more recently, in Guatemala and Friuli (1976). Less intense earthquakes may be caused by the collapse of subterranean cavities, which are frequent in karstic zones, or to the fragmentation of large rock-masses. In the 1963 Vajont disaster, the collapse of Mt. Toc was accompanied by earthquake shocks and tremors which, though not intense, were picked up by instruments at a considerable distance. The areas of the world with the highest seismicity generally occur in the most recently formed mountain chains in the Alpine-Himalayan system, and in particular in the vicinity of the so-called *fire-belt* in the Pacific Ocean. This fire-belt, characterized by numerous active and extinct volcanoes, includes the Andes, the mountains of Central America, the Rockies, the Aleutian Islands, the Japanese archipelago, the Tonga archipelago and New Zealand. This area accounts for 80 per cent of earthquakes, whether *deep*, of tectonic origin, or

superficial, of volcanic origin. The importance of seismology lies in the fact that this discipline represents the only means – and they are indirect means – of coming up with the information which enables us to obtain at least an approximate reconstruction of the inside of the Earth.

Sensitive instruments called seismographs enable us to record on a graph, called a *seismogram*, earthquake shocks and tremors, and to localize the *focus* and the *epicentre*.

The focus is the point within the Earth's crust, where an upheaval has originated, i.e. a laceration of the structures deep down, from which the 'elastic' waves depart. On the basis of the focal depth, earthquakes are divided up into *normal*, *intermediate* and *profound* types, when the respective depths are less than 60 km (40 miles), between 60–300 km (40–200 miles) and more than 300 km (200 miles). The epicentre is the point vertically above the focus, along the same Earth's radius. There are two types of elastic waves produced by earthquakes: longitudinal and transverse. The longitudinal waves spread by expansion and compression, successively, as if the particles forming the interior of the Earth were vibrating in the same direction. Vice versa, the transverse waves cause vibrations perpendicular to the direction of

Left: the summit of the volcano Fujiyama. Top right: aerial view of the crater of the volcano Villarica at Cautin in Chile. Below: geysers and fumaroles in Yellowstone National Park, Wyoming, USA. These phenomena are associated with delayed volcanic activities. Geysers are intermittent jets of very hot water and steam, caused by the water beneath the earth overheating. Fumaroles are gaseous emissions at great heat, with steam as the predominant product.

Collapsed buildings after the 1964 earthquake in Alaska. The destructive consequences are caused mainly by the undulatory vibrations, which fan out radially from the epicentre.

propagation. The speed at which the seismic waves spread depends on two factors: the density of the medium crossed, and the coefficient of elasticity, which is different for the two types of wave. The longitudinal waves are faster; they start from the focus at the same time as the transverse waves, but reach the seismograph first and are registered on the seismogram as *primary waves*. After a certain time-gap, which increases with the distance from the focus, the transverse waves also reach the same seismograph, and are registered as *secondary* or *shear-waves*. After a further time-lapse the seismograph registers a third type of wave, the *long* or *L-waves*. These leave the focus and spread by means of concentric rings, with the same mechanism by which the ripples in a pool of water spread outwards when one throws a stone into the water. It is in fact this third wave-type which is felt as the actual earthquake shock, even though, compared with the other two types, they are relatively slow and purely superficial.

If these L-waves are important because of the destructive effects which they have in the areas around the epicentre, it is, on the other hand, only by analyzing the primary and secondary waves that one can discover important data about the strata crossed, i.e. about the internal composition of the Earth.

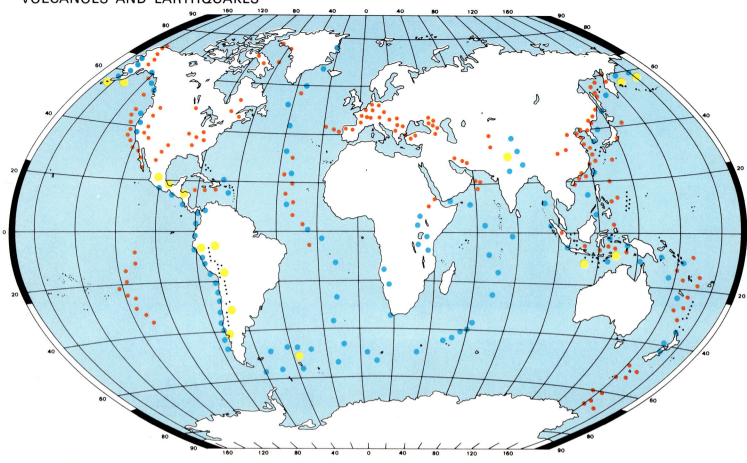

- • strong in intermediate strata
- ⏐ normal
- • strong at surface
- ● strong at depth

The intensity of earthquakes is expressed by means of the Mercalli scale, which is based principally on the empirical assessment of the effects produced by them on the Earth's surface. In 1935 Richter introduced the concept of *Magnitude*, based on the measurement of the breadth of given elastic waves on the basis of their distance from the epicentre.

In addition to the only too well-known earthquake disasters occurring in continental areas, one of the effects produced by earthquakes is the highly dangerous *tidal wave*, which, in its course, may swamp coastlines with breakers up to 30 metres (100 feet) in height. Commonly known as *tsunami* waves as well, they spread at very high speed, sometimes in excess of 600 kph (400 mph) for thousands of kilometres. In 1960 the Arica earthquake in Chile caused a tidal wave that spread as far as Alaska and Japan.

Considered, by and large, to be inevitable and uncontrollable calamities, earthquakes are the topic of specific research aimed at predicting them in both time and space. This is in order to limit the damage caused and safeguard the lives of those likely to be affected. Although it is hard to determine with precision the moment and the place in which an earthquake will strike, an assessment of the earthquake risk can be reached by knowledge of statistically

The geographical distribution of earthquakes. The greatest seismic groups tally with the tectonically youngest areas, which have recently undergone folding. One may note, in this respect, the close association with the great mountain ranges of the Alpine orogenetic cycle and with the distribution of volcanoes.

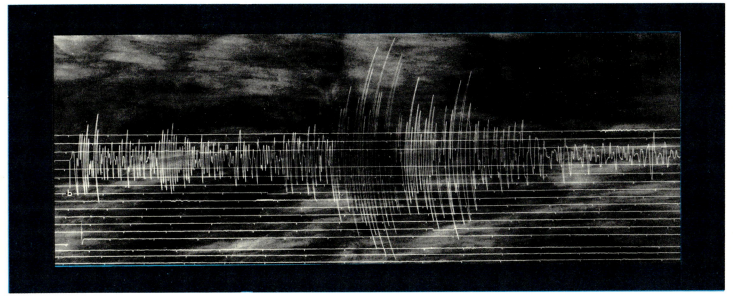

Top right: results of the 1964 earthquake in Alaska. Top left: Managua, a year after the earthquake. Below: an earthquake occurring on 28 February 1969, with the epicentre 200 km (125 miles) south-west of the Portuguese coast. The seismogram shows the vertical component obtained with a Wiechert seismograph; by comparing seismograms from different stations it is possible to localize the epicentre fairly accurately.

possible seismic occurrences. Once this first assessment has been made, it is possible to look for the best remedies to limit the destructive effects of the earthquake. In various countries such as the USA, Japan and the USSR *Seismic Risk Maps* have been drawn up on which, taking into account past history and the geological and tectonic structure of the various regions, the territory is divided into various classes of earthquake risk. For a certain zone, each of these classes indicates the probability of an earthquake occurring and the maximum intensity which it may reach.

Other studies have the task of examining the possibility of controlling earthquakes. Researches at the National Center for Earthquake Research in San Francisco have recently come up with new control methodologies, starting from the assertion that it is possible to cause 'artificial earthquakes'.

FROM ROCKS TO PRECIOUS STONES

The use of natural stones for ornamental purposes was certainly one of the first manifestations of man's aesthetic sense. Necklaces, pendants and bracelets made of coloured stones have been found among the remains of the most ancient of palaeolithic cultures. As early as protohistoric times, the civilizations of Asia and the Middle East, Egypt, and the central and southern Americas expressed their highest artistic achievements in the production of splendid gems and refined jewellery. In addition, the unique beauty and magnificence of precious stones must have aroused, in primitive peoples, a sense of mystic veneration which caused them to attribute to them (as powerful talismans) astral origins and extraordinary magical and therapeutic powers.

So what are these precious stones? And how did they come to be in the rocks in which they are discovered?

The reply must start from present-day knowledge relating to the nature and genesis of the rocks themselves. These appear, on the whole, as aggregates of two or more minerals, or as chemically homogeneous bodies forming part of the lithosphere. The strange geometric shapes assumed by many minerals are the expression of their internal structure, resulting from the specific spatial matching of infinitely tiny elementary polyhedra; this regular arrangement of matter produces the solid crystalline state. Of the crystals which thus come into being, many are considered as precious and it is from these that we obtain beautiful gems. The Aristotelian theories about the origin of minerals referred to the influence of rays issuing from heavenly bodies, which gave rise to a sort of spontaneous generation in the bowels of the Earth. Other theories attributed to minerals a particular life of their own and the possibility of growing by means of assimilation of substances by special circulatory systems. The presence of sometimes liquid inclusions, which are frequent in crystals, in fact appeared to justify this hypothesis. In maintaining that minerals were capable of reproducing themselves, Pliny made a distinction between male and female varieties, usually on the basis of their different colours.

Such beliefs survived until the end of the Middle Ages when Georgius Agricola, a German physician and distinguished naturalist, undertook his systematic study of minerals on the basis of their objective properties. He realized that some of them could be formed by the precipitation of substances contained in watery solutions. We know now that the genesis of minerals can be brought about from a liquid state, a solid state or a gaseous state, by variations in the chemical composition of the environment in which they are formed, in relation to the pressures and temperatures at work.

During the solidifying of the magma within the Earth's crust, the slow cooling process allows the formation of crystals with more or less equal dimensions which give the rock a granular, granite-type structure. Crystallization is helped by the presence of volatile substances and occurs in successive stages, each one of which corresponds with the separation of typical minerals. The first to form are the crystals of accessory minerals such as zircon, apatite and titanite. Subsequently, the so-called *mafic* minerals, silicates combined with iron and magnesium such as olivine, the pyroxene group, amphiboles and biotite, are separated, simultaneously with the plagioclastic minerals, chalco-sodic silicates combined with aluminium; lastly, orthoclase, muscovite and

Left: an aggregate of muscovite leaf-crystals. Muscovite is a fairly common silicate in various types of rocks, hallmarked by very easy cleavage and a fairly high level of transparency. Other minerals belonging to the same class of micas are olivine and tormaline, as well as the amphiboles and the pyroxene group. The micas are found fairly frequently in a large number of intrusive rocks. Right: an agglomeration of wulfenite crystals. This is a lead molybdate, a rare mineral belonging to the same class as the tellurates, chromates, wolframites and sulphates.

quartz crystallize. In the various phases of solidification, the magmatic residue is enriched with volatile components, and when it fills up fissures in the surrounding rock, large crystals form from it, including the highly-prized corundum, beryl, topaz, tormaline and chrysoberyl. In a later phase, an exclusively gaseous residue remains, which gives rise to minerals such as cassiterite, fluorite, tormaline and sulphides.

When the magma reaches the surface, the rapid cooling in the atmosphere prevents the chemical elements from organizing themselves into a crystalline structure. Thus we find the formation of volcanic glass or obsidian.

A crystallization by the precipitation of substances from watery solutions corresponds with both the hydro-thermal depositing of sulphides, fluorides, silicates and quartz, at relatively high temperatures and pressures, and with the genesis of evaporitic rocks, formed essentially by gypsum, anhydride, rock-salt and other chlorides and sulphates, at lower temperature and pressure. Lastly, in the course of the metamorphic processes, we find a re-crystallization in the solid state of pre-existing minerals; typical here are talcum, asbestos, graphite and garnets.

Minerals are often concentrated in the actual rock in which they originated,

forming primary deposits. Following the disintegration of these rocks by outside physical processes, the more resistant minerals accumulate in sedimentary deposits before being carried away; we thus find the formation of these deposits of secondary origin in which most of the more precious gems are found. Since earliest times the gem-bearing alluvia (deposits left by flooding) of Burma, Thailand and Ceylon have been world-famous, producing the finest varieties of rubies and sapphires. In the laboratory one can observe the formation and growth of crystals, by the precipitation of substances contained in watery solutions. In conditions of over-saturation minute little *crystalline nuclei* appear which already have a polyhedral shape and the chemical and physical features of the final crystal. Growth comes about by the successive addition of layers parallel to the original faces; a consequence of this pattern is the well-known law of the *constancy of the dihedral angle*. The observations made in this respect by the Danish naturalist Nicholas Stanone at the end of the seventeenth century led to the theoretical proposal of the existence of a clear-cut internal structure of crystals.

Likewise, the cleavage which, in some crystals, entails the splitting-up into similar, and smaller and smaller, polyhedra, leads to a consideration of the

Left: fibrous, thread-like native silver on calcite crystals. This is found in deposits of hydro-thermal origin generally associated with blende, pyrite and galena. Right: pyrite, iron sulphide in pentagonal-dodecahedral crystals. This belongs to the group of sulphides and is used industrially for the mining of sulphur. The main deposits, where pyrite is in general associated with cuprite, are of the hydro-thermal type, i.e. linked with recent volcanism. In the case of the gold-bearing pyrites, these are mined so that the precious metal can be extracted. Large pyrite deposits exist in Elba.

Top left: large cubic rock-salt crystals (or sodium chloride). It is possible to find large crystals in the natural state, although it is more common to find crystalline masses, formed by tiny crystals with an obvious cubic cleavage. Below: amethyst quartz crystals belonging to the class of oxides and hydroxides, along with cuprite (copper oxide), uranite (uranium oxide), cassiterite (tin oxide) and rutile (titanium oxide). Top right: concretions of malachite, due to a mixture of carbonate and copper hydroxide. Below: massive tabular apatite crystals (a mixture of phosphate, fluoride and calcium chloride).

crystalline structure originating from the matching of minute *elementary cells*, which, when they come into contact with each other, fill up the space like the bricks in a wall. Each one of these cells, which Haüg called *integrating molecules*, expresses the chemical composition of the substance, and results from the orderly spatial arrangement of atoms, ions and molecules. The elementary cells of minerals can produce fourteen fundamental Bravais types.

Important progress in our knowledge of crystalline structures was made in 1912 by Laue, who for the first time used crystals as diffraction gratings for X-rays. The effects, impressed on photographic plates, gave rise to a series of markings, the arrangement of which represents the innermost symmetry of the structure. Two or more crystalline types in whose similar structures we find particles of almost the same size in equivalent positions can give rise to *isomorphism*. This phenomenon results in a mixing in the solid state, of the extreme ends or terms, with the formation of minerals with an intermediate chemical composition. This is the case with the isomorphic families of garnets, spinels and olivine. In addition, a compound may crystallize with different structures. The carbon from which, in different conditions, diamonds and graphite are formed, offers a good example of *polymorphism*. Every crystalline

type has morphological features which simplify identification. The various degrees of symmetry defined by the presence of axes, planes and centre give rise to 32 classes grouped in seven crystallographic systems.

Often, however, crystals do not have a regular polyhedral form because, by growing in aggregates, contact between them restricts development. In the natural state, the partial concretion of crystals on a rocky substratum causes particular aggregates known as *druses*, if the substratum is tabular, and *geodes*, if it corresponds with the inner surface of a subspherical cavity. Fine geodes of amethyst quartz come from Brazil and Madagascar. Crystals with an elongated prismatic constitution, caused by subparallel or radial association, form strange aggregates which, because of their appearance, are called bacilliform, acicular (needle-shaped), fibrous, coralloid, globular etc.

The identification of crystals is helped not only by morphological examination but also by knowledge of their optical features and other physical properties such as their specific weight, cleavage and hardness.

Hardness consists in the resistance put up by a surface to scratching. The German physicist Friedrich Mohs observed in 1816 for various pairs of minerals which of the two could scratch the other. As a result he proposed a

Top left: delicate prismatic crocoisite crystals, dawn-red in colour with an adamantine lustre. This fairly rare mineral is a lead chromate and is found in oxidation areas of lead-and zinc-bearing deposits. Below: an aggregate of rare orpiment (arsenic sulphide) crystals. This mineral, so called from the Latin auri pigmentum *because of its golden yellow colour, originates from low-temperature hydro-thermal processes. It is found in Yellowstone Park, in deposits connected with geyser activity. Right: large amethyst quartz crystals.*

Top left: granular olivine crystals, from Germany. These transparent, clear crystals can be used to extract precious gems. Below: a fibrous-radial aggregate of hedenbergite from Yugoslavia. Top centre: prismatic polychrome tormaline crystals from Elba. Typical of pegmatitic gangues, this is used in gemmology. Below: Italian serpentine, a mineral produced by the alteration of peridotitic rocks. Top right: prismatic epidote crystals, frequently found in metamorphic rocks. Below: aggregate of amazonite from Colorado, a variety of microcline, used as a gem.

scale of ten degrees of hardness. Although only of relative importance, this scale is nevertheless a useful guide to crystals. The lowest values refer, in ascending order, to talcum, gypsum, calcite, and rise to corundum and diamond, with values of nine and ten respectively.

Some of the optical features of minerals depend on their degree of light-absorption. We thus find opaque minerals, in cases of high absorption, and transparent ones, where there is no absorption, with translucid types between the two. Furthermore, the selective absorption of the colour spectrum of white light gives rise to the colour of minerals. In some, known as *idiochromatic*, this relies exclusively on the chemical composition and the physical structure.

Other minerals are *allochromatic*, that is, they take on a colour because of impurities existing in the crystalline structure; thus, if pure, corundum is colourless, whereas traces of chromium determine the red coloration of rubies, iron and titanium, and the blue of sapphires. The presence of chromium is also responsible for the green of emeralds, and the presence of iron causes the sky-blue of aquamarines.

However, the elements responsible for the different colours of minerals have not been identified in all cases.

As it crosses through crystals, light slows down and changes direction to some extent, both processes depending on the nature of the crystal; when it is a question of white light, this is broken down into the colours of the rainbow. We talk, respectively, of *refraction* and *dispersion*. The values of the refraction index which are peculiar to every crystal type, are used for the purposes of identification.

The amount of light reflected by the surface of minerals determines the lustre, which may be metallic, adamantine, vitreous, mother-of-pearl (nacreous) or silken (sericeous). Further, phenomena of reflection are responsible for pretty optical effects in certain crystals containing inclusions. One such is *asterism* which occurs in certain sapphires and rubies which incorporate tiny needle-shaped crystals of rutile arranged in accordance with the pseudo-hexagonal symmetry of the host crystal. Cutting to achieve a curved surface helps the play of reflections which gives rise to the appearance of a six-pointed star. In a particular variety of quartz formed by fibrous crystals we find the distinctive *glinting reflection*. Depending on the hue, we thus have *tiger's eye*, *cat's eye* and *hawk's eye quartz*. A subtle play of light, due mainly to phenomena of diffraction, causes *iridescence* in opals.

Left: calcite crystals with hexagonal symmetry. One of the commonest minerals, found in carbonated rocks, it originates from watery solutions, sometimes at ambient temperatures. Top right: wulfenite crystals with a tetragonal structure like those seen earlier. Below: scolecite, a tectosilicate for aluminium and calcium, in fibrous-radial crystals. It forms on the surface as a result of the hydro-thermal transformation of feldspars from basic rocks.

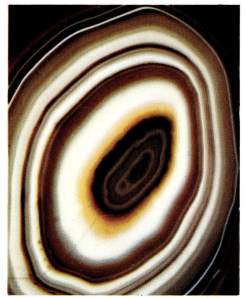

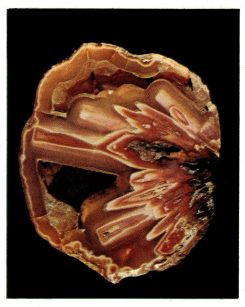

Left: agate, a semi-precious stone generally formed by the depositing of chalcedony quartz by watery solutions of the same in consecutive and differently coloured strata. The depositing occurs mainly on the inner walls of geodes. In cross-section, agates have a fibrous or fibrous-radial structure, fairly regular concentric strata, and shades of colour, varying from brownish red to bluish-grey and milky-white. Right: two examples of agate; from central Mexico (top) and the Rio Grande do Sul in Brazil (below).

If subjected to irradiation, many minerals emit light. This effect, which is fairly conspicuous in fluorite, is known as *fluorescence*. The fluorescent colour is often a useful diagnostic feature.

Gems are obtained by cutting from crystals which have particular qualities of purity, transparency, hardness, lustre, colour and rarity. Top of the list come diamond, ruby, sapphire, emerald and topaz; the rest, some of which are also extremely beautiful, are usually considered to be semi-precious: these include aquamarine, garnets, amethyst and opal. Lastly we have what are known as hard stones, such as turquoise, malachite, lapis lazuli, jade, agate and onyx, used both ornamentally, and for the manufacture of artistic objects.

Extremely beautiful vases and statuettes are evidence of the high artistic level achieved by peoples in the Orient many centuries ago.

In the remote past, precious stones were generally used in the natural state; at most the faces of the crystal were polished, the yardstick being to reduce the weight of the stone as little as possible. Nevertheless, the art of intaglio or incision was practised by Egyptians, Chaldeans, Babylonians, Persians and Assyrians.

The carvings of cylindrical seals by the Mesopotamian civilizations, and the

251

scarabs of malachite, emerald, obsidian, agate and amethyst found in Egyptian mummies as ornaments and amulets are impressively made. Working with hard stones with strata of different colours such as agate and onyx (to obtain a better contrast), the Greeks and then the Romans sculpted human figures and faces in relief, and produced magnificent cameos. Technically speaking, the incision was made with metal chisels, or chisels with corundum tips; today diamond points are also used with drills; these are lubricated with a mixture of oil and very fine diamond or corundum powder. Polishing is then carried out with fine emery.

Carving quartz is done with the use of hydrofluoric acid, once the crystal has been covered with a layer of wax on which the design has been made.

The art of facetting precious stones is still carried on today. The stone-cutter must have extensive knowledge of the optical and crystallographic properties of the mineral in question, if he is to create shifting plays of light by combining shades of colour with the effects of dispersion and reflection.

A good combination between the refractiveness of the mineral and the angulation of the facets will produce, in gems, a whole series of inner reflections visible to the observer, the result of which is the gem's *brilliance*.

Various aspects of quartz. Top left: amethyst quartz. Below: a druse of clear and colourless quartz crystals from the Dauphiné in France. Right: a clear quartz crystal with inclusions of needle-shaped rutile. Druses are formed by a nearly-parallel association of crystals arranged on a different type of substratum; they are formed in fissures in rocks from moving solutions or vapours. The small needle-shaped spikes of rutile – titanium dioxide – are formed by prismatic crystals due to the hydro-thermal alteration of different titanium-bearing minerals.

Top left: verdite or Transvaal jade. This rock is formed by clay and a variety of chromium-bearing mica called fuchsite. Below: a piece of fish-shaped Australian jade jewellery, made in New Zealand. Centre: detail from a seventeenth-century Chinese jade screen showing silk processing. Top right: a Chinese gold disc, with jade figures, from the third century B.C. Below: eighteenth-century jade jewellery with pearls. The term 'jade' indicates various minerals characterized by similar properties such as hardness, opacity and oiliness to the touch. Jades are found in the natural state in alluvial deposits.

Cutting differs with the type of gem; it ranges from the brilliant cut, typical of diamonds, to the table cut and step cut found in emeralds.

For the opaque, semi-transparent and transparent semi-precious stones, with bright coloration, the *cabochon* cut is usual, that is, where the surface is curved.

In any event, in the initial stages of the cutting process, possible cleavage planes are used. Alternatively, cutters use circular saws. Next, for the cutting of facets, the stone is fixed to some appropriate mounting, and worked with grinding-wheels of steel, copper, brass, tin, or lead, depending on the particular degree of hardness of the stone.

The facetting process, which is usually preceded by 'rough-hewing', is then followed by polishing, to remove any surface scratches caused by working the stone. When the cutting is finished, the gems are put on the market; the value is proportionate to the weight measured in *metric carats*. The metric carat (cm), equal to 0.200 grammes, is subdivided into 100 parts called *points*. For more valuable stones the *metric grain* (one quarter of a carat) is still in use, being the equivalent of 0.050 grammes. The word 'carat' appears to derive from *Keration*, the Greek term for the carob seed. In ancient times, the weight of

gems was established by comparison with the weight of the seeds of various plants.

The value of precious stones is assessed not only by their weight but also by their colour, transparency, purity, cut and, last but by no means least, their rarity. The most valuable gems are undoubtedly diamonds, rubies, sapphires and emeralds. But high prices are also fetched by other stones like padparasca, tanzanite, opal and alexandrite, in the case of specimens of rare beauty.

The study of gems, with the purpose of establishing their genuineness, type and, thus, their value, borrows many methods used in mineralogical analysis. Chemical tests can obviously not be used, and the same goes for hardness testing. Diagnoses are thus made essentially on the basis of the optical or visible features. Observation under the microscope, particularly in the case of coloured stones, often reveals the presence of tiny liquid, gaseous and solid inclusions with distinctive arrangements. Solid inclusions are usually formed by microscopic crystals with characteristic form. It is more difficult to identify and interpret shapeless inclusions, which are often of a carbon nature, and bands of colour with different intensities. Close examination of these features makes it possible to detect, in many instances, the deposit from which the gems

Top left: topaz – aluminium fluoro-silicate – in which a certain number of hydroxide ions can replace the fluoride. The chemical composition of topaz is like that of andalusite – aluminium silicate – which is also used in jewellery. Top centre and top right: tormalines: these stones are formed mainly by aluminium silicates, containing various other elements and mainly boron, hydrogen, sodium and lithium. Below: various types of cut, in different precious stones.

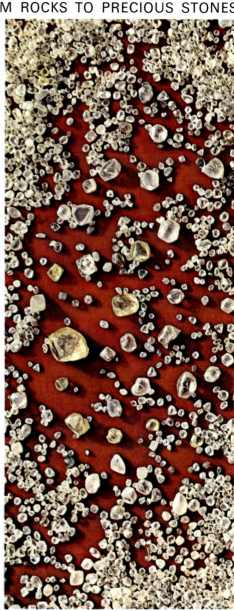

Top left: looking for diamonds. The primary deposits occur in kimberlite-filled volcanic chimneys; the secondary deposits, on the other hand, are formed by alluvial deposits. Below: crude diamond still in the host rock. This valuable mineral is formed by slow crystallization from volcanic magma, in specific temperatures and at specific pressures. Centre: head-dress brooch of diamonds mounted round pink brilliants. Right: rough diamonds from Pretoria (South Africa).

originate, and is useful in distinguishing between natural and synthetic stones.

The diamond is the queen of precious stones. Formed by pure carbon, it frequently crystallizes octahedrally. Its genesis is still uncertain, but it is thought that, having formed deep-down in the Earth's crust, it is then brought to the surface by the magma. The greatest primary deposits are found in South Africa; diamond-bearing rock, known locally as *kimberlite*, fills explosive, funnel-shaped volcanic chimneys known as *necks*. Mining is open-cast or underground.

A diamond is considered pure if, when magnified ten times, it does not show any inclusions. It is rare to find a pure, colourless diamond, and in fact only ten per cent satisfy this criterion; diamonds often have faint blue and pink shadings, which are much sought-after, but in the majority of cases the coloration is more marked.

THE TWO OCEANS: AIR AND WATER

In the solar system, the Earth is the only planet which, because of its distance from the Sun, has the temperatures and pressures which allow water to be present in three forms: solid, liquid and gaseous.

The world, which started out white-hot, extremely fluid and capable of emitting gaseous substances, then tended to envelop itself in a covering of steam, as a result of the gradual cooling process; this steam, together with other magmatic gases, was released from the interior of the earth as a result of volcanic eruptions. When the temperature and pressure of this atmosphere reached levels which allowed the steam to condense, *water* was produced. As water accumulated, it then gave rise to the first seas. Here the surface was horizontal, i.e. perpendicular to the direction of the force of gravity.

As far as the origin of the salinity of sea-water is concerned, some maintain that the salts accumulated when the Earth was still in the fluid state, by leaching from the original atmosphere, while the volatile substances, in which the Earth was particularly rich, were undergoing condensation. Others, however, maintain that the salts were carried to the sea by rivers and thus represented the product of leaching from the land. On the basis of these hypotheses, and taking into consideration the fact that the lithosphere has a predominantly sialic composition, this initial leaching would only have given the sea those compounds which are currently found there in small quantities. On the contrary, the chlorides, sulphates, and carbonates, which are more abundant, would have formed in a later, second stage, with the leaching of evaporitic rocks produced by successive climatic variations. An objection that one can raise about this hypothesis is that these evaporitic rocks are, nevertheless, sedimentary rocks which originated in an already salty marine environment.

The problem of the origin of salinity can be cleared up if we accept that the salts were not carried to the sea just by rivers, but that they originated in other ways, for example, from submarine volcanoes which were very active in the past and capable of producing large quantities of chlorine, sulphur and carbonic acid. With regard to the variations in salinity down the ages, many maintain that in the original seas it was at present-day levels, and that later amounts carried down to the sea by rivers did not alter it. In any event, sea-water is a balanced solution, governed by a complex series of reactions, affected in the main by pressure and temperature.

The vast expanses of salt-water on the Earth's surface are divided into *seas* and *oceans*; these differ from each other on the basis of area, origin, form of the basin, chemico-physical and dynamic features such as density, temperature, tides, and so on. Seas can be considered as comparatively small offshoots of an ocean; they are situated either at the edge of an ocean – like the Baltic Sea and the Mediterranean, or within a continent – like the Caspian Sea and the Aral Sea. Seas are thus affected by the adjacent ocean or continent. The origin of the inland basins can be attributed to the Earth's crust sinking, or to the emergence of mountain chains, which cause the type of land-locking we see in the case of the Caspian Sea. The oceans occupy vast, deep depressions; these have evolved very gradually through various geological periods, as a result, principally, of *continental drift* and of the geodynamic processes of the Earth's crust. There has been much controversy about the classification of the oceans, which were

Left: a schematic picture of the water-cycle, indicating the various processes of evaporation and precipitation. Top centre: the geo-thermal circuit. Water on fault planes descends, heats up on contact with hot rocks, and re-emerges warm. Top right: the percolation of water into the subsoil with the formation of a water-bearing stratum. Below: diagram of the circulation of underground water. Above: there is a free fold, known as a water-table, while, below, because of the presence of two impermeable levels, we have artesian water.

initially divided into five; Atlantic, Indian, Pacific, Arctic and Antarctic then into four; Atlantic, Indian, Pacific and Southern and later into three; Atlantic, Pacific and Indian, with the Arctic and Antarctic Oceans being considered, respectively, a branch of the Atlantic Ocean and a 'sea' joining the three main oceans together.

Modern technology, and especially observations by man-made satellites, has made it possible to make an accurate calculation of the area of the various seas: they occupy $\frac{7}{10}$ths of the Earth's surface – land occupying just $\frac{3}{10}$ths – which is in the region of 149 million sq km (93 million sq miles). In the course of various geological periods the surface area of the oceans has been subject to change both because of the folding of the Earth's crust and of the sedimentary evolution of the ocean basins. There have been fluctuations in the sea-level, which is subject to movements brought about by major climatic shifts, linked with astronomical phenomena which have been responsible, among other things, for the varying extent of the ice-caps on the poles. The ice-caps of the Quaternary Period, during their massive expansion, caused a conspicuous lowering of the sea-level, sometimes involving drops of up to 150 metres (500 feet). These same, present-day ice-caps have a mass which could, if they thawed

257

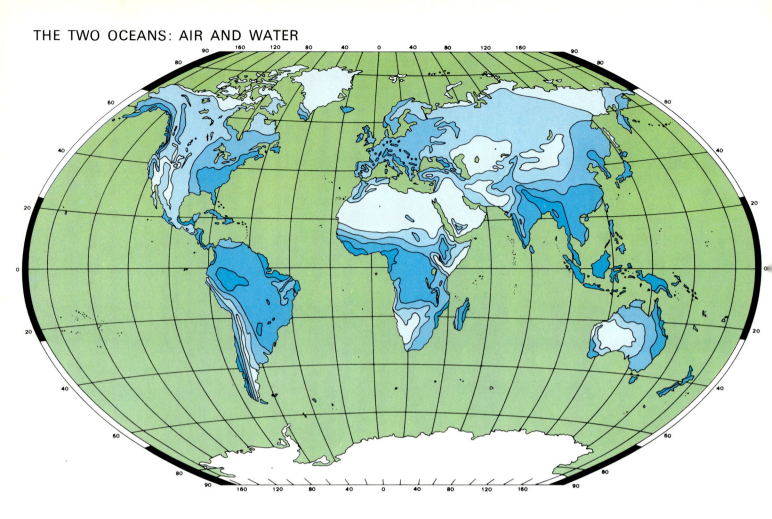

- [ ] less than 250 mm (10 ins)
- [ ] from 250–500 mm (10–20 ins)
- [ ] from 500–1000 mm (20–38 ins)
- [ ] from 1000–2000 mm (38–77 ins)
- [ ] more than 2000 mm (77 ins)

completely, raise the level of the sea by 30 metres (100 feet).

The temperature of the sea is caused by the heat transmitted by the Sun's rays which heat the upper layers. The warmest ocean appears to be the Pacific, because of its large area lying between the tropics, although as a whole the water in the northern hemisphere is warmer than that in the southern hemisphere. This is because in their southerly regions the oceans do not have continental barriers capable of preventing the inflow of deep, cold currents circulating around Antarctica. Likewise because of the relative absence of continental land-masses, the isotherms (the lines which join points having the same temperature) in the southern hemisphere tend to run with the parallels because of only slight deviations caused by the continental land-masses in these regions. Similarly, the temperature of the ocean waters fluctuates less daily, and over the year, than do temperatures on land because of the greater thermal capacities of water. In the sea, the temperature drops with depth, rapidly at first, and then more slowly, the most conspicuous variations being in the first 300 metres (1000 feet). In the very lowest ocean-depths the temperature tends to increase slightly because of the strong pressure exercised by the water above. In the marginal basins, on the contrary, the temperature decreases gradually

A map showing the distribution of precipitation. In relation to the various latitudes and altitudes, we can see the various rainfalls, i.e. the different distribution of precipitation in the various months. The highest rainfall occurs in the equatorial, monsoon and tropical areas and is closely linked to the vegetation and climate.

Left: a rainy day on the river Skrang in Borneo. Given the hot humid climate, rainfall is very plentiful and favours agriculture and the spread of endless forests. Right: rain-laden clouds announcing a storm.

down to a certain depth, beneath which it remains virtually constant. These basins are usually separated from the oceans by sills, intrusions of rock on the sea-bed, which prevent the deep currents from carrying out their compensatory action.

Another fundamental feature of sea-water is the salinity, which is expressed, in general, as the quantity, in grammes, of salt present in one kilogramme of water. In the open sea and on the surface, the average salinity is 35 per 1000, which means that if 1000 grammes of sea water are evaporated, there will be a residue of 35 grammes of salt. The most plentiful salts are the sodium and magnesium chlorides, the magnesium, calcium and potassium sulphates, and calcium bicarbonate. The salt content of sea-water may vary from place to place, depending on the climatic conditions. For example, high temperatures cause considerable evaporation which, if not compensated for by rain or additional water from other sources, makes the sea very salty. The inflow of fresh-water, or melted ice or snow, produces dilution and thus a reduction of the salinity level, and hence of the density of the water.

Salinity is thus lower at the mouths of large rivers such as the Amazon or Congo because of the large influx of fresh water. In the marginal basins (seas),

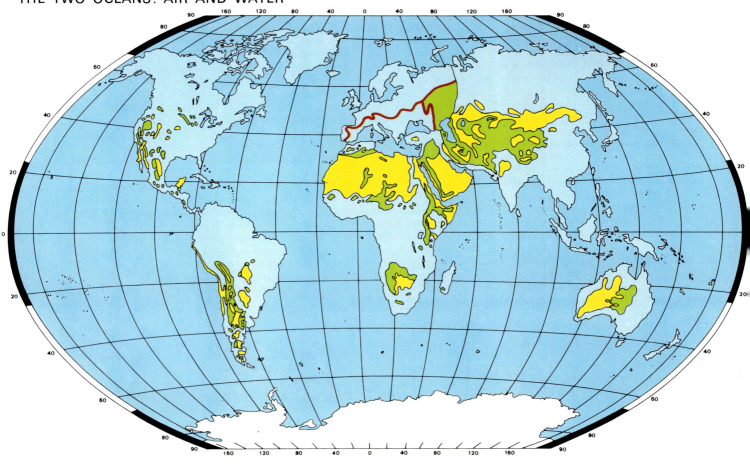

☐ exhorheic zones
☐ endorheic zones
☐ arheic zones
☐ watersheds

where the exchange is limited, there are larger variations of salinity than in the open sea. In the Mediterranean, the salinity varies with the latitude; in the northern areas the inflow from the rivers in the Pyrenees and Alps lowers the salinity by dilution; in the southern areas, on the other hand, the arid climate and very high condensation increase the salinity level; in the Baltic Sea it hovers around seven per 100 and goes as low as one per 1000 in the Gulf of Bothnia, because of the wet climate, low evaporation and large influx of river-waters. In the Red Sea the salinity level reaches more than 40 per 1000 because its entire area lies in an arid region, fed by no river. Then we have completely anomalous land-locked basins such as the Dead Sea, where salinity reaches very high levels in the region of 275 per 1000. In this case the high saline content is due not only to the aridity of the region but also to the influx of the River Jordan which winds its way through saline rocks which are extremely soluble. Salinity varies both on the surface and with depth, in the sense that, normally, it increases from top to bottom because the saltier water, which is denser and heavier, tends to make its way towards the bottom.

The variations in temperature and density between adjacent masses of water cause the ocean currents. In fact, the world's major water-masses are subject to

The hydro-graphic distribution and features of the various continents. Zones having no surface hydro-graphy correspond to the arheic zones, generally typified by desert environments such as the Sahara, Arabia, Mexico, the Kalahari and Turkestan; the zones in which rivers run out into an enclosed basin, for example, Lake Eyre (in Australia), Lake Chad and the Caspian Sea, correspond to the endorheic areas. Lastly, the zones where rivers run out into the sea are called exorheic zones.

An aerial view of a tributary of the Amazon. Covering 7 million sq km (4.5 million sq miles), Amazonia is the largest hydro-graphic system in the world with 50,000 km (30,000 miles) of navigable waterways. Because of its sub-equatorial, hot and humid climate, overcast skies and daily rainfall, Amazonia is covered with dense virgin forests.

fairly regular and constant movements. We have surface currents and deep currents, which are either ascendant or descendant. The main factor determining the swifter surface currents is the action of the wind, whereas, deeper down, the wind has no effect, and the water moves along very slowly indeed, from the denser to the not so dense zones, by convection.

The surface currents are warm when they come from the equatorial zones, and cold when they come from the polar – both Arctic and Antarctic – regions. They affect the upper layers of the sea, to an average depth of 300–400 metres (1000–1500 feet), and the water carried along in them differs from the surrounding water in temperature, salinity, density and colour.

In the Atlantic and Pacific Oceans the warm currents are symmetrically distributed, north and south of the equator, and flow always in the same direction because they derive from the respective trade-winds; these are constant winds which blow from the north-east in the northern hemisphere and from the south-east in the southern hemisphere. These winds cause the north- and south-equatorial currents, which are separated by a counter-current flowing along the equator. These currents cross the oceans from east to west, and then rise towards the central latitudes, i.e. from where the water is less

dense, as a result of plentiful precipitation in the equatorial belt, to where the water is denser because of the arid climate and considerable evaporation in the Tropics. Once at the central latitudes, they once again flow back across the oceans in the opposite direction, i.e. from west to east, and return to the more extreme latitudes, thus making a full-circle. Within these circuits anti-cyclone areas are often formed, i.e. high-pressure areas, where the water remains perfectly calm. In the north Atlantic one area of this type is the well-known Sargasso Sea, situated inside the Gulf Stream, where the calm water favours the development of specific algae, known as gulf-weed (*Sargassum*). In the Indian Ocean the warm currents are not as a rule distributed symmetrically north and south of the equator; in fact, to the north, because they are caused by the monsoons – periodic winds which blow for six months in one direction and six months in the opposite direction – the currents change direction every six months.

Among the cold currents coming from the polar regions we find, in the Atlantic Ocean, the *Labrador* current and, in the Pacific, the *Oja Shivo*, coming from the Bering Sea.

In the Atlantic, Pacific and Indian Oceans there are also cold currents in the

Left: an ice-floe crowded with penguins. Top right: a cave carved out of compact ice by a sub-glacial river. Below: ice sculpted by staggered thawing during summer. This phenomenon is linked with the constant direction of warm winds.

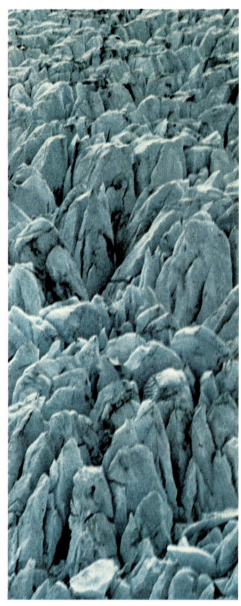

Left: a massive and impressive iceberg off the coast of Greenland. To have an idea of its size, imagine that the tip visible is just one-ninth of the entire ice-mass. As they thaw, icebergs can capsize because of a shift in their centre of gravity; a serious hazard for anyone in the vicinity. Right: these large rugged pinnacles are evidence of the constant movement of the ice and the considerable stresses which develop in it.

southern hemisphere, caused mainly by the influx of the current around the Antarctic.

The warm currents are climatically very important because they tend to soften the climate; the steam contained in the air in fact retains the temperature of its place of origin for some time. In the North Atlantic, for example, the *Gulf Stream*, despite its northerly latitude, makes the climate mild in southern Greenland, Iceland, Great Britain and Scandinavia. A branch of it enters the White Sea, and leaves the port of Archangel free of ice for more than six months a year (at a latitude of 65°). Among other things, the *Gulf Stream* is responsible for the fog which is common in these parts, caused by the condensation of the warm, damp air coming off the sea. In the South Atlantic, a branch of the south-equatorial current carries warm water as far as the River Plate. In the Pacific Ocean, the counterpart to the *Gulf Stream* is the *Kuroshivo*; this current creates a mild climate on the eastern coast of Japan, flows back across the ocean in the opposite direction, and thus also benefits the coasts of Alaska and Canada.

The cold ocean currents are responsible for carrying the icebergs from where they are formed to lower latitudes. In the Atlantic, for example, icebergs

carried along by the cold *Labrador* current reach latitudes of 40°, on a level with Newfoundland, where they are melted by the warm water of the Gulf Stream. When they melt, the detritus contained in the ice sinks to the bottom and forms the well-known *Newfoundland Bank*. The deep movement is caused by the different density of the water and is also affected by the morphology of the ocean-bed. The deep currents, which are generally slower than the surface currents, originate in the Arctic and, above all, Antarctic seas where, because of the formation of ice, the denser and thus heavier water tends to sink and move towards lower latitudes. Their polar origin explains the fact that even on the equator the water in the ocean depths is cold.

In the southern hemisphere the deep currents spread out more freely because their route is less impeded by the smaller land-masses.

Biologically speaking, the ascendant currents are very important, because they carry to the surface not only cold water but also important mineral substances which are transformed by the vegetable matter forming the *phytoplankton* into organic substances.

The seas near Peru, western Africa and California are particularly fertile. The tropical seas, on the other hand, are barren, because the absence of

A photograph in polarized light of melting snow-crystals. The iridescence is due to the various orientations of the ice-crystals.

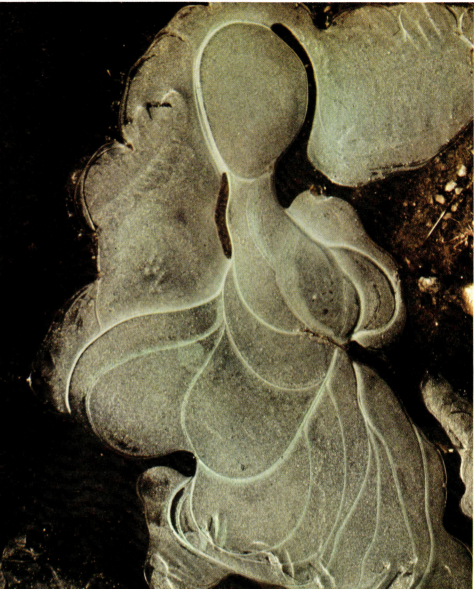

Left: a detail of bubbly ice. The small bubbles are caused by tiny gas inclusions, which have been imprisoned in the ice as it thickens. In time, and very gradually, these bubbles tend to disappear, giving rise to compact ice with a density almost equal to that of water. Right: ice can sometimes assume amazing shapes, as in the case of the 'manacle' pictured here; these are temporary and local, due to partial thawing and then immediate refreezing. These forms disappear altogether when summer arrives.

ascendant currents prevents the influx of nutritive salts.

In the marginal seas, which are usually separated from the oceans by a submarine sill, the surface currents are associated with the difference of levels of the water, whereas the deep-sea currents are caused by the difference in density between the sea-water and the water of the adjacent ocean.

In basins with an arid climate, like the Mediterranean and the Red Sea, evaporation exceeds rainfall, the sea-water is more salty, and its level is thus lower than the level of the ocean. At the Straits of Gibraltar, for example, water enters the Mediterranean on the surface, but deeper down the current flows in the opposite direction and the denser water of the Mediterranean crosses over the sill, and pours into the Atlantic Ocean. In wet climates, like that in the Baltic Sea, rainfall exceeds evaporation; as a result the sea-water, which is more diluted and at a higher level, flows towards the ocean. In the Black Sea the movement is quite specific in as much as the Dardanelles sill is so high that it prevents the renewal of deep water. There is thus a lack of oxygen exchange, hence, below 200 metres (700 feet) we find an environment with a shortage of oxygen with the result that there is the formation of sulphidic acid by bacteria.

The marine and ocean currents are also important from a biological point of

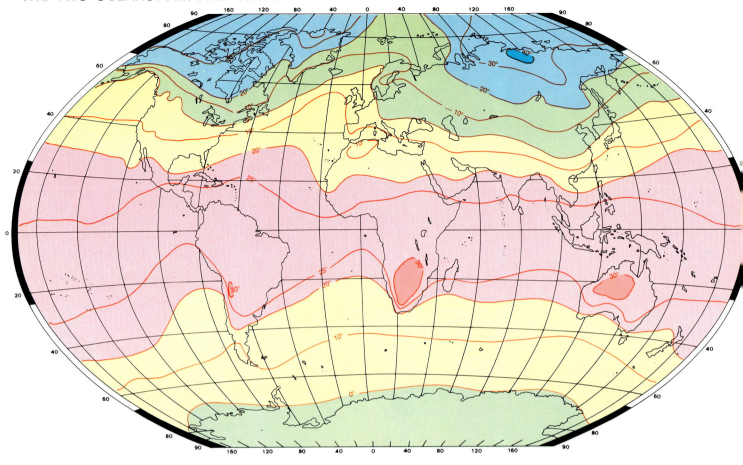

view because as well as making the physical and chemical properties of the water uniform, they have the effect of softening the climate, as in the case of the intertropical waters which, when carried to high latitudes, make it possible for life to flourish in otherwise uninhabitable regions. Equally important the cold water from the polar regions moving away from the ocean-bed reaches the equatorial zone where it mixes with the warm water and lowers the temperature.

If we consider the continental waters, we can see that on the Earth's surface a considerable quantity of water is retained within morphological depressions which form lakes, both large and small, and which number several thousand. On the basis of their origin, the most numerous are undoubtedly those of glacial origin, which are distributed mainly in high latitudes and at high altitudes, areas most affected by the Quaternary glaciation. In Scandinavia alone there are more than 3000 lakes, while in North America the small number is compensated by the presence of the Great Lakes – Lakes Superior, Erie, Huron, Michigan and Ontario.

Lakes of tectonic origin are also important, i.e. those lakes situated in depressions in zones shaped by the lowering of the Earth's crust. In the case of

An isothermal chart for January. It is evident from the individual charts that the behaviour of isotherms is influenced by the distribution of the oceans and continents and, above all, by the arrangement of the principal mountain ranges. It should be pointed out that the January isotherms correspond with winter in the northern hemisphere and summer in the southern hemisphere. A chart for July is shown on page 268.

Left: the Boîte river at Cortina after a fall of snow in December. Right: the (Alpine) ibex, a true relic of the fauna of the Quaternary Period.

tectonic types with folding, we find the synclinal lakes of Neuchâtel and Brienne in the Swiss Jura; the rigid tectonic type, as in the Red Sea rift valley, is represented by Lakes Nyasa, Tanganyika, Albert and Rudolf in East Africa, Lake Tiberias and the Dead Sea in the Middle East, and Lake Baikal in Russia.

There are also lakes situated in volcanic craters or calderas, like the lakes in Latium (Lazio) – Bolsena, Vico, Bracciano and Albano – or explosive chimneys (conduits) like the *Maaren* in the Eifel in western Germany.

Another great fresh-water reserve is the glacier. Even today, glaciers cover more than 15 million sq km (9.3 million sq miles) of the Earth's surface and are the remnants of the very extensive Quaternary ice-caps. In that period a thick covering of ice lay over the great Alpine chain from the Pyrenees to the Alps, and from the Balkans to the Himalayas, with enormous glacial tongues reaching right to the foot of these mountain ranges.

At the present time the largest glaciers are to be found in Antarctica and Greenland, where the ice can reach thicknesses of more than 3000 metres (10,000 feet). The peripheral areas of these caps reach the sea where the motion of the waves and the tides break down the fronts and create *icebergs*. The form of these icebergs ranges from massive in the northern hemisphere to tabular

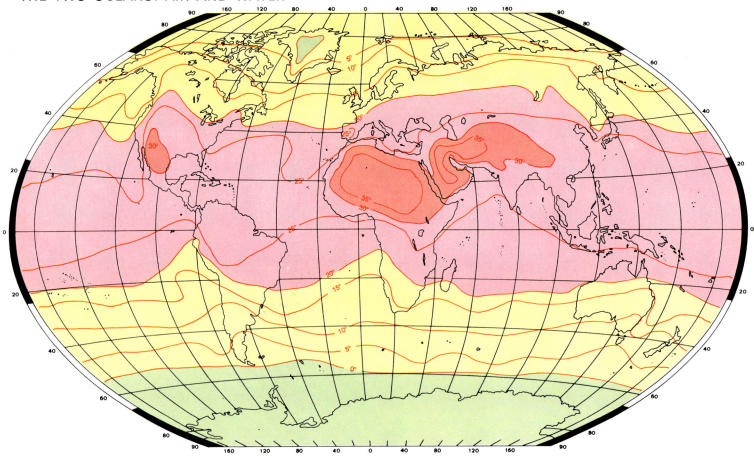

around the Antarctic, where single icebergs can reach dimensions of a few score square kilometres. The glaciers originate in most cases above the *persistent snow-line*, where the basin that feeds them is situated, and reach to below this limit through what is called the *ablation* process. The persistent snow-line marks the transition between zones where the ground is always snow-covered and zones where the snow melts completely when summer arrives. The altitude of this limit depends on various factors such as latitude, local climatic conditions, and the angle and shape of the land on which the snow falls and accumulates. At the present time it lies at sea-level at the Poles, at 600 metres (2000 feet) in Greenland, at 2700 metres (9000 feet) in the Alps and at 4900 metres (16,000 feet) in East Africa (Mt. Kilimanjaro) 5895 metres (19,340 feet).

Unlike lakes and glaciers, rivers are only temporary water-containers. In the world we have rivers and water-courses which flow in areas with a uniform climate, like the Amazon or Congo, and others which rise in areas with considerably different climates from the areas through which they flow, like the Nile or Mississippi. To the water which collects in lakes and rivers should be added the meteoric water which is absorbed by the ground. This filters its way into the soil, and makes it way downward until it reaches a level of

An isothermal chart for July. Here, too, we can see how the distribution of air-temperatures on the continents is closely linked with geographical factors. It follows that there is frequently little coincidence between thermal zones and astronomical zones. The highest absolute temperatures have been recorded in Death Valley, California, and in the Libyan Desert (67°C) (153°F), i.e. in inland continental areas.

Two photographs depicting the hot African summer: a Tunisian boy diving into a pool at Gafsa (left) and peasants gathering the harvest on a Moroccan plateau (right).

impermeable rock. Here it stops and forms a water-bearing stratum. This water is important for man because it provides him with his drinking-water.

Let us now turn to the other 'ocean': air.

A gaseous envelope, called the atmosphere, surrounds and protects our planet, to which it is attracted by the force of gravity. It does not have a homogeneous composition, but has different strata which are called, from bottom to top, the *troposphere*, the *stratosphere*, the *ionosphere* and the *exosphere*. The greatest concentration of gases occurs in the troposphere where, in the order of their respective quantities, we find nitrogen, oxygen and argon, followed by steam and carbon dioxide.

In the troposphere we find the main meterological and optical phenomena such as clouds, precipitation, sunsets, rainbows and mirages. Its instability is due to the reduction of the temperature as altitude increases.

In the next 'wrapper', the stratosphere, the temperature tends to increase and reaches its highest point in the ozonosphere, where, because of the action of ultra-violet rays, oxygen is transformed into ozone. Beyond this layer the temperature decreases up to an altitude of 80 kilometres (50 miles), where the ionosphere begins and the temperature starts once more to increase gradually

and steadily. In the ionosphere, where the upper limit touches on an altitude of more than 500 kilometres (300 miles), the fairly rarefied gases are ionized by the photoelectric effect of the Sun's rays. Here, the refraction of the electromagnetic waves used in radio-communications takes place. Beyond the ionosphere lies the exosphere, the last layer of the Earth's atmosphere, which gradually merges with the solar plasma.

The atmosphere protects the Earth from ultra-violet radiation which is a hazard to biological life, and regulates the irradiation of those infra-red rays by steadying the temperature of the air which would otherwise be too high during the day and during the night, too low, which is what happens on the Moon, which has no atmosphere. The amount of the Sun's rays arriving in one minute on to an area one centimetre square, at the upper limit of the atmosphere is known as the *solar* constant. Equivalent to about 1.94 calories, it varies according to the activity of sun-spots, which follows an eleven-year cycle. All the phenomena observed in the atmosphere are thus related to the activity and relative position of the Sun and depend also on temperature, pressure and humidity which are created locally.

The distribution of temperatures across the globe is made with the use of

The centre of a typhoon 480 kilometres (300 miles) south-east of Tokyo. The photograph was taken by the American meteorological satellite Tiros 5. These violent whirlwind storms, of tropical origin, are frequent in the west Pacific and in the China Sea. They can produce exceptionally strong winds with speeds of 160 kph (100 mph), with torrential rain.

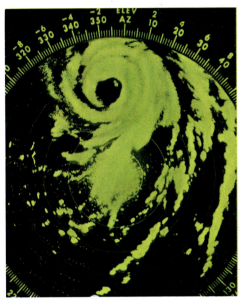

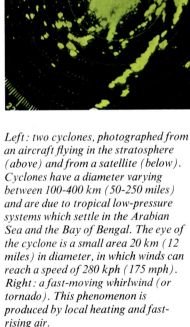

Left: two cyclones, photographed from an aircraft flying in the stratosphere (above) and from a satellite (below). Cyclones have a diameter varying between 100-400 km (50-250 miles) and are due to tropical low-pressure systems which settle in the Arabian Sea and the Bay of Bengal. The eye of the cyclone is a small area 20 km (12 miles) in diameter, in which winds can reach a speed of 280 kph (175 mph). Right: a fast-moving whirlwind (or tornado). This phenomenon is produced by local heating and fast-rising air.

isotherms, which are the lines joining points having the same temperature.

The highest temperatures occur in the inter-tropical regions which receive a larger amount of solar rays because those regions lie where they receive the Sun's rays at a greater angle. As latitudes increase, so the temperatures decrease because of the reduction of the angle of the Sun's rays, until a minimum point is reached which corresponds with the polar ice-caps, where the Sun's rays hit the Earth's surface like tangents. The atmospheric pressure represents the weight of the atmosphere over a given area; at sea-level and 0 °C the average pressure is 1013 millibars and equals the weight exercised by a column of mercury 760 mm tall. The pressure decreases with altitude, with the increase in temperature and with the increase in relative humidity. Pressures are indicated with *isobars*, which are lines joining points having the same pressure.

In the world, as a result of the different amount of solar radiation received, we have areas of low pressure, or *cyclonic* areas, corresponding with the hotter regions, and high-pressure or *anticyclonic* areas in the colder regions. The cyclones and anticyclones may be permanent, seasonal or temporary.

The Azores anticyclone, for example, is a good example of a permanent one;

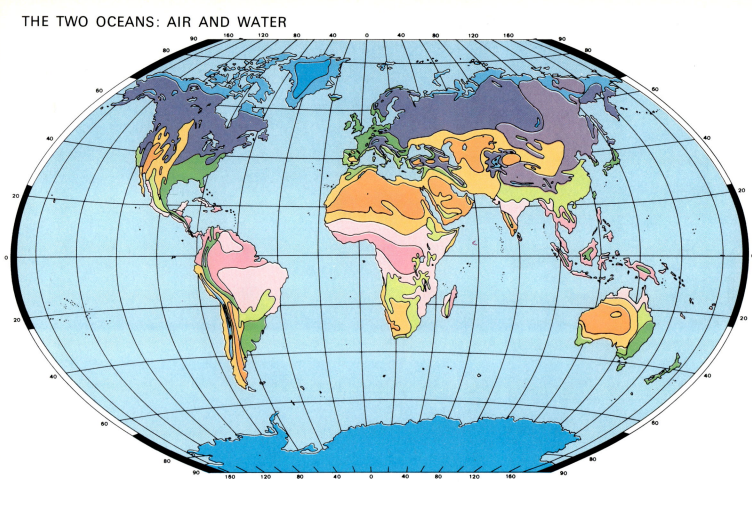

- ▢ Equatorial climate with constant precipitation (forest)
- ▢ Tropical climate with periodic precipitation (savannah)
- ▢ Arid climate: mainly subtropical (deserts)
- ▢ Semi-arid climate: subtropical or hot-temperate (steppe)
- ▢ Hot temperate climate with dry winter
- ▢ Hot temperate climate with dry summer

- ▢ Hot temperate climate with constant humidity
- ▢ Subarctic climate with dry cold winter
- ▢ Subarctic climate with humid, cold winter
- ▢ Semi-snow (tundra)
- ▢ Snow (ground permanently frozen)

A chart showing the major climatic zones of the world.

seasonal examples are the anticyclone which in winter settles in Siberia, and those which install themselves in the monsoon regions. The amount of steam produced by evaporation of the liquid surfaces represents the humidity in the air.

A volume of air cannot contain an unlimited quantity of steam; once a given amount has been reached, known as the *dew-point*, the excess quantity of steam condenses. In this way the heat stored as *latent heat* in regions where evaporation is more intense is released in the form of *sensitive heat*. Condensation may occur when the air-mass cools or when the amount of steam in it increases. Air cools by irradiation, aeration or movement upwards. Usually, air rises vertically by convection when it is heated by the Earth's surface or, when hotter, when it flows above strata of cold air, or when it is forced to rise up beside a mountain slope.

In most cases, rising air denotes instability and represents the major cause of rainfall. Air rises because it is warmer (or hotter) and less dense than the air above it, and continues to rise until it finds strata of air which have the same temperature and pressure. It should be said, however, that for condensation to occur it is not enough to reach saturation point; there must also be solid

An important factor influencing the climate of a given area is the duration of the seasons, which establish the vital rhythm for the local fauna and flora. Here, from left to right and top to bottom: the changing look of birch trees as season follows season – spring, summer, autumn and winter respectively.

particles in the air, usually dust and fumes, which act as *condensation nuclei*, which prevent the droplets which form on them from evaporating immediately. When there are no such particles condensation does not occur, and the air becomes over-saturated in as much as steam continues to be there even above the saturation point. In general, condensation occurs at temperatures above zero; but water can exist in the liquid state at temperatures way below freezing-point when the air does not contain enough *freezing nuclei*, which are less plentiful than condensation nuclei. The presence of droplets of water in the atmosphere is responsible for picturesque optical phenomena such as rainbows, which are visible when the air is very humid, for example, after summer storms. Another feature of the atmosphere surrounding us is that it is in constant motion. The movement of air is sometimes so rapid that it whisks away anything in its path, and sometimes so slight that highly sensitive instruments are needed to detect it. Vertical movements produce hailstorms, hurricanes and tornadoes, and the horizontal movement of air produces wind.

The mass of tiny droplets of water and ice gives rise to the clouds which can spread across the entire troposphere. On the basis of their height we can divide them into *low-cloud*, up to about 2500 metres (8000 feet) from ground-level,

intermediate cloud, from 2500–6000 metres (8000–20,000 feet) and lastly *high cloud*, up to about 13,000 metres (40,000 feet). On the basis of the shape, however, we can distinguish three main types of cloud: *cirrus, cumulus* and *stratus*. Another particular type of cloud is *fog* or *mist*, produced by the condensation of steam around solid nuclei, in the lowest levels of the atmosphere.

The particles forming clouds are very small. As a result, before precipitation, they must reach an appreciable diameter to overcome the resistance of the ascendant (rising) currents which have produced them. The particles then join together by a process of coalescence and, once sufficiently large, fall to the ground in the form of rain, snow or hail: this precipitation always occurs in low-pressure areas.

In the equatorial zones we find high and uniformly distributed precipitation, with an average of about 2000 mm, and the same goes for western continental coastal areas, because of the arrival of moist air from the sea. In the monsoon and tropical regions rainy seasons alternate with dry seasons. The hot tropical deserts are characterized by an almost total absence of precipitation, as are the cold continents and the polar regions, where high-pressure systems persist.

The disastrous aftermath of the typhoon Hurricane in Australia. The exceptional wind-speed can rip off roofs, and even blow down entire buildings, as shown here, as well as uproot trees and bring down telegraph poles.

Running for safety during the arrival of a typhoon in Australia. These people are looking for some refuge from the blast of the wind, aware all the while of the catastrophic damage it can cause.

Wind is produced by the movement of air-masses from high-pressure regions to low-pressure regions. These masses do not move along a straight line, but are deflected by the Earth's rotation, according to Ferrel's Law, and as a result of the friction of the surface. The degree of movement, i.e. the wind-speed, depends on the *barometric gradient*, which is given by the ratio between the difference in pressure between two points, and their distance. Winds may be constant, periodic or local. The most typical constant winds, which are also among the strongest, are the trade-winds, which originate in the anticyclonic zones in the Tropics, and head towards the equatorial low-pressure areas, blowing from the north-east in the northern hemisphere and from the south-east in the southern hemisphere.

Periodic winds change direction at regular intervals, because of the different thermal conditions in two adjacent areas. The most typical and important periodic winds are undoubtedly the monsoons, which occur between the Indian Ocean and India. These change direction every six months. In winter the monsoon blows from the cold continental masses, where there is a high-pressure system, towards the sea, and in summer from the sea towards the continent, bringing with it a great deal of rain.

Among the periodic winds we can also include *breezes*, caused by the different daily thermal behaviour of the sea in relation to the land, and the barometric imbalances thus produced. There are sea- and land-breezes: the former blow from the sea towards the land during the afternoon, when the land's warmth causes a low-pressure area to form, which draws air from the sea. Then the opposite occurs, during the hours of darkness, when because of the greater thermal capacity of water, the sea retains more warmth than the land, and becomes the site of a low-pressure area, as a result of which the movement of air changes direction and the breeze blows off the land. Similar breezes occurring on lakes are caused in the same way. A similar phenomenon, though not linked with the different thermal capacity of land and water, occurs in mountain areas where mountain- and valley-breezes form. These blow, respectively, upwards by day and downwards by night.

Along with constant and periodic winds we have local winds. These are movement of air-masses connected with specific barometric situations. One of the most notorious local European winds is the *Föhn*, a Tyrolean-Swiss word, referring to a spring or autumn wind which is produced by considerable barometric imbalances between the opposite slopes of a mountain.

Top left: cirrus fibratus, *a delicate, transparent, silken-looking type of cloud, frequent in the forefront of atmospheric disturbance. Top right:* altocumulus undulatus. *Below left:* cirrostratus fibratus. *Like fine veils, these clouds often form haloes round the sun and moon. Centre:* cirrocumulus undulatus radiatus. *This forms between 6000-7500 metres up (20,000-25,000 feet) in undulating banks, and indicates good weather. Right:* altostratus, *cloud of varying thickness present above banks of* altocumulus, *which denotes growing atmospheric disturbance.*

Top left: stratocumulus stratiformis. *This is a grey, stratified cloud formation with the bottom lower than 600–1000 metres (2000–3500 feet) which entirely covers the sky. It occurs often in the winter months and may cause light or moderate rain. Top right: fleece-like* cirrus. *Below left: scattered, roundish* cumulus, *caused by masses of hot rising air. Centre:* stratus, *lower, layered clouds which give the sky a leaden colour. Right:* cumulonimbus, *storm-cloud with considerable vertical height.*

Climatically speaking, the *Föhn* softens the climate, and indicates a thaw or melting snows; as a result agriculture is possible in areas where it would otherwise not be feasible. Unfortunately, variations in temperature can produce sudden avalanches and flooding because of the rapidly melting snow, as well as forest-fires. A hot, dry wind, like the *Föhn*, occurs in the Rocky Mountains, where it is called the Chinook; this wind blows from the West in the tributary valleys of Missouri, USA, and makes it possible to grow cereal crops at relatively high latitudes.

The exaggeration of normal atmospheric phenomena, like those described above, can give rise to fairly violent occurrences which are sometimes responsible for many casualties and enormous damage. One of the commonest of such phenomena is the thunder storm, followed by hailstorms and hurricanes. Thunder storms are fairly localized occurrences which originate from rising currents in low-pressure areas. They are characterized by electric discharges, thunder, squalls and torrential rain.

The formation of hail is associated with the existence of vortices in thunderclouds. In these the droplets of water are thrust by the ascendant currents to very high altitudes where, because of the very low temperature, they turn into

ice. They then fall, by means of gravity, to the bottom of the cloud, are surrounded by another liquid layer, rise once more, the liquid layer freezes, and in this way the ice gradually swells concentrically until it becomes large enough to overcome the thrust of the rising currents, and falls to the ground.

A more violent phenomenon than the thunder storm is the hurricane.

A hurricane consists of three zones: a central area known as the *eye*, a peripheral area known as the *vortex*, and an outermost area known as the *margin*. The vortex forms where trade-winds converge; because of the effects of the Earth's rotation, these winds start to move in a circular path, anticlockwise in the northern hemisphere and clockwise in the southern hemisphere, at a speed which increases progressively from the outer to the inner area. Here, in the innermost area, the air tends to rise very quickly and, once it has reached the level where it becomes saturated, condenses, and produces, because of the high temperature of the region from which it originates, a strong increase in heat which constantly draws new air from below. At the same time, at ground-level, more air is drawn from the outside towards the central depression to replace the rising air. In this way very strong winds are produced which can reach speeds of more than 150 kph (100 mph). The rain stops in the eye of the

Top: fog in a mountain-pass. Below left: mist forming at dusk in the Po Valley. The conditions necessary for the formation of this mist are provided by a total absence of wind, a calm sky and high degree of humidity in the air. Below right: mist forming at sundown over a marsh. This mist is caused by evaporation, i.e. produced by the influx of steam from the surface of the marsh and by subsequent condensation.

Examples of rainbows. This phenomenon is due to the Sun's rays striking droplets of steam, passing through them, and being reflected by their other face and diffracted into the colours of the spectrum, with red outermost and violet innermost. The colours of rainbows may vary in brightness, depending on the size of the water droplets struck by the sun's rays. Sometimes, the rainbow described (known as primary) *may be followed by a* secondary rainbow *which is thinner, with the colours arranged in the opposite sense, i.e. with red innermost and violet outermost.*

hurricane, the clouds vanish, calm returns, and oppressive heat follows.

Hurricanes are most destructive in coastal areas where the high wind-speed destroys everything in its path: houses, trees, power-lines and so on. A similar, more violent but more short-lived phenomenon, is the *tornado,* a vortex which forms on land which is of limited diameter – not more than one kilometre (half a mile). Tornadoes occur above all in the south-eastern United States, and the various types – trunk-shaped or funnel-shaped, and cylindrical – which they can assume, are formed by droplets of condensed water. A tornado extends from the bottom of cumulo-nimbus cloud to the ground, and its movement follows the host-cloud. It is caused by strong rising currents, which can reach speeds of 300 kph (150 mph), which are produced by the difference in pressure between the periphery and the centre of the vortex, where a marked depression occurs. The disastrous effects of tornadoes are due to the rapid lowering of pressure, which happens in a matter of minutes and causes houses (if shut up) and vehicles to explode. This phenomenon is due to the internal pressure which is higher than the outside pressure, and to the wind which whisks away anything in its path.

In places where tornadoes are most frequent underground shelters have been

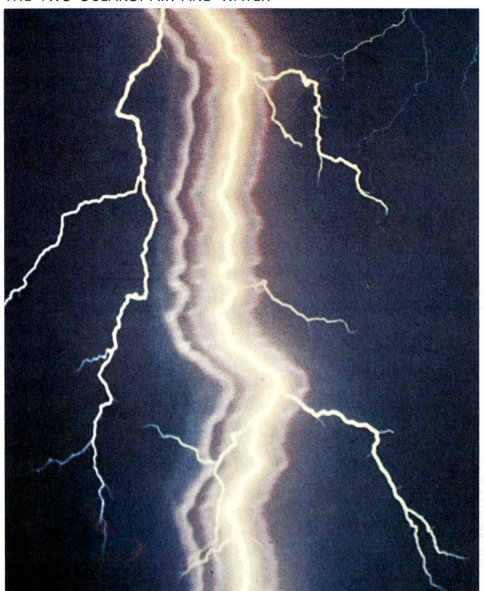

built to protect people. If no such shelter is at hand, and a tornado is imminent, people are advised to find a relatively protected place like a hollow or a hole, or flatten themselves as much as possible on the ground and cover their heads. In the case of a hurricane, on the other hand, although there is a drop of pressure similar to that witnessed in tornadoes, the lowering occurs in a period of time sufficient to even out the pressure.

Waterspouts, which occur at sea, are also tornadoes. At the centre of the vortex of these spouts the rising forces of the air is such that water is sucked up into it.

Temperature, pressure and humidity are the principal factors which determine, day by day, the weather in a certain area, and the sequence of different weather conditions in an area defines its climate. Astronomical and terrestrial factors such as the variation of solar radiation, the distribution of land and sea, and the presence of hot or cold ocean currents can have a considerable effect on the climate. These are currently accompanied by anthropical factors, such as the construction of man-made lakes, land-reclamation, reforestation or deforestation which can sometimes produce very significant changes in climate. The closeness of the sea, for example, has the

Various types of lightning. This is produced by electric discharges produced either inside a storm-cloud or between a cloud and the ground. This phenomenon, which is invariably accompanied by thunder, is due to the strong electric charges originating from the friction between the solid or liquid particles moving inside the cloud. Lightning occurs when the difference in potential between the two points at which the discharge occurs, which in some cases can reach 30 million Volts, is sufficiently high in relation to the distance.

Top: mother-of-pearl-like clouds occasionally visible in the ozonosphere, between 20 kilometres (12 miles) at one p.m. and 25–28 kilometres (15–17 miles) in the late afternoon after a winter sunset and at high latitudes. Delicate in shape and colour, they reveal the presence of steam in the stratosphere. Below: higher than the above clouds, at more than 80 kilometres (50 miles) we can observe the so-called noctilucent clouds. Bluish or silvery-white in colour, they appear as parallel bands with a thread–like texture.

effect of softening a climate, mainly by reducing fluctuations in temperature.

The amount of energy absorbed in summer by sea-water is stored and returned during the following winter. In fact the sea can be considered a vast reservoir filled with heat, because the heat spreads deep down, contrary to what happens on the Earth's surface where the heat is absorbed only by a stratum of limited thickness. Regions farthest from the sea in fact have a continental climate, with marked temperature variations.

Much help in identifying climates is provided by vegetable life in as much as it expresses accurately the environmental conditions of the areas concerned.

In the equatorial zones, like the Amazon and Congo basins, the very high humidity and temperature promote rain-forests which have their dinstinctively lush and dense vegetation, known as *three-layer* vegetation, based on the various heights of the dominant plant species and their light requirements.

Lastly, let us take a look at how our weather forecasts are made. First and foremost, there must be a picture, put together as accurately as possible, of the state of the atmosphere at a given moment over a sufficiently wide region. This means that we must know the distribution of temperatures, pressures and relative humidities. These factors already make it possible to predict the

development of atmospheric disturbances, especially their direction and intensity. As a result of this knowledge it is then possible to assess the movement of cyclones and anticyclones, and the dynamic pattern – incidentally always fairly complex – of these movements. It is generally a question of variations which are so rapid that fairly precise forecasts can be made for periods of not more than 36 or at most 48 hours. Depending on the circumstances in fact, cyclonic and anticyclonic areas tend to gather momentum or to dwindle so rapidly that predictions about their development is problematic even for experts.

Because weather-forecasting also requires knowledge of temperatures, pressures and humidity at altitude, the *Land Chart* is accompanied by the *Atmospheric Chart*, the function of which is to represent large-scale phenomena, without there being any interference to them by mountains. Conversely, this constitutes a modifying element which affects the lower atmosphere. These far-reaching disturbances are related to the flow of solar energy coming from space, and to the Earth's rotation. In detecting the state of the atmosphere at altitude, recourse is made to meterological satellites and radio-probes capable of continual transmission of the data recorded. Based on

In the ionosphere one can observe variations in the electronic density, related to the varying activity and position of the sun. Among the most eye-catching effects are the Northern Lights which generally occur at sunset, before the long nights of the polar winter. This phenomenon, localized around the poles of the Earth's magnetic field where it becomes dazzlingly bright especially at night-time, is attributed to the bombardment of the higher strata of the atmosphere by high-energy protons and electrons.

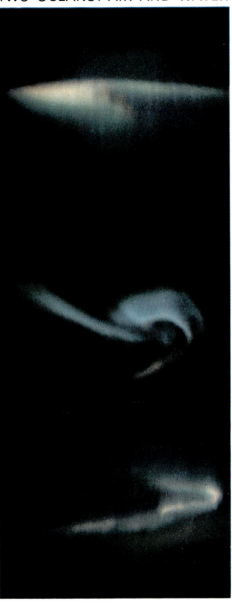

Left: radar antennae installed at Thule in Greenland. The whole polar sky is constantly scanned by these highly sensitive screens. Right: various forms of Northern Lights, drapery-like, ribbon-like and radial. The different gases present in the ionosphere, when excited by the Sun's rays, take on various colorations. Oxygen, for example, diffuses red or green light, nitrogen and blue light. In this way it is possible to analyze the composition of the ionospheric gases.

international agreements, there are some 600 high-altitude meteorological stations, all over the world, in which measurements are taken every twelve hours. The information gathered is immediately transmitted to the computer and relayed to meteorological charts within a maximum of three hours. In this way it is possible to analyze the formation and disappearance of major atmospheric vortices which, by means of the repercussions that can be predicted for the lower atmosphere, enable us to make medium-range forecasts.

SCIENCE FOR MAN

THE DEVELOPMENT OF TECHNOLOGY

In the history of scientific thought it has been a rare occurrence for research investigating natural phenomena to be an end in itself, to spring from nothing more than a straightforward desire for knowledge: other factors have usually been present. Men have always felt the need to know about the natural world about them, and often a practical reason has been found to justify curiosity: to defend themselves from the dangers and threats inherent in nature herself, in the first instance; and, in the second, to dominate nature. To achieve these goals, the means which has to all appearances been most effective down the ages has, until now, been the use of things which man has made with his own hands and toil.

It is of course true that different paths have also been tried in certain phases of the development of human civilization. In particular, attempts have been made to establish a relationship with nature by means of magic, and this approach has certainly not necessarily resulted in any major technological progress. However, this path did make it possible to set other positive activities in motion, it has led step-by-step to the development of areas of research and the elaboration of doctrines which, as in the case of alchemy, astrology and primitive medical practices, have then advanced and have been radically transformed. And the end result, which is scarcely identifiable now, has been their eventual inclusion in the mainstream of rational scientific thought.

What is more, the relationship between science and technology can certainly not be defined in simplistic terms, and limited to the bland statement that the latter advances systematically as a result of progress made by the former. This would be a gross error, just as it would be grossly misleading to say that all scientists, in every stage of their work, have their eye on some kind of technological application. The bonds between scientific research and technological development are, on the contrary, complex and interwoven. Many practical achievements have presumably had no back-up in terms of theoretical elaboration, except for whatever happens to be in the minds of those directly responsible for it. Similarly, many scientific discoveries have never had any subsequent practical application. In other respects we find the case of scientific discoveries which have remained untapped on a practical level for fairly long periods of time, only then to be put to use at some later stage, when they have become of prime importance. And if it is true that the most sophisticated products of technology are often backed up by major advances in theoretical knowledge, it is no less true that a great many scientific discoveries have been made possible only after technology has been able to provide appropriate research tools. All this emphasizes the extreme complexity of the relationships between science and technology, relationships which defy any simple attempt to define them but which are reciprocal and continuous.

This close association also makes it difficult to define the scientist and the technologist and to distinguish clearly between them. If one tries to do this one runs the risk of coming up with artificial and unbalanced definitions. In different ages we come across individuals who, in different degrees, match present-day criteria for defining a scientist or a technician, but they were hardly thought of in those terms by their contemporaries. This is true of surveyors in ancient Egypt, the architects of Greece and Rome, the various craftsmen in the

Middle Ages, and present-day engineers and scientific researchers. It is hazardous and possibly even unfair to insist that two functions, that of the scientist and the technologist, so closely related are yet absolutely distinct, but this was certainly so in the case of the philosophers of ancient Greece, who were perhaps the purest scientists who ever lived. These men required their slaves to undertake virtually every conceivable task of a practical nature while they concentrated on pure theory; but their own scientific achievements proved to be totally ineffectual, in concrete terms, when it came to combatting the collapse of that civilization. Leonardo da Vinci, on the other hand, appears in quite a different light, as an unsurpassed example of all-round genius, a scientist and technician in one and the same breath, and the complete antithesis to the Greek philosophers in his manner of tackling practical problems. Leonardo embraced every fundamental aspect of human activity. Living as he did at the dawn of the modern historical era, he witnessed mankind's major triumphs over various natural phenomena, and he can in fact be rightly considered as a symbol of the scientific age.

Whether subordinate to one another, or friendly allies advancing in step, the fact remains that human science and human technology have changed the face of the world, and more and more decisively conditioned the circumstances of human life itself. In the pages to come we shall look at certain of the more significant aspects of this vast and many-sided achievement. We shall see how man has gradually made himself master of the Earth's surface, master of the sea, and master of the skies, by means of complex systems of communications and transport. We shall see how he has been able to harness and make use of the various forms of energy which exist in the natural world to work his more and more powerful machines, which have taken on the role of nothing less than mechanical slaves, always at his beck and call. And we shall see how and with what means man has often combatted the ills which menace the human body.

There can be no doubt that in all these instances we are faced with positive achievements. But now, when mankind sees its domination over the forces of nature being consolidated, when the problems posed by distance, climatic hardship and all manner of environmental hazards can easily be overcome, just as diseases which were once lethal can now be eradicated, other more threatening problems face mankind.

If we look at the face of the Earth, we see that it is dotted with huge metropolitan centres, sliced through by canals, highways and railways, flown over by supersonic aircraft and cultivated by extremely up-to-date methods. But we can also put our finger on scourges and menaces which have never reared their head or been present at all, even in a much lesser degree, in other eras. The land, the sea and the sky itself, which is becoming more and more accessible, are now removed from normal natural conditions, and hostile to life, because of the various forms of pollution. Entire species of creatures have vanished, or are threatened with extinction; and, even more tragically, as civilization pushes ahead the last primitive human peoples are also being wiped out. Huge sections of mankind, more numerous than ever before, live today in conditions of extreme poverty and degradation. And in the atomic arsenals in the hands of the superpowers there are enough bombs and warheads to sweep our planet clean of every single form of life.

Can this fearsome situation be categorically attributed to the development of technology? There is no doubt that technological development has been necessary and instrumental in bringing this situation about; but it would be hazardous to maintain that it is an inevitable consequence of technological development. It is extremely hard to pass a fair verdict. If we strike an optimistic note, it seems reasonable to say that the damage caused by our present mode of development should not be attributed to science and technology *per se*, but rather to the use that has been made of science and technology by men, and in particular by men who have wielded power. Thus in the first place it is in the development of new and more equitable human relationships that slender hopes for the survival of our species and perhaps of life itself must reside.

COMMUNICATIONS AND TRANSPORT

There is no life without movement and expansion. On the individual level every single living being must undertake a series of moves to find food, and in most cases to avoid becoming in turn the food eaten by others. Movement is vital to all those activities which are part of survival itself, and of the satisfaction of the requirements laid down by the process of survival. On the collective level, every living species tends to occupy the largest possible area allowed it by the respective environmental and climatic conditions and limits, and by the competition for that area shown by other species. These observations enable us to consider the capacity of movement and expansion as one of the most significant signposts for the success of a given species in its struggle for survival.

In this respect modern man is unrivalled. He can move around on the ground better than any stronger and swifter quadruped; he can move through water better than any fish, and through air better than any bird, at least as far as distance and speed are concerned. And he has managed to thrust his way to places where no other form of life, except perhaps certain microscopic spores, has ever been, beyond the limits of the Earth's atmosphere. In addition, whether fixed or mobile, man's settlements are to be found in virtually every latitude and longitude, and, broadly speaking, at more or less any altitude or depth. And in every such settlement man is capable of transporting cargoes which are quite out of proportion to the weight of his own body. Lastly, he can convey sounds and images over any distance, across an enormous variety of material obstacles, and, of course, across the void.

All these abilities are not the direct result of man's physical power. Indeed, from a strictly anatomical viewpoint, men are not particularly well equipped for running, swimming, or carrying heavy loads; and flying is literally beyond their reach. But other animals and creatures can only rely on the strength and nimbleness of their limbs. Man, on the other hand, thanks to faculties which have taken thousands and thousands of years to mature and develop, is capable of building himself, artificially, the means with which to move, and these turn out to be vastly superior to those with which nature has supplied him, and to those with which any other animal has been provided. If the truth be told, the first step taken along this road were not particulary glorious, because they followed in the footsteps of other creatures. In fact, the first great step forward as far as the use of our own legs for walking, and the use of our own shoulders for carrying things, were concerned, came as a result of the use of beasts of burden and animal pulling power.

The decisive advance, the revolutionary importance of which has in fact become proverbial, was nevertheless the invention of the wheel. We have no way of knowing where, when or by whom this invention was made, nor how often (perhaps extremely often) the invention was made anew in different places and at different times. All we can do is to speculate about its origins. It is plausible, for example, that the wheel came about as a result of the gradual improvement and shaping of a cylindrical tree trunk, placed beneath a load as a form of roller. Or, given that the idea of rotatory movement is already inherent in the primitive tools used to make fire by rubbing them together (as it is in the first rudimentary drills) it is possible that in certain situations the wheel was inspired by these tools. Historically speaking, the first models of wheels,

A caravan in the Tassili in Algeria. The use of animals to transport goods and people is still widespread over much of the earth's surface and can in some cases be considered competitive with modern means of transportation in those regions, such as deserts, where there are absolutely no roads or modern machinery.

fastened to carts, are encountered towards the latter half of the fourth millennium B.C. in the lower reaches of the Tigris-Euphrates valley. Later examples occur in Crete (2000 B.C.), Egypt and Palestine (1600 B.C.) and China (1300 B.C.). The importance of the wheel for human civilization is not limited to transport. On land it represents the means of coping with friction caused when a load is being dragged along. More generally, though, the wheel offers the possibility of translating a circular movement into a straight movement, and vice versa. Thanks to this invaluable feature, the wheels crops up in a vast number of variant forms among the basic component parts of a huge number of mechanisms. It has been used in the past in, for example, pulleys, mills, ships' tillers and the gearing in clocks; today it is part of every machine which involves rotating movements.

The spread of this crucial invention, which has been used for various means of transport, be it two- or four-wheel vehicles, such as coaches and chariots, princely hearses, engines of war, carts used for transporting victuals, did not, however, bring with it, in those ancient times, a corresponding development on a wide scale of transport systems. . . . Many factors stood in the way of this development, such as, first and foremost, the scarcity of easily passable

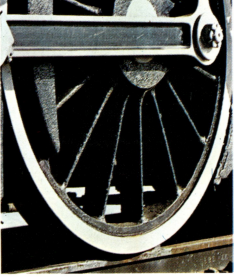

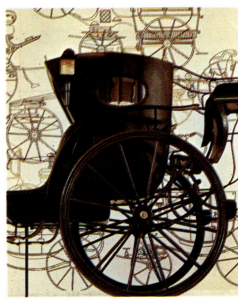

thoroughfares, the greater convenience (as a consequence) of using beasts of burden, and the concentration of the more heavily-populated centres near navigable water-courses.

In fact, in ancient times people used a type of harness for beasts of burden and animals used for pulling loads which was not attached in any way to the animal's shoulders. It was placed tightly round the neck, thus hampering the animal's breathing, especially under conditions of stress and strain. In addition, man has still not discovered the moveable forecarriage for carts, and it is this shortcoming which accounts, among other things, for the largely straight routes of the great Roman highways. It would also appear that man had not yet adopted the use of the horseshoe, nor the practice of using a team of animals to pull one vehicle (although we should mention, for example, the Roman *quadriga* with its team of four horses abreast).

It was not until the Middle Ages that these problems were solved. The primitive harness used for beasts of burden was replaced by a more rational version. This method, which had in fact been known in China since about the first century A.D. and was probably introduced to the West from Asia in the eighth century, took a hold and became widespread in Europe in the tenth and

Top left: an Etruscan wooden wheel; the hub and spokes are made of bronze and the rim of iron. Below: the wheel as it was applied to the earliest bicycles. Centre: a locomotive wheel. Top right: the wheel of a coach made in 1890, from the Transport Museum in Lucerne. Below: a modern car-wheel.

Top: the triumphal chariot of Maximilian I in an engraving dated 1518; such ceremonial chariots and carriages were much used during the Renaissance for royal weddings, victory celebrations and festivals. Below: Madagascan farm-carts, showing how, at the present time, means of transport of very different technological levels co-exist on Earth, depending on the various stages of development of the countries concerned.

eleventh centuries. The essential improvement consisted in the adoption of a type of collar which left the animal's throat free, and thus enabled it to use its energy fully without its breathing being obstructed. Likewise the iron shoe became generally used, and this enabled the animal's hooves to get a better grip on the ground. Teams of animals also started to be used, making it possible to make the most effective use of their pulling power.

These advances which we have mentioned, and which appear modest but are in fact crucial – plus the introduction of the moveable forecarriage in carts – together with the advent of a socio-economic framework which had changed radically when compared with ancient times, opened the way to the staggering development of transport and communications networks which have gone on expanding from the Middle Ages through the Renaissance right up to the present day. Different types of vehicles, with ingenious features built into them, became more and more numerous, keeping pace with the improvement and enlargement of the highway network. The capacity and strength of carts increased, stage-coaches came into use for passenger traffic, and so did state-run and privately-owned haulage companies. At the same time the carriages used by the aristocracy and the wealthy reached the last word in refinement.

But from the strictly technological point of view the most decisive steps in this historical phase, which led right up to the end of the eighteenth century, were still the initial ones. By the end of the eighteenth century, in fact, the possibility of making any substantial step forward was hampered by an insurmountable obstacle, the limitations of speed and strength of the only available means of traction: the animal. If man was to move forward, it was necessary to find a substitute for the animal, a substitute which, compared with a beast of burden, would be both stronger and more robust. Man found the necessary replacement in the steam-engine.

The first vehicle for use on land which did not have animals pulling it – the steam-engine – saw the light of day and started to be used in the early nineteenth century in Great Britain, thanks to the work of R. Trevithick. The steam boiler was already in evidence, and in other respects the use of tracks or rails can be traced back to the second half of the sixteenth century. In 1797 Trevithick made a model steam-engine and after several unsuccessful experiments went to London in 1808 to demonstrate a locomotive with carriages. But it was taken for some kind of toy. Nevertheless the efforts of this pioneer had brought the right moment that much closer. In 1812, in Leeds,

Top: two illustrations of trains used on the Liverpool-Manchester line, inaugurated in 1830, in which G. Stephenson played a decisive part. Below left: an old steam-locomotive. Below right: an Erie-Railway locomotive (United States).

Il primo automobile a vapore costrutto nel 1765 da Cugnot

Vettura a vapore di Galsworthy Gurney   1831

Omnibus a vapore di William Church 1832

Vettura a vapore di Squire e Macerone   1833

Fabric showing the history of the automobile (Milan, Bertorelli Municipal Collection). It is not hard to see that the first self-propelled vehicles paid little heed to aesthetics, or aerodynamics. For all this, they represented an extraordinary step forward when compared with animal-drawn carts and carriages, which had been used by man almost exclusively up until the end of the eighteenth century.

under the guidance of J. Blenkinsop, the first practical locomotive started to operate. This was followed by three more locomotives, and their task was to link the city of Leeds with the nearby coal-mines. In 1813 W. Hedley put his locomotive on the rails, and he was followed a year later by G. Stephenson.

It was then necessary to resolve a few technical details. The first locomotives were in fact slow and costly and for some years the steam-driven train found itself competing with the horse-drawn train, which had existed in England since 1803.

The contest was finally won by virtue of the continual improvements that were made both to the structure of the locomotives and the strength of the rails. On 15 September 1830 the Liverpool-Manchester line was opened – the first public railway in England in which all the traffic was steam-powered. This new invention spread like wild-fire, confirming the existence of a technological situation capable of supporting it and ready to exploit it. Regular passenger lines were established in the United States (Charleston-Hamburg S.C., 1830), in France (St. Etienne-Andrézieux, 1832), in Germany (Nuremberg-Fürt, 1835), in Italy (Naples-Portici, 1839).

In this way man had at his disposal the means to expand his traffic and

FIAT 1902 - *Modello 12 HP*

communications systems to a degree which had been inconceivable in the days of the horse-drawn cart; means without which the development of industrial societies would never have been able to forge ahead.

Another decisive step was the motor-car. The fast and powerful train was cumbersome, and restricted to its two tracks. It was thus no use as an individual form of transport, which could use a less inflexible highway than the railway line. With the establishment of the principle of mechanical traction, the next step was to apply it to vehicles which were not bound to follow fixed tracks.

Long before the arrival of the locomotive experiments had been carried out in Holland in the seventeenth century with carts driven by the wind. The following century saw attempts to adapt clockwork mechanisms and experiments with compressed air and steam. It should by now be evident enough that when we talk of the motor-car, or the locomotive, or any of mankind's major discoveries and inventions, it is wrong to talk of a single inventor. Major advances have come about rather as a result of the work of whole generations of inventors and technicians. The official laurels for the invention of the first 'true' motor-car are nevertheless awarded by some experts to N. J. Cugnot from Lorraine. In 1769 he constructed a large, heavy tricycle

The 12 hp. FIAT 1902. At the beginning of this century, and although the spread of the motor-car on a mass scale was still some way off, the automobile had reached a technically remarkable stage of development, even though, to our eyes, accustomed to the aerodynamic shapes of modern vehicles, the aesthetic aspect leaves a lot to be desired. What is more, what we tend to consider as jalopies nowadays, were in fact capable of remarkable performances and even made some concessions to comfort.

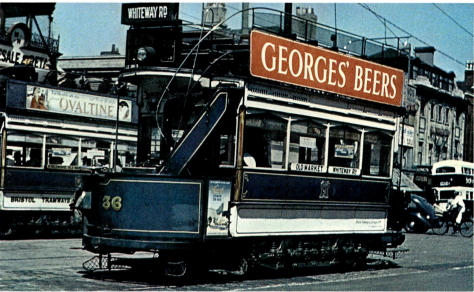

Top left: a tram in Prague, showing the trolley pole which picks up the continuous supply of current. Top right: a tram built in Bristol in 1900 which remained in service until 1941. Below left: the New York subway. Below right: a modern tram in Vienna, consisting of an articulated carriage with two compartments.

powered by steam which, it would appear, kept in motion for no less than twenty minutes.

In Cugnot's wake, over the decades, came a host of inventors, among them C. Dallery in France, N. Read and A. Kinsley in the United States, and W. Murdock in Great Britain. The steam-powered motor-car was thus developed. But the fact that it was not independent, and was difficult to handle, stopped it from taking the place of its counterpart, the horse-drawn vehicle, which was not ousted until the advent of the rail-bound locomotive. Success finally came with the development of a more versatile and sophisticated type of engine: the internal combustion engine. Although the Belgian Lenoir and the Austrian Marcus had constructed (though not developed) successful models twenty years earlier, it was in 1885 that the German C. Benz triumphantly presented to the world his three-wheeled motor-car powered by a petrol engine. The following year it was the turn of his fellow-countryman G. Daimler, pioneer of the automobile industry, to produce a four-wheeled car. From that moment the development and spread of the motor-car across the globe was uninterrupted. Improvements came thick and fast, many of them brought about by skilled technical experts whose names are still in many cases unknown

to the public at large. In their place fame went to the heads of the automobile industry, such as, among others, the American Henry Ford.

As far as public transport is concerned, a crucial contribution was made by the introduction of electric traction. From the 1840s onwards there were scores of attempts – most of them fruitless – to use the energy generated by electric batteries for transport. It was only after 1870, following the invention of the dynamo, that electricity made its real debut in the field of transportation. The first electric locomotive to operate without batteries was presented at the industrial exhibition in Berlin in 1879 by W. von Siemens. The energy for it was supplied by a third (live) rail. Two years later the first commercial electric track came into service in Lichterfelde, near Berlin, using the current supplied directly by the two tracks. About four years later the American L. Daft introduced the 'trolley' and replaced the use of live tracks with overhead wires. Simultaneous and similar solutions were also successfully introduced by another American, F. J. Sprague, and the Belgian C. J. van Depoele. 1888 saw the first installation of an electric tram service in a large city, Richmond, and before long this type of vehicle became a part of urban traffic in both America and Europe, replacing the horse-drawn tram which had been in service for

Top left: a Parisian bus with its now rather antiquated design. Top right: the famous red double-decker buses used in London. Below left: a bus for tourist excursions. Below right: an American bus station; in the United States in particular there is a whole network of passenger transport which uses buses, even over very long distances, and is a strong competitor to the railroad system.

An Austrian rack-geared train in the Innsbruck region. Overland means of transport have taken on a wide variety of forms to match the features of the terrain covered: for example, funicular railways and cableways, tractors, vehicles mounted on skids or skis and tracked vehicles for steep or icy terrain.

more than 50 years. Within a few decades the electric tram was in turn faced with further fierce rivals. With the improvements made to the motor-car and the introduction of the bus to transport passengers, the tram started to decline in importance in the 1920s and was gradually replaced, especially in the United States and England, by taxi services. Together with its close relative, the trolley-bus, the tram has nevertheless survived up until now, though competing with other forms of urban transport, in much of the rest of Europe and elsewhere.

One area in which electric traction plays a vital role is that of rapid urban transport, which is basic in easing the problems of traffic congestion in the world's largest cities. The first underground railway came into service in London as far back as 1863; and the first overhead railway appeared in New York in 1871. In both cases, as in all the other railway lines of the day, the trains were steam-powered and this caused a whole series of problems. From 1895 onwards, with the introduction of the western metropolitan line in Chicago, electric traction started to be used in this field as well.

The heaviest use of electric traction has occurred above all in inter-urban railway lines, both for passenger and goods services. The first railroad of this

type was the Baltimore-Ohio line, which came into service in 1895, and was followed by a huge development in this sector. There was a lull, however, after 1930, when the rival Diesel locomotive came on the scene. But today whole countries, and large areas of other countries using Diesel locomotives as well, are still served by electrified railways.

If the last century has seen advances which are unique in the history of overland transport systems, the current situation looks problematic. Major technical advances on the one hand are paralleled by difficulties, economic and social, which prevent some of the advances being fully exploited. On the one hand, the surge forward shows no signs of slackening off, be it in terms of road or rail transport. Automobiles have been built which are capable of speeds of more than 1000 kph (625 mph), as demonstrated by the four-wheeled model powered by natural liquid gas driven by G. Gabelich on the Bonneville salt flats in Utah on 23 October 1970, when he reached a speed of 1016 kph (635 mph). In 1973 the Soviet Union announced that it had built a similar racing-model capable of reaching a top speed of 1200 kph (750 mph). Lower speeds, but nonetheless within the 600–1000 kph (375–625 mph) range, have also been recorded by vehicles driven by jet, turbine, and piston engines. In other

Top: the Tokaido line in Japan, served by high-speed trains. Below left: the turbo-train in service between Boston and New York, fitted with gas turbine engines, capable of covering the 530 kilometres journey (330 miles) in 3 hours and 15 minutes, with top speeds of 270 kph (170 mph). Below right: a monorail in Tokyo; the monorail is used mainly in preference to the traditional railway for transporting freight over short distances in industrial complexes.

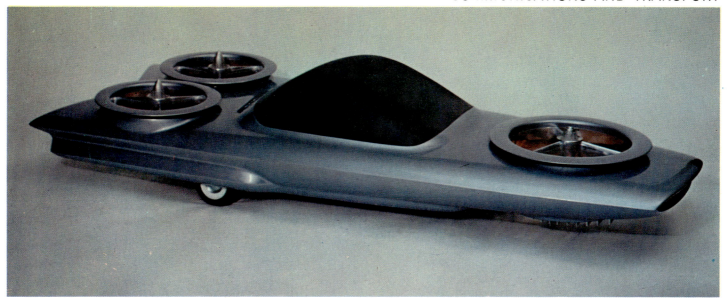

Top: a Ford-designed motor car which may be manufactured in the future, capable of travelling along highways on wheels and taking off and flying at the flick of a switch; it would be kept airborne by three turbines, while the power would be supplied to the driving wheels by an ordinary engine. Below: another futuristic design, elaborated by Fiat.

respects extremely large vehicles have been produced, as for example the 8-track Marion tractor, used to transport the Saturn V rocket, which weighs 8165 tons when fully loaded; and there are trucks with load capacities in excess of 200 tons. As far as railways are concerned, the traditional steam, electric and Diesel trains have been joined by extremely modern turbo-trains, which have clocked up speeds of around 400 kph (250 mph). The Osaka-Okayama line was inaugurated in Japan on 15 March 1972 with an average speed of 180 kph (112 mph), and such speeds are on the way to becoming the norm in other countries. In France, for example, the Paris-Lyons line is scheduled to start operating in 1980, with the 421 kilometres (263 miles) of track being covered in less than two hours. Another area in which interesting results have been achieved in recent years is the 'air cushion' design for vehicles, which are driven by a new power-unit, the linear induction engine. This engine is friction-less, silent, does not pollute and can reach extremely high speeds. Similar interest has also been aroused by completely automatic trains, which were one of the highlights at the Montreal Expo in 1967. As far as goods transport is concerned, a major step forward has been the use of containers which can be swiftly transferred from one means of transportation to another.

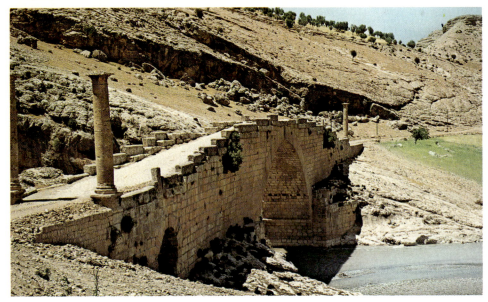

All these important steps forward have found themselves confronted lately by possibly crucial limitations: the rationalization of traffic in general, the fight against pollution, the saving of energy, all these are goals which are nowadays starting to be considered as major priorities, and which are certainly ill-attuned to the creation of new and spectacular records in speed of transportation. Consequently, in the area of passenger transport, plans have been drawn up which aim at making public transport competitive with private transport. For example, types of buses for dual use – on roads and on tracks – have been designed. So also have systems incorporating separate thoroughfares for public transport, and request services for buses, controlled via radio by computers. In other respects, the need to fight pollution and the need to reduce the consumption of energy have forced planners and designers to consider new solutions involving engines which cause less pollution and are more economical.

The topic of overland transportation would be incomplete without at least a brief mention of the communication routes that man has opened up in the face of every conceivable sort of natural obstacle.

As early as prehistoric times there were major thoroughfares linking one

Top left: a suspension bridge in the Baciuma region in Ethiopia. Below: a Roman bridge in Turkey. Right: the Etrusco-Roman bridge at Vulci abbey on the River Flora.

Top left: an impressive example of a freeway intersection in the city of Toronto in Canada. Below: a section of a motorway with a rest area in the background. Right: a complex cloverleaf intersection on the freeway linking Brooklyn to the majestic Verrazzano Bridge in New York. Built with specific objectives, freeways and motorways are intended mainly to link important centres both rapidly and conveniently.

distant region with another. We have already referred to the great caravan-routes in the East; another highway of major importance was the European one which went in a north-south direction. Roads built during the modern era often retrace the routes taken by these ancient highways. In general terms it can be said that throughout history the types of roads used, and the routes taken by them, have to some extent dictated the sorts of vehicles intended to use them, but that they have in fact themselves been conditioned, to an even greater extent, by the various types of vehicles that needed them. We have mentioned the straight roads built by the Romans, who were the greatest highway architects of all, in ancient times, and have noted that the straightness of the road was partly because there was no moveable forecarriage in the carts they used. Similarly, at the end of the nineteenth century, the demand for better roads by cyclists and then by motorists gave rise to the spread of the asphalt surface. It is evident that where roads are concerned, and also where the accessory construction of tunnels and bridges is concerned, the progress made by mankind down the centuries has been quantitative rather than qualitative. The length, breadth and number of roads has increased, and the strength of the materials used and the smoothness of the surface have been improved, and the

same can be said of bridges and tunnels. But, conversely, it is impossible to single out any innovations in these structures which really make them essentially different from those built in the past.

If man, in some cases, had to use enormous amounts of energy to build roads on the Earth's surface, roads of another type have come ready-made by nature. Water-courses, lakes and the sea have from earliest times constituted excellent if inflexible communication routes. To ply these routes man has constantly created appropriate means of transport which have also undergone continual improvement down the ages.

The first boats of all to be used were, as far as we know, tree trunks, or tangled masses of branches found floating on the water by the first unpractised navigators. Later, trunks and rudimentary rafts were made for use as craft. The first type of craft to show signs of methodical human application was the canoe. Either as a dugout made from a hollowed out tree trunk, or as a more complex structure consisting of a framework covered with bark or hides, the canoe is fairly akin in its concept not only to every small boat that has sailed the seas in every corner of the world but also to actual ships.

Historically speaking, the first mention of boats and ships comes from

Above: an aerial view of the docks and the Bay Bridge in San Francisco. Opposite: the highest railway in the world – the Peruvian State Railway. A train at the Limetree Pass, altitude: 4850 metres (15,900 feet).

ancient Egypt where, as early as 4000 B.C. people used fairly large boats and craft. These are considered relatively primitive, however, given that they had no keel, poop or prow, no internal timberwork, but were made by superimposing, overlaying or tightly wedging together various layers of timber. In fact, they were more like rafts than ships. We do not have much information about the types of craft used by the Cretans, who in about 1500 B.C. created a powerful fleet, or by the Phoenicians, their successors as lords of the sea. But two major innovations are nevertheless attributed to the Pheonicians; the use of a double-decker system for the oars, and the decisive introduction of the bow ram for their warships.

The ancient Greeks achieved a remarkable technical competence in the construction of their ships, which had many modern features – keels, poops, prows and an inner framework; and the same can be said of the ships built by the ancient Romans. Mention should be made of the galleys, which were propelled by oars, usually wielded by slaves, although they were also usually equipped with masts and sails. Together with oar-propelled vessels, but mainly used for trading purposes, there were the various types of sailing ship; the sails were generally square or rectangular, although the Egyptians, for example,

A fisherman in his outrigger canoe on the Mahawali river at Puttalam, Sri Lanka. The canoe can be considered the most primitive type of craft to display the systematic and rational work of man's hand; in its conception it is a forbear of our modern boats and ships.

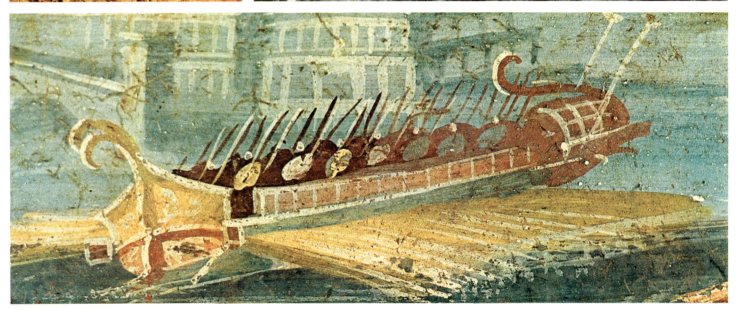

Top left: taking supplies on board a Roman ship: detail from the Trajan Column. Top right: a fishing boat in a detail from a fresco from Pompeii. Below: a Roman warship with rostra, also from Pompeii.

used triangular shaped sails with the point at the bottom. From perhaps the second century A.D., through the Middle Ages and up until about the thirteenth century, the lateen sail was widely used in the Mediterranean; this was triangular in shape with the lower edge horizontal with the deck, and it replaced the square sail, being lighter and easier to handle. Its widespread use was due in particular to the Arabs. The typical lateen-rigged mediaeval craft was two-masted. As far as the hull was concerned no major innovations were made beyond those known to the Greek and Roman ship-builders.

At the same time, though somewhat slow to mature, the seafaring, ship-building technology of northern Europe was making its own advances. A boat which was discovered at Als in Denmark in 1921, some 14 metres (46 feet) in length, can be dated about 300 B.C., and is possibly the oldest to come from those parts – at least the oldest about which we have any definite information. It is still of a primitive design, somewhere between a large canoe and an improved version of a raft. The Nydam ship, found in 1869 near Flensburg, on the Baltic coast of Germany, is a more developed type of craft, which represents a blueprint on the basis of which Nordic vessels were built for about 1000 years. This ship, which dates back to about 250 A.D. and is 23 metres (75

feet) in length, has certain modern features, such, for example, a prow and a poop. In addition it has a tiller, situated on the starboard side. Like the craft found at Sutton Hoo in Suffolk, England in 1939, which dates back to about the seventh century A.D., it was powered solely by oars. Other ships found in Norway, dating back to the ninth or tenth centuries have a square-shaped sail as well as oars.

In about 1200, and probably in Holland or in a nearby region, there was an extremely important invention: the stern rudder, which brought with it major improvements in the construction of ships in general.

As the Middle Ages drew to a close and the modern era approached, further progress was made both in the structure of ships and in their dimensions. The number of masts was, as a rule, increased to three, and the compass, now in widespread use, created safer conditions and enabled voyages to become longer. Thus the fifteenth century caravel came and went, as did the galleons of the sixteenth and seventeenth centuries, the frigates of the latter half of the seventeenth and of the eighteenth century, and the swift nineteenth-century sailing ships and schooners. Improvements were made both to the hull, which became increasingly slender and manoeuvrable, to the masts and rigging,

Above: laden with tea, the American clippers Teaping *and* Ariel *head into the English Channel on 6 September 1866. In the nineteenth century, great sailing-ships such as these were highly developed and continued to ply the high seas, despite the competition of the steamship. Below: the first steamship in the Mediterranean, off Naples.*

Left: the training ship Polinuro.
Right: the Amerigo Vespucci.

which became increasingly complicated and refined, and to the dimensions, which reached the point where some ships measured more than 100 metres (328 feet) in length. Likewise tonnage and displacement were increased to more than 5000 tons in some cases and the speed of some ships was also increased.

But, as in the case of overland transport, the revolutionary development for maritime transport was the introduction of steam propulsion. With this development ships were no longer at the mercy of the wind, but could make headway independently of the atmospheric conditions. The first hapless attempt to build a steam-powered boat was made by D. Papin in about 1690. In 1736 an Englishman, J. Hulls, took out a patent for a steam-powered tug, but it never materialized. After another unsuccessful attempt by the Frenchman J. B. d'Auxiron in 1774, the Frenchman J. C. Périer managed to propel a boat by steam for the first time in history in the Seine at Paris. The first really functional steam-powered vessel was nevertheless the *Pyroscaphe* (steamship) built by the French Marquis C. de Jouffroy d'Abbans. This was a small paddle-steamer with a displacement of 182 tons which plied up a branch of the river Saône near Lyons in 1783.

The French were quickly joined by the English, and then in turn by the other

major countries of the world. The method of propulsion used in these early years of steam-powered ships was the paddle-wheel, which appeared in several forms, some more effective than others. Thus the new type of ship found itself competing with the sailing ships, which were doomed to die out. Before long the steamship was capable of crossing the Atlantic. The first to do so was the *Savannah*, which made the voyage between May and June in 1819. She was in fact a sailing-ship equipped with an auxiliary engine. The first crossing made by a thoroughbred steamship was made in April 1827 by the *Curaçao*, which took 22 days to sail from Rotterdam to the West Indies. Regular passenger services were subsequently introduced with increasing frequency: in 1838 the first transatlantic crossing made by the packet-boat *Sirius* (714 tons) with 40 passengers on board took eighteen days and ten hours. And in 1840 Samuel Cunard founded the first fully-fledged transatlantic shipping line.

But at the same time the other shipping routes had in no way been overlooked. In 1825 the steamship *Enterprise* sailed to India, but the introduction of regular services, apart from a mail service, proved to be uneconomical. And, yet, ships began to ply the routes between England and the Iberian Peninsula and Egypt, and Europe and South America.

The liner Cristoforo Colombo *built in Italy after the Second World War. The term 'transatlantic liner' which applies mainly to ships used for passenger services across the North Atlantic, came into use in 1907, when the* Lusitania *and* Mauretania, *both launched in Scotland, came into service.*

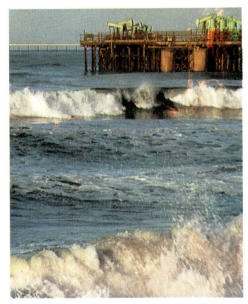

The two top illustrations show the oil-tanker Universal Ireland *(326,000 tons) at sea and under construction. Below left: an example of marine pollution, which has become a grave problem as a result of the huge increase in the number of oil-tankers. Below right: a 270,000 ton oil-tanker.*

In 1836 the Englishman F. Pettit Smith and the Swede J. Ericsson patented simultaneously designs for engine-driving propellers. Simultaneously again, both men in 1839 built screw-driven ships, the former's called the *Archimedes* and the latter's the *Robert F. Stockton*. These ships represented a considerable step forward from the paddle-wheel, and, together with the establishment of the use of iron for ships' hulls, allowed better and better performances.

Where engines were concerned a significant advance came about by the adoption of an engine based on compound expansion; this was more economical than the earlier engine based on the single expansion principle. In this new engine, having functioned at high pressure in the first cylinder, the steam was used again at a lower pressure in a second cylinder. The first ship in which this type of engine was used was the British ship, the *Brandon*, in 1854. The *Holland* was the first transatlantic ship to be equipped with such an engine in 1869. Later, with the addition of a third pressure stage, a type of triple-expansion motor was eventually introduced.

An even more significant step forward was the development of the turbine engine. The first ship in which this was fitted was the *Turbinia*, built in 1894 at Wallsend-on-Tyne in England, and designed by C. A. Parsons. With about

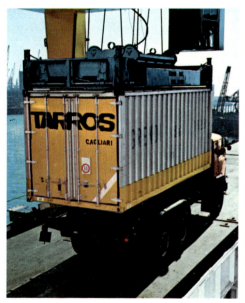

2000 horse-power at its disposal this ship reached the speed of 43.5 knots (63.9 kph – 40 mph) during her first official trials in 1897.

At the same time, during the latter decades of the nineteenth century, major progress was also made with regard to the structure of ships. With the successful introduction of steel into the shipbuilding industry, ships gradually took on the appearance which is so familiar to us nowadays. The increased power of the engines being used also allowed the size, burden and speed of vessels to be increased.

Naturally enough the internal combustion engine was also introduced in the field of navigation. In 1903 the *SS Wandal*, in service on the Caspian Sea, was fitted with this type of engine. It was then widely adopted running at first off diesel oil, which was particularly suitable, first for freight, and then for passenger services. These engines took up less room and were cheaper to operate than the steam-engine. One specific use to which internal combustion engines were put, as in the case of the *Wandal* itself, was to generate electricity which was then used for an electric type of propulsion system.

We have now come to the present. And it can be said that today in the field of maritime transport we find the same variety among ships as we do among

Top left: a container depot. Containers are special crates used for transporting freight; they are of a specific shape and size, designed to simplify the loading and transhipment of goods from railway trucks to lorries, ships and in some cases even aircraft, and vice versa. The other photographs show containers being loaded and unloaded.

Top left: a British SRN 6 Hovercraft. The hovercraft is a particular form of wheel-less transport which uses an air-cushion to maintain its lift and forward movement, situated between the liquid or solid surface over which it moves and the appropriately designed lower section of the craft. Below: a hovercraft built by Bell Aerosystems for the US Navy. Right: a British SRN 4.

vehicles for overland transport. And here too we can only give some rough indication of the dimensions and performances on record. We should mention, where liners are concerned, the French ship *France* which has an overall length of 315.5 metres (1035 feet). In 1973 the world saw the arrival of oil-tankers with a tonnage of some 500,000 tons, and an overall length in the region of 400 metres (1300 feet). But as a result of the energy crisis there has been a decline in the construction of super-tankers, now, not least because of the re-opening of the Suez Canal, we are seeing smaller tankers being built. Where freighters and cargo ships are concerned, we should mention the Japanese *Usa Maru*, with a displacement of 264,000 tons, an overall length of 337.71 metres (1120 feet) and a beam of 53.25 metres (173 feet).

As far as the future development of maritime transport is concerned the *Hovercraft* has aroused a great deal of interest, operating as it does on the air-cushion principle. It is already quite widespread and can reach speeds of up to 150 kph (90 mph). These craft (one of the variants of which can be considered to be boats working on the air-cushion basis) are capable of moving above the surface of the water by means of a layer of air between them and the water surface. Atomic power might well be an even more significant factor in the

years to come. It has already been experimented with in the Russian ice-breaker *Lenin*, in the American vessel *Savannah*, the German ship *Otto Hahn* and the Japanese *Mutsu*.

But now we come to what is perhaps mankind's most extraordinary exploit of all: flight. If man's imagination and fantasies have nurtured illusions and created legends about flight in every era, the obstacles which stood in the way were such that it was less than two centuries ago that men first managed to become airborne. This slow start, when compared with overland vehicles and ships, was not due to some quirk of fate, but rather to a specific practical fact. In developing his means to move about on land and on water man had to some extent merely to improve upon nature. But in order to fly, he had to conquer nature. In fact today we know this is scientifically true. Unlike winged creatures, man is too heavy to become airborne by his own strength alone, in as much as the power developed by his muscles is insufficient in relation to his body-weight. And the same goes for all those creatures used by man to pull and haul his vehicles. For this reason it was not enough for man simply to imitate the flight of birds. To take off from the ground he had no option but to resort to carefully designed mechanical devices, and these could only be developed when

A rare picture of the US atomic submarine Queenfish at sea in the Atlantic Ocean.

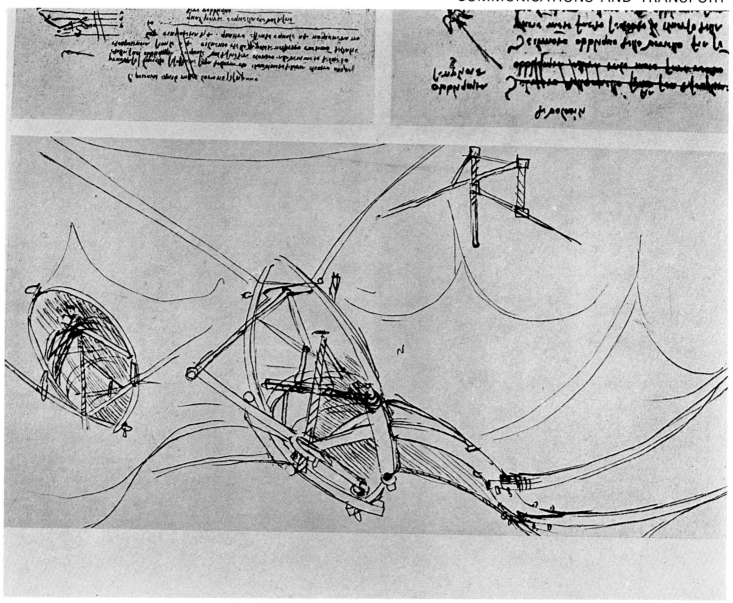

A drawing of a machine with wings, manually operated by means of screws, by Leonardo da Vinci. This famous scholar had the idea of solving the problem of human flight by using structures with flapping wings, but later studies showed that such a system would not work.

scientific and technological knowledge had reached the extremely advanced stage it reached in recent times.

Scientific study of the problem of flight nevertheless had a famous precursor: Leonardo da Vinci. He in fact tackled this extremely challenging area in an age when theoretical and practical resources were very poor, but his brilliant mind managed to shed light on more than one fundamental aspect of the problem.

It was nevertheless not until the end of the eighteenth century that a practical solution was found, and this solution was the simplest and most feasible of all the theoretical possibilities then being discussed. This was the montgolfier or fire-balloon (named after the Frenchmen who built it, the Montgolfier brothers), a large balloon filled with hot air which, by virtue of its lightness, could rise into the air as a result of the thrust given it by the air itself, based on Archimedes' principle – like a piece of cork which, when immersed in water, tends to rise to the surface because of the thrust of the water. On 19 September 1783 the first men in history, the Frenchmen De Rosier and D'Arlandes, sailed up into the air, and finished their flight with a happy landing. At almost the same time the Frenchman Charles was building an actual air-balloon, which was filled with hydrogen, a gas which is lighter than air. In practical terms,

almost as soon as this new means of transport came into being it quickly reached technical perfection. However, it never managed to gain the status of a development of major practical importance, because it was impossible to fly and steer and also because of the dangers to which it was exposed in adverse atmospheric conditions.

The air-ship had a more glorious, though somewhat tragic career. This was a combination of an air-balloon and an engine which drove a propellor. In 1852 the Frenchman Giffard built the first air-ship (or dirigible balloon) which was completely limp and powered by an engine which developed 3 hp. Various improvements were made to the concept in later decades. In the first decades of the twentieth century either semi-rigid or wholly rigid air-ships were built. These were elongated in shape and capable of speeds in excess of 100 kph (60 mph); and they were used for regular passenger and freight services across the Atlantic. But air-ships were costly, hard to handle and too cumbersome. What is more they were dangerous in bad weather conditions, because they too were filled with hydrogen, which is highly inflammable. Suffice it to recall the disasters that befell the *Italia* which was smashed to pieces on ice in 1928 and the *Hindenburg* which went up in flames in 1937, probably because of an

Left: the first hot-air balloon, built by the Montgolfier brothers, which, on 5 June 1783 reached an altitude of 1500 metres (4920 feet) in about ten minutes. Top right: an airship flying over a large city. Below right: a sports meeting with lighter-than-air craft or aerostats.

Above: the first biplane built by the Wright brothers. Below: the Fairey Swordfish.

electrical fault, both with considerable loss of human life. Today the air-ship is obsolete.

The final triumph of human flight was not due to a device lighter than air, but rather, as we well know, to the aeroplane, which is heavier than air. The physical laws on which the flight of an aircraft is based are much more complex than those which govern the way in which a craft which is lighter-than-air rises into the sky. And next to treating the subject in a strictly theoretical way, it is perhaps most helpful to give just a few informative details. The reason which allows the aircraft to rise up into the air is a dynamic one, rather than a static one, as in the case of the balloon. The secret is the wings which, because of their particular shape, generate in the air as they move forward, a resistance, and part of this takes the form of an upward thrust known as the 'lift'. In order to rise up into the air and stay there, the aircraft must have a speed which will maintain the lift above the aircraft's overall weight. The first means used to overcome air resistance and keep the aircraft moving forward was the propellor. This was driven by the engine, and thrust a column of air backwards. On the principle of action and reaction, as it thrust the air backwards, so it was thrust forwards, taking with it the aircraft to which it is mounted.

The aeroplane made its debut very late in history, as compared with other means of transport. In addition it took a very long time for the transition from theory to practice to take place. As early as 1809 G. Cayley designed an aeroplane which had all the main characteristic features of this form of transport: two rigid wings, a powered propellor and a fuselage, and rudders. But the aeroplane never actually materialized. For almost a whole century Cayley's attempts were followed by a series of ill-fated and sometimes tragic efforts by a host of pioneers: F. du Temple, A. Pénaud, V. Tatin, C. Ader, F. Philips, H. Maxim and H. Langley. Special mention should be made of the German O. Lilienthal who, between 1891 and 1896, was the first man actually to fly with the help of gliders.

Fortune (which is the *mot juste* in this case) was eventually to smile on the American brothers W. and O. Wright. After an unsuccessful trial on 14 December 1903, on 17 December they achieved the first powered flight, which lasted twelve seconds. The Wrights' aircraft was a biplane equipped with landing skids, and a 16 hp engine.

From that moment aviation went ahead by leaps and bounds. L. Blériot flew at an average speed of 41 kph (26 mph) in 1906, and 75 kph (47 mph) in 1909. In

Top left: a wooden aeroplane propellor. The propellor is a mechanical device consisting of a special hub firmly fixed to the rotating shaft of an engine, on which the blades are mounted. The propellor (or screw) was used in ships before being adapted to aircraft. Below: the screw of a very large ship. Right: the propellor of a turbo-prop aircraft.

Top: American military helicopters. Below left: a Sikorsky helicopter, used for carrying large cargo. Below centre: the controls of a helicopter. Below right: a twin-engined helicopter.

the same year W. Wright flew for 62 kilometres (39 miles) without touching down, and then the longer distance of 124 kilometres (77.5 miles). Later in 1909 H. Farman flew a distance of 200 kilometres (125 miles), and Blériot crossed the Channel in 27 minutes (a feat which had been preceded by H. Latham's attempt, which ended up in the sea).

The first aircraft were hand-built with timber and canvas, and usually somewhat crude and primitive, with no fuselage or cockpit. After Blériot's flight, and after some longer flights such as the flight across the Alps, the Leghorn-Bastia crossing, the Tunis-Rome flight, and many others which took place in the years immediately leading up to the First World War, aircraft began to be taken seriously by industry.

During the four years of the First World War the aeroplane developed extremely rapidly. Improvements were made to the types of propellor, the structures were strengthened and refined, and new engines were introduced, particularly of the radial type. Speeds increased to more than 200 kph (125 mph) and flights of more than 1000 kilometres (625 miles) were made. 1919 saw the introduction of the first transatlantic flights. The first such flight, covering 6800 kilometres (4250 miles) with various stop-off points, was made

in May 1919 with a four-engined military Curtis seaplane which took off from Newfoundland and landed in Plymouth. In the following decade civil aviation developed considerably. Aircraft were now constructed entirely of metal (keeping pace with the expanding light metal industry) and became larger, more comfortable and safer. Regular freight and passenger services were established.

Further daring and spectacular flights made good publicity for the performances of this new means of transport. In May 1927 C. A. Lindberg made the first nonstop flight from New York to Paris. In the summer of that same year R. E. Byrd flew over the North Pole. At the same time flights over extremely long distances with various staging-posts became more and more common. From the technical point of view, there was a shift from one- and three-engined aircraft to two- and four-engined models, both for safety and constructional reasons. The engines themselves underwent a whole series of improvements too.

In this same period a highly successful relative of the aeroplane arrived on the scene. This was the helicopter, and because of its ability to take off and land vertically, and stay virtually motionless in mid-air it has become more and

Top: a glider just about to be released from the aircraft towing it. The glider is an aeroplane without an engine, kept airborne by the dynamic reaction of the air on the wings. Below left: a towing aircraft seen from the cockpit of a glider. Below centre: two gliders being towed by a single cable. Below right: two gliders being towed by two cables, seen from the towing aircraft.

Top: the Proteus jet engine. The jet engine is a drive-unit which generates the propulsive thrust necessary to move an aircraft forward, using the acceleration of a fluid which is expelled in the opposite direction to that of the desired movement. Below left: cross-section of a turbo-jet. Below centre: the air-intake of a Rolls Royce turbojet. Below right: one of the four jet engines of the supersonic Concorde.

more important in this modern day and age. I. Sikorsky experimented with this concept as early as 1909, and the helicopter started to be actually used for practical purposes in 1935. It obtains the power needed to keep it airborne from a large screw with long blades on a vertical axle.

If the progress this far in the field of aviation was spectacular, the advances witnessed during the Second World War, and in the last three decades, have been even more so. On 27 August 1939 the first jet aircraft took to the air. This was the German Heinkel He 178. In May 1941 the first English jet followed suit, and on 1 October 1942 it was the turn of the Americans. These historic flights opened up a whole new era in the history of aviation. In fact, jet-powered aircraft had truly revolutionary features when compared with propellor-driven models, and they have produced performances which the latter could never have attained. The principle of the jet-engine is in effect quite simple: we all know that if some of the liquid contained in a receptacle is released though a hole, a direct force is applied to the receptacle in the opposite direction to that in which the fluid flows. The same principle applies to the jet-powered aircraft. Gas is sucked in from the atmosphere at the front of the jet-engine and thrust out backwards at high speed by a turbine engine. These

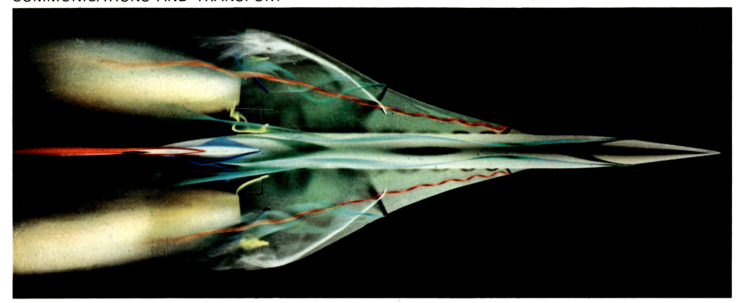

gases, when so expelled, thrust the aircraft forwards.

In the period of time between the end of the Second World War and the present day there have been still further developments in both military and civil aviation. The dimensions, capacities, altitudes and speeds have all increased, and so has the comfort of passenger aircraft and the carrying capacity in aircraft used for freight. An important step forward, which has been almost forgotten today but which is of interest even from a strictly scientific point of view, was the first time that an aircraft broke the 'sound barrier', in other words the speed of sound as it travels through the air, which is about 1200 kph (750 mph). Physics show that an aircraft capable only of speeds below the 'sound barrier' flies in certain given conditions, which differ markedly from the conditions experienced by an aircraft flying at speeds higher than it. What is more, there is a narrow speed range around the speed of sound in which a totally anomalous and dangerous situation occurs, with the likelihood of unforseen effects such as vibrations, depressions and vortices. As a result this critical speed was for a long time considered to be a barrier hard to overcome. But on 14 October 1947 C. E. Yeager made the first supersonic flight in the United States, in a *U.S. Bell SX-1* rocket-powered aircraft. In so doing he

Top: visualization, using the coloured thread method, of the turbulence produced in flight by a supersonic aircraft. Below: the French Concorde *which started making regular scheduled flights on 21 January 1976.*

Top left: a Boeing 747 (Jumbo Jet).
Below: A Boeing 727. Right: an
aircraft seen through a 'fish-eye' lens.

showed that man's technology had reached a level where it was capable of providing the materials and structures to overcome even this major obstacle. Since that day this critical speed has been exceeded on innumerable occasions, and has in fact become quite commonplace for a whole range of aircraft. In fact the speed of sound, which is called Mach, has even become a unit of measure, and, if we disregard space-flights which are not discussed in this chapter, aircraft have obtained speeds of Mach-7. And where altitudes are concerned, aerial flights seem to have overlapped with space-flights when we talk about altitudes of some 35 kilometres (22 miles).

Today the age of supersonic aircraft seems to be really upon us with the Russian *TU 144* and the Anglo-French *Concorde*, with the impending introduction of many regular passenger services (the first schedule flight by the *Concorde* took place on 21 January 1976). But the future of this means of transport is somewhat debatable because of the extremely high operating costs. A rosier future seems to lie in store for the *STOL* (short take-off and landing) aircraft and *VTOL* (vertical take-off and landing) aircraft. Both these types of aircraft have been undergoing extensive trials for several years now. In Canada an airline using *STOL* aircraft capable of carrying 50 passengers flies regular

passenger services between Ottawa and Montreal. The far-reaching development in the last few years in the field of aviation has brought with it new and
unforeseen problems. The major ones are atmospheric, and in some regions
acoustic pollution, and to some extent thermic pollution as well; and the
excessive consumption of energy resources. It is thus forseeable that in the near
future we shall see this sector being rationalized and in all probability cut down
in size.

A need that man has always felt, and which technology has been able to meet
in an extremely satisfactory way, is that of communicating with other men; to
express his own thoughts and feelings and to give news and in turn receive
news. Articulate language can indeed be considered the first major technique
developed by mankind for this purpose. In fact other living species often
communicate with sounds which in some cases have different tones and forms.
These in turn have fairly definite meanings. But there is nothing in nature to
compare with the structure of man's languages, with all their clear-cut rules
which can be passed on from individual to individual, and their extraordinary
range of expression. The next step, obviously enough, was writing. Here too we
could quote other living species which in some cases use signs to communicate

*A V-STOL (Vertical Short Take-Off
and Landing) aircraft. This type of
aircraft which has been developed in
recent years, is capable of taking off
from and landing on very short
runways. It goes some way towards
meeting the need to link large cities
without using long runways in airports
which are some way from the city-
centre.*

Top left: a Morse transmitter dating back to 1898. Top centre: the telephone built by A. Meucci. Top right: an old-fashioned wireless. Below left: a modern telephone. Below centre: a recently designed teleprinter. Below right: a colour video-cassette machine with monitor.

with, but once again there is no comparison with the degree of complexity and expression achieved by our own written languages.

Spoken words enable us to communicate instantly, but at close quarters. The written word enables us to communicate in time and space, but does not enable us to communicate instantly over long distances. This latter was thus the next goal to be reached, as far as the means of communication were concerned. In this respect various civilizations have in the past devised ingenious systems – from the tom-tom to smoke signals, to flashing mirrors, to methods of signalling at sea. But all these primitive methods had severe limitations: acoustic signals become quickly much less clear over longer distances and eventually become completely inaudible; visual signals could be picked up easily, even a long way off, in perfect conditions, but could only be transmitted in a straight line, could not go round obstacles, and could not negotiate the curve of the Earth's surface. The problem at hand was thus to direct the desired signal towards its destination and amplify it once it had become too faint to be clearly heard, and to be heard directly. The decisive turning-point came through the exploitation of electromagnetism. The telegraph, which was the first really effective means for transmitting long-distance messages, is an

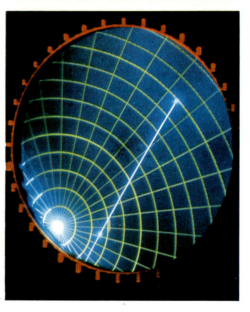

electromagnetic device. The message, which consists of electric impulses, is sent with the help of conducting leads or wires which form the telegraph line. Together with an appropriate transmitting code, the telegraph was developed in the United States by S. F. B. Morse in 1837. In its wake came the telephone. This was first invented almost simultaneously by the Italian Meucci and the American A. G. Bell who had the good fortune to patent the invention in 1876. The telephone is an electromagnetic device used for the long-distance transmission and reception of the voice, based on the principle of converting sounds into electric impulses and vice versa.

The experimental determination of the existence of electromagnetic waves, carried out by the German physicist H. R. Hertz, between 1885 and 1889, brought about the liberation of human communication systems from the last material shackle – wires. In 1895 the Russian scientist A. S. Popov devised an apparatus capable of recording electromagnetic signals produced a long way off. Making the most of extremely fortunate circumstances, in the following year the Italian G. Marconi patented his wireless telegraph, and the road to radio-communication was thus opened. Radio-communications are based on the emission, from the aerial or antenna of a transmitter, of electromagnetic

Left: radar aerials with a parabolic reflector (or dish). Top right: radar being applied to meterology. Below: detail of a sighting radar screen.

Top left and right: the Daily Express, the first transmission of a page of a daily newspaper via satellite. Below left: the INTELSAT IV satellite for space telecommunications.

waves which spread out into the surrounding space, with different patterns depending on their frequency, until they are picked up by the antenna of a receiver.

During the twentieth century, telecommunications have developed in parallel with, or as a consequence of, the development of electrification; and with the development of electronics whole new areas have also been opened. In about 1925 the Englishman J. L. Baird and the American C. F. Jenkins carried out various rudimentary experiments concerned with the transmission of images. Within a decade television in its turn became a *fait accompli*. The first public television service came into operation in London, in 1936, marking the whole new area and era in the field of human communications. This invention has, as we know, been developed to the point where it can allow man's eye to alight at close quarters even on the surface of distant planets.

Along with the telegraph, the telephone, the radio and the television, modern man has at his disposal, or will shortly have, a whole host of devices which either incorporate some of the features of the above, or else present completely new features. To list them would require a great deal of space. A science has even been developed, called cybernetics, which deals specifically

325

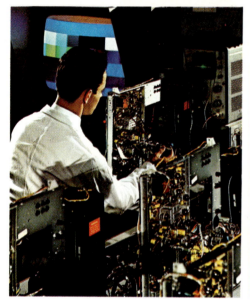

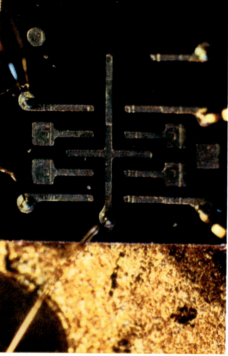

with monitoring and communication in both machines and living beings. Teleprinters represent an improvement on the telegraph. These enable messages to be transmitted immediately 'in clear', i.e. with letters of the alphabet. An important future role will probably be played by the video-telephone, which is something of a cross between a telephone and television set, capable of 'uniting' several people in different places, and thus saving a considerable amount of time and energy which would otherwise be wasted in terms of travel.

Special mention should be made of radar, a device used for localizing and detecting the presence of fixed or moving objects by means of radio or Hertzian waves. The principle on which radar works is based on the reflection of electro-magnetic waves. Radar, which was first put to practical use in Great Britain and the United States on the eve of the Second World War, has become a vital element in aeronautical engineering, and has become essential for night-flying and navigation in conditions of poor visibility.

A further step forward in the development of telecommunications has come about as a result of space flights, where, as we well know, major importance is attached on the one hand to the transmission of as much information and data

Top left: part of a computer used to control a nuclear reactor. Below: electronic telecommunications equipment. Centre: a detail of an electronic molecular circuit. Top right: on a TV screen the computer produces the modifications made by the draughtsman. Below: magnetic disc memory (or brain).

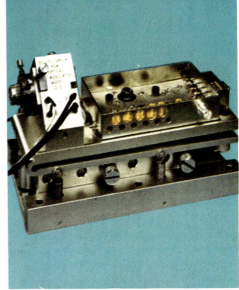

Top right : the mechanical assembly of an optical modulator and pilot amplifier used in the telecommunications system and based on the modulation of impulses in code. Below : an optical receiver for demodulating optical carrier waves modulated in phase. Left : a TV channel transmitted via laser, of the argon type. The green strip on the left of the photograph represents the diffusion of the beam in the atmosphere. Note the telescopic system which forms the receiving antenna on which the laser beam acts.

as possible and, on the other, to the use of apparatus which is as light and as small as possible. This has encouraged a continuous process of miniaturization.

Among the equipment which has contributed to the development of telecommunications, or which seems to guarantee a bright future to the technology are the transistor, a tiny device which conducts, modulates and amplifies electric signals by means of the use of a solid substance, and which is about to celebrate its thirtieth birthday, the laser, a device capable of generating coherent light waves, whose effectiveness has been experimented with for example in the amplification of very weak signals, and optic fibres, which can be used as wave carriers and are thus capable of transmitting information with light rays.

MAN'S USE OF ENERGY

Between man's capacities for work and those of any other creature there lies a radical and crucial difference. The nest-building bird, the beaver and the bee all carry out their tasks on principally instinctive bases. In other words the things that they can do rely only in the second instance on learning, and are far more dependent on innate factors. For example, no one has ever taught a swallow how it should build its nest; from birth the swallow possesses the physiological abilities which enable it to acquire this capacity, under normal conditions of development. But all swallows will build their nests in more or less the same way, and similarly there is no chance, not even as a result of training, of a swallow ever managing to build anything other than a swallow's nest. The same applies, broadly speaking, to every other species of animal, even bearing in mind the very wide variety of existing situations – with the exception, that is, of man. In fact while the respective activities of every other species have remained closely bound by clearly circumscribed forms of behaviour, in man – down the ages and with the help of decidedly fortunate circumstances – there has been a series of exceptional structural modifications of extraordinary importance: the upright position on two legs, the opposability of the thumb to the other fingers, and linguistic articulation. Thanks to these modifications man has been able to initiate new and more complex forms of activity, using the nervous system as a point of departure: and these modifications have in turn resulted in the development of an intelligence to levels never attained by any other known species. All these acquisitions have enabled man not only to assume new forms of behaviour but also to transmit these forms of behaviour to other men, and hence to successive generations, by means of education and upbringing. And here we come to the way man's specific attitude to work manifests itself. This is a faculty which allows him to undertake, if necessary, completely new activities and tasks, in relation to the requirements which may from time to time occur, without being excessively bound by instinct. And thanks to this special gift, man has been in a position to tread his path towards a state of dominion over nature, using the fruit of his own labours.

The first form of energy available to man was certainly that supplied him by his own muscles. As there was progressive improvement in the nervous coordination between hands, eyes and brain, our remote forbears became gradually capable of making the first decisive step towards the development of human technology: the use of tools forming no part of the human body. It should be born in mind that the first tools of this type were of the simplest conceivable kind: stones or branches picked off the ground. This hypothesis, which is the most likely to be true, is backed up by the fact that our closest relatives, the monkeys, clearly display a capacity to use simple objects which come to hand, such as sticks, to bring their food to them. Unlike present-day monkeys, however, prehistoric man systematically developed his use of these first tools, and learnt to make artificial modifications to them. In this way there emerged the first techniques for making tools and working various materials. We are in possession of much archaeological evidence concerning the use of stone and bone. The same, unfortunately, cannot be said of early man's use of wood, which stands up less well to the rigours of time but which was

This ancient Egyptian papyrus scroll shows various tasks being carried out by the only source of energy then available: muscular energy, supplied by man and animals.

undoubtedly used as widely as stone and bone or even more widely.

Man thus slowly discovered that his own energy could be supplemented by the use of appropriate tools and devices – for example, a large boulder, too heavy to be moved by a man's arm alone, could be shifted by that same arm wielding a sturdy branch as a lever; animals which were too strong to be overcome, or too nimble to be caught as prey by unarmed human hunters, could be overcome or caught by blows dealt with stout cudgels or pointed spears. And a branch which a man's bare hand could not wrench from a tree could be cut off with a sharpened stone in the grasp of that same hand. At the same time man started to learn that work and tasks could be carried out not only with muscle but also by external objects appropriately situated or arranged. For example, a heavy object situated at a certain height had a disproportionately greater effect, if dropped on a given target, than the effort used to cause its fall. In other respects, hunting tools such as the bow or snare soon showed that they incorporated the inherent principle of the use of previously stored amounts of energy, on the basis of elastic distortion.

Still in the prehistoric era, man learnt how to produce and make full use of fire. This was possibly the most significant discovery ever made, both because

Top left: ploughing with dromedaries in Tunisia. Top right: ploughing in the Andes in Peru (in the Cordillera Negra). Below: threshing corn in the Darri-i-Shikar valley in Afghanistan.

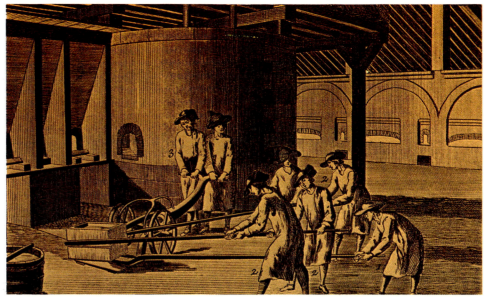

Left: a rudimentary tool for hoisting water and irrigating fields. Right: two illustrations of an eighteenth-century glass works (from Diderot's Encyclopédie*).*

of its crucial consequences in the field of technology, and because of the enormous influence it was to have on the psychology and customs of men, as a miraculous weapon against the cold and darkness. In addition fire represented the first instance of the large-scale use of energy above and beyond the human muscles. Ancient man had wood, principally, and occasionally coal to thank for fire. The use of the energy developed by fire brought with it the development of metal-working, ceramics, and the working of other materials.

In the prehistoric community, therefore, extremely important techniques had emerged and had been developed. The great state-units which then came into being, one after the other, in ancient times did not, however, for their part make further proportionately significant advances in the technological field. Of course there were major achievements in the field of building, navigation, transportation and the development of waterworks, but overall these were quantitative rather than qualitative advances. The reason lay in the structure of society, which was based on slavery. Thus the Egyptian pyramids, Babylon, and the grand architectural achievements of Greece and Rome were all built by slaves. The slaves represented an especially convenient and cheap source of energy, from the point of view of the ruling classes. Within the latter there also

emerged – as a result of the scorn with which slaves were regarded – an attitude which disdained manual work in general, and this in turn led to the exploitation of free craftsmen, where the latter existed.

Animal energy began to be used not only for the purposes of transportation but also in other sectors, and particularly in agriculture. Likewise, where navigation was concerned, flowing water and the winds started to be used as sources of energy. But here too, slaves were used on a massive scale. Even the water-wheel, which made its appearance among various ancient peoples, did not become very widespread because it was considered less economical than the slaves who would be used in its stead, and, even more important because of a matter of principle (although this is to some extent supposition). It was not in the interests of the ruling classes to entertain solutions which represented a dangerous substitute for the slave labour on which their power was founded.

It was rather in the Middle Ages that the roots of present-day industrial civilization can be detected. With the collapse of the structure of the slave-based societies, and the emergence of a new feudal type of society on its ruins, the problems of discovering ways of harnessing energy came once more to the fore.

Human labour was no longer available on a large scale, but the court

Top: windmills. Below left: a water-mill. Below right: a wind-pump for irrigation near Mallia in Greece. Nowadays the energy contained in running water and the wind is exploited to produce electric energy in plants which, from a technological viewpoint, can be considered to be the direct descendants of those pictured here.

An eighteenth-century steam pump. The first steam-driven units gave very low performances, in as much as they needed a large consumption of fuel for the power they produced; they were originally used in the mining industry.

economy on the one hand, and the establishment of relatively free urban centres on the other, led in fact to a reappraisal of manpower. It was thus gradually found to be more economical to exploit animal energy, and this development spread far and wide during the Middle Ages. For his part, the working man, whether a servant or a free craftsman, found encouragement in the partial improvement of his social status and was stimulated to look for further improvements in his living and working conditions, and particularly through the progressive reduction in the muscular effort required of him.

In addition, with the extraordinary expansion of both the economy and industry which was taking place, animal energy also started to appear inadequate. It was necessary to find other sources of energy which were more powerful and more constant. As a result hydraulic power was exploited on a huge scale, using the water-wheel, which had been known since antiquity and which was now used not only in mills but also to carry out mechanical tasks of various sorts in many workshops. From the twelfth century onwards, with the appearance of the windmill (which had already existed in primitive forms in Persia for many centuries) aeolian or wind-power started to be used as well. Metal-working was still carried out with wood or charcoal for some time to

Left: an open-cast coal mine in Bohemia. Right: miners at work. Coal was the first source of energy to be used in large quantities in modern industry, and the existence of vast coalfields was crucial to the development of the first major industrial powers.

come, but between the twelfth and thirteenth centuries coal from the first English mines also came on the scene. With the Renaissance and the dawn of the modern era the process of industrialisation, which had started in the Middle Ages, became more and more intense. Simultaneously, major advances in the cultural and scientific fields prepared the way for the explosion of modern technology which was given great impetus at the end of the eighteenth century. From the sixteenth century, in many parts of Europe, there was an enormous increase in the demand for coal, destined for furnaces and other industrial uses, as well as for domestic purposes. The demand for mechanical energy increased proportionately, both for industry and for the extremely taxing work entailed in mines. In the meantime the mediaeval craftsman was gradually transformed into nothing less than a modern type of technician. With these conditions prevailing, the continual observation of the powerful effects of heat led inevitably to the idea of deriving from heat itself the much-needed mechanical energy. The time was now ripe for the advent of heat-powered machinery.

First it was the turn of the steam-engine. Blueprints for this resounding invention were already to be found in the works of Heron of Alexandria, in the

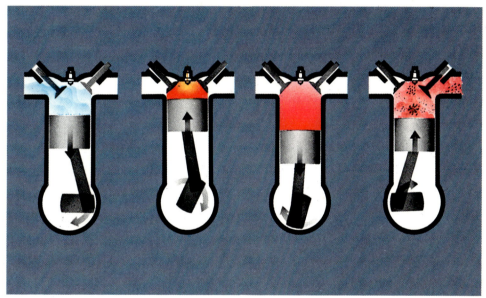

Top: a large explosion engine. The term 'explosion' applies to those internal combustion engines in which the explosion of a mixture of carbonized air is produced by an electric spark. Below left: the engine of the Ferrari 312 B. Below right: a diagram showing the four-stroke principle of a combustion engine (induction, compression, combustion and exhaust).

first century B.C., in the works of Leonardo da Vinci, and later, in the sixteenth and seventeenth centuries, in those of G. Cardano and G. B. della Porta, of S. de Caus and of E. Somerset. The first practical steam-engine was not built, however, until 1698 by the Englishman T. Savery. This was a steam-powered pump, in which the water was sucked in by the vacuum obtained by the condensation of steam in a tank. A less rudimentary type of steam-engine was constructed by T. Newcomen in 1705. This was also a pump, but here, on the basis of previous studies and experiments made by O. von Guerricke, C. Huygens and D. Papin, a cylinder fitted with a moving piston was used. This type of engine became widely used in England and Holland, where it remained in use for pumping out mines and draining land, and for irrigation purposes, until about 1830. In fact it was with this machine that steam became a source of energy used on a huge scale for industrial purposes. The problem of improving on Newcomen's still rather crude machine was tackled with determination by the Scottish engineer J. Watt who, from 1763 onwards, managed to make a series of crucial modifications and improvements to it, including the introduction of the condenser, the use of steam-pressure rather than atmospheric pressure to move the piston, and the use of linings made of

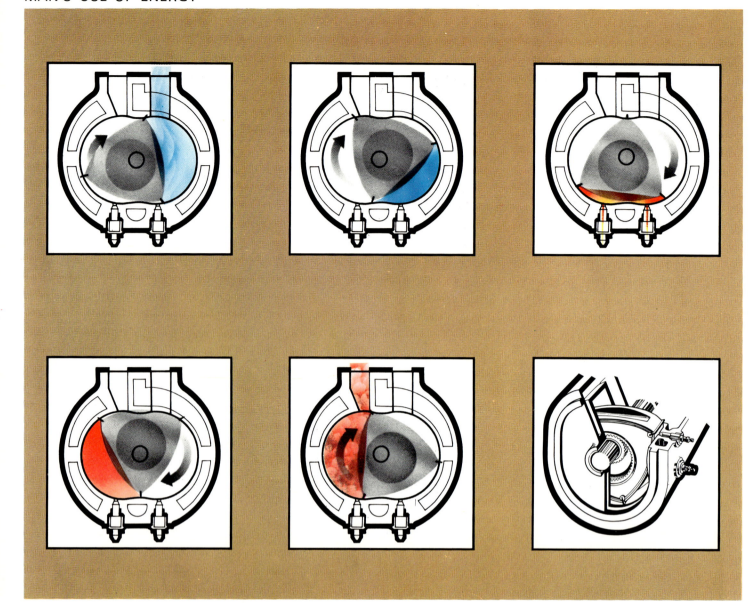

refractory material. This new type of engine, which had an alternate movement in which each stroke was motive, and was thus capable of easily transforming the straight motion into a rotating motion, could now be adapted to any type of mechanical task.

The improvements subsequently made in the nineteenth century were aimed above all at reducing both cost and weight, and increasing power. Among the most significant we should mention the use of progressively increasing expansion and fractional expansion, the use of high pressure and, connected with this, the process of overheating steam and the use of high operating speeds. The use of increasingly high pressures and piston speeds led not only to increased efficiency, but also to more powerful engines, some of which could develop several thousand horsepower. The triple-expansion engine, introduced in 1874 by A. C. Kirk, is still used today in a few steamships. But today, if large amounts of power are to be produced with steam, another type of engine is preferable. This is the turbine engine, which we shall deal with separately at a later stage.

As far as heat-powered machines are concerned, the steam-engine was joined in the nineteenth century by the internal combustion engine. In the

Diagram of a Wankel-type engine, in various phases of operation. The Wankel is an internal combustion engine which has a rotor in the form of an equilateral triangle instead of pistons which move alternately.

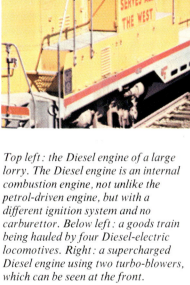

Top left: the Diesel engine of a large lorry. The Diesel engine is an internal combustion engine, not unlike the petrol-driven engine, but with a different ignition system and no carburettor. Below left: a goods train being hauled by four Diesel-electric locomotives. Right: a supercharged Diesel engine using two turbo-blowers, which can be seen at the front.

steam-engine, the fuel on which it runs is enclosed in a furnace outside the engine, and the engine itself contains a secondary liquid, which is usually water, situated between the combustion chamber and the moving parts, and which expands by means of evaporation. In the internal combustion engine, however, the products of the combustion process, in the high-pressure gas state, act directly on the moving parts.

If the steam-engine was the invention not of any one man but the creation, rather, of whole generations of technicians, the same can also be said of the internal combustion engine, which was also made possible by a gradual accumulation of knowledge and experience. The incentive behind its creation lay nevertheless in the general acknowledgement of the drawbacks of the steam-engine: low performance, excessive weight and low functional capacity. Among the precursors of the internal combustion engine we might mention: W. W. Cecil and S. Brown who, in about 1820, experimented in England with engines operating on a mixture of hydrogen and air; S. Carnot, who made certain advances in the theory of the engine; W. Barnett who, in 1838, designed a rudimentary engine; J. J. E. Lenoir who, in 1860, built an illuminating gas engine. In 1862, in Paris, A. Beau de Rochas published a theory containing the

operative principles on which the modern car engine is based. In 1867 the German N. Otto, following the ideas of the unsuccessful Italians Barsanti and Matteucci, embarked on the production of a piston engine. In 1876 Otto and Langen built the first type of four-stroke internal combustion (or explosion) engine. Here, in a cycle arranged in four phases, the movement of the piston was produced by the explosion or combustion of a mixture of air and fuel in the cylinder, the explosion being caused by an electric spark. It is impossible, with the space available, to give even an inkling of the countless improvements which this type of engine has undergone to date: the basic principles on which it is founded have nevertheless remained substantially the same. Together with the four-stroke combustion engine, the two-stroke model has also become widespread in recent decades. Although this latter gives a slightly lower performance, with the same revolutions and cubic capacity it develops considerably more power than an equivalent four-stroke engine. Another type of internal combustion engine which has become of major importance is the Diesel engine, invented in Germany in 1897 by R. Diesel. In this type of engine, which has a considerably higher thermal performance than that of the combustion engine, and which becomes worn less quickly, the fuel burns rather

Left: the assembly of an alternator with a reaction turbine. The turbine is a drive-unit which uses the mechanical energy supplied by a fluid (water, gas or steam). Top right: the Sondrio power station with Francis turbines. Below: a steam turbine.

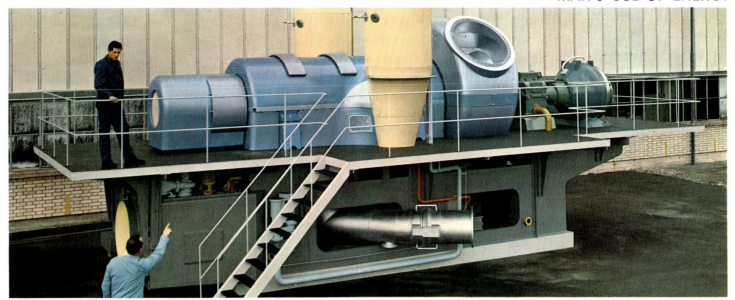

Top: an auxiliary generator, designed to give an additional supply of electric energy in peak periods. The generator is a device designed to transform any form of energy into electric energy. Below: inspecting the conductors inside a generator built by a Russian firm.

than explodes: in other words the combustion is not quick to occur, as in the combustion engine, but quite a lot slower, and the piston is moved by the high pressure of the hot gases thrown off by the burnt fuel. Diesel engines are particularly valuable wherever there is needed a source of energy of average power, reasonable size, and capable of operating for a long time non-stop. A type of combustion engine which may rightly be called revolutionary is that invented by F. Wankel, and produced by the NSU company from 1956 onwards: instead of pistons giving an alternate movement, this engine has a rotor in the shape of an equilateral triangle. Although it is still not completely competitive with its piston-driving counterpart, it has conspicuous advantages, such as the smaller space and lower weight required per horsepower, the almost total absence of vibrations, and the low maintenance costs.

The advantages of a drive unit with a rotating and continuous action rather than a straight, alternate motion have been seized upon by a category of engines which have become more and more important in the exploitation not only of thermal but also of hydraulic energy: the turbine engines. These are drive units whose main parts consist of a paddle-wheel driven by a fluid. In the so-called 'action' turbine the fluid strikes the blades when emitted from

external pipes; in the 'reaction' turbine the fluid is emitted from pipes placed between the blades; as it leaves the pipes it exercises a thrust in the opposite direction to that of its own flow. The source of energy used to drive a turbine varies with the type of turbine in question. Thus we find steam turbines, which use a rise in the temperature of the steam and direct the subsequent thrust on to a series of blades fitted to the rotor. Then we find gas turbines, where the blades are rotated by a current of gas at both high temperature and high pressure, obtained for example by the combustion of kerosene or methane. Finally, there are hydraulic turbines which transform the energy of a waterfall into mechanical energy produced by a turning part; these latter are usually connected to an electric generator which is driven in this way.

After the development of the dynamo in about 1870, plans were swiftly drawn up to spread the general use of electrification. The first public electricity distribution plant was put into operation in London on 12 January 1882, followed that same September by one in New York. These plants supplied direct current and thus were of fairly limited use. In the years immediately following alternate current started to be used. From then on the spread of the use of electricity has been such that our very day-to-day life, with all its most

Top left: a transformer. The transformer is a static electrical device for transforming alternate-current electric energy (single- or multi-phase) into electric energy of the same frequency but with different tensions. Below left: the nucleus of a 400 MVA transformer. Right: a transformer built by the E. Marelli company.

The electric power station and dam on the Colorado River at Davis Dam in Nevada, USA. Hydroelectric power stations are of vital importance as inexhaustible sources of cheap and non-pollutive energy. Unfortunately there is little hope of their numbers increasing much more, because the most suitable and accessible sites have by now been almost all occupied, at least in the highly industrialized countries.

commonplace features, would be inconceivable without it. Today the production of electricity in the various countries around the world has reached a high degree of efficiency and automation, and many different problems have been tackled and solved to make this possible. As far as generating current is concerned, it was first and foremost a matter of using large sources of energy which, in the original state, came in a form that differed from electricity: for electricity is not found in nature in a form in which it can be harnessed for man's use, and it is no coincidence that its use has come about relatively late in history. To this end engines, and usually turbine engines, for the production of electric energy and the respective generators are concentrated in large plants in just a few sites. These are power stations, and can, essentially, be divided into two categories: thermoelectric power stations and hydroelectric power stations. The former can in their turn be subdivided into power stations operating on steam or Diesel, geothermal power stations and nuclear power stations. The hydroelectric plants can for their part be subdivided into power stations operating with running water and plants linked up to reservoirs. The former incorporate the creation of an artificial lake obtained by damming up a river and on a constant basis use a sometimes fairly small drop in the level of

The two illustrations show the cooling towers of the Neurath thermoelectric power station.

the water. In this way they can exploit a constant rate of flow which is higher than the minimum rate of flow of the river in the periods when the river is low. The latter, on the contrary, only let the water flow down from the dam when the basis is full, thus exploiting very considerable differences in levels with not very large rates of flow. They are used intermittently, and preferably in peak periods, when the demand for energy is higher. The thermoelectric power stations, on the other hand, consist solely of huge steam-boilers, operated on gas, coal or other types of fuel. They provide a constant supply of energy, or one that at most only varies slightly.

Another problem which has had to be solved has been that of distributing electric energy. This is carried out by switching plants, where the energy received at the high tensions necessary to obtain long-distance transmission of it is converted by means of transformers into lower tensions, so that it can be distributed to primary distribution lines. Next, because the tension is still too high for ordinary use, it is reduced still further, and the current is then distributed to the consumer along service lines.

From the brief list of the means generally and currently employed to utilize energy, the reader may already have become aware of an increasingly vital and

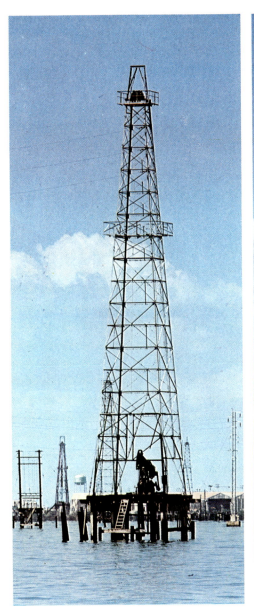

Left: oil-wells on Lake Maracaibo in Venezuela. Right: an oil-well in Maracaibo. Oil (or petroleum) is another important source of energy. Unfortunately, however, despite its advantages as far as profit and easy use are concerned, it also involves major drawbacks; it pollutes (at least as it is used in engines at the present time) and its reserves are likely to be exhausted sooner or later.

dramatic problem: the problem of the sources of energy. Even in the apparently most favourable case, i.e. hydroelectric energy, which can be considered to be inexhaustible, there are in fact serious limitations: the number of water-courses in the world is not infinite, the most accessible are now for the most part being used to the full, and every new dam and power station that is built involves higher, almost prohibitive costs. For this reason the maximum amount of energy produced in a given time by such and such a means cannot easily be increased. And the amount of energy in question is now clearly inadequate for the overall needs of the world. Things look even bleaker for the other sources of energy in use, at least as far as those currently most commonly used are concerned, such as coal, oil and natural gas. In fact the reserves of all these are being depleted and even though the discovery of new deposits and fields makes it harder and harder to give accurate predictions about the probable life of the various natural supplies of fuels available to mankind, the fact remains that this life, based on present-day rates of consumption, can certainly not last for ever. It may last more than a few decades, but it will not last more than a few centuries. What is more, the problem of the depletion of our natural resources is further aggravated by other problems which may

Left: petroleum plant at Hassi-Messaud in Algeria. Right: a large oil-pipeline. The exploitation of oil-energy has brought with it major technological problems, such as the processing (refining) of the crude product and transportation of it from producer to consumer.

perhaps be incidental, but which nevertheless entail dangers which are no less real. The concentration and distribution of energy are subject to geographical, historical, political and social influences which complicate the current situation in at times very grave ways, and to such an extent that solutions of a technological nature run the risk of being inadequate, if not futile. Suffice it to recall the facts that emerged at the end of 1973 after the Arab-Israeli war when there was a partial embargo on oil-products by the oil-producing Arab countries. This example is more sensational than substantial, in as much as the world was already in the throes of the energy crisis and wrestling with the difficulties of the political conditions being imposed by rich countries, in terms of raw materials and technology on, and to the detriment of, other less fortunate or less developed countries. But for all this it is a crystal-clear example which brought to light certain truths more clearly than anything which had happened before. Thus the general awareness of energy problems has increased considerably. To illustrate the point that we have reached, probably better than any other data or consideration, we therefore include the following figures. In 1860 the world consumption of energy was 1,100,000 million kilowatt-hours. In 1900 this figure rose to 6,100,000 million kilowatt-

New York blacked-out: on the night of the 9–10 November 1965 a breakdown in the world's most complex electrical junction system caused the electrical grid serving the New York area to stop working. The city was plunged into chaos and the people living there into a state of panic. Perhaps this is an inkling of where the present-day model of technological development might be leading us in the not-too-distant future?

hours, and by 1950 the consumption of the world stood at 21,000,000 million kilowatt-hours. In recent years the annual rate of consumption has soared to around 50,000,000 million kilowatt-hours. Now, a fairly recent estimate has revealed that the so-called under-developed countries, which account for more than 70 per cent of the world's population, consume only one fifth of all the energy consumed throughout the world, and less than one sixth of all the electricity produced. It would seem that this imbalance is destined to increase. Finally, we should mention that the pollution produced by the traditional fuels runs the risk of increasing the amount of carbon dioxide in the atmosphere to a dangerous level.

Once more mankind thus finds itself having to tackle the challenge posed by nature, but this time the situation is complicated. The need to find new sources of energy has now become urgent.

One such source, which has already been in use for a while, and is now being used on an ever larger scale, is one which would currently seem to offer the most immediate and tangible prospects: nuclear energy. On 2 December 1942, in Chicago, E. Fermi and his colleagues developed the first 'atomic pile', consisting of a large, appropriately structured block of uranium and graphite.

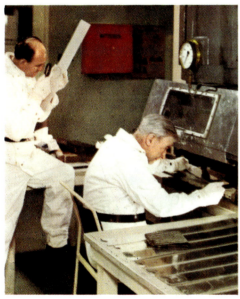
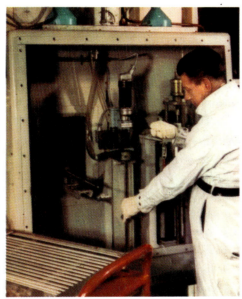

Opposite: the methane-pipeline (Cortemaggiore-Caviaga) across the river Po, near Cremona. Methane is used industrially, domestically and as a fuel for automobile engines. Above left: a concentration of uranium. Above right: plutonium being prepared. Below left: nuclear fuel in the form of uranium-and-zirconium bars. Below centre: nuclear fuel in the form of cylindrical 'tablets'. Below right: the purification of mineral containing radioactive substances with solvents.

This was the primitive model for the electro-nuclear plants using natural uranium and graphite which not many years later were to spring up and be put to practical use. This 'pile' was a nuclear reactor based on the 'fission-process'. We should point out that nuclear fission consists in the splitting of atoms of heavy elements (uranium, actinium, thorium) into lighter elements, thus producing energy. Because the energy developed by a reactor assumes the form of heat, this heat can be transformed into electric energy by means of appropriate plant.

This spread of the use of nuclear energy has gone hand in hand with the controversy about whether or not this energy is in effect competitive with the other forms of energy in use today. This dispute is by no means over: on the other hand, there are certain definite advantages; for example, thermo-nuclear power stations give relatively low-cost energy, do not pollute the atmosphere (except as a result of radiation 'leakage', but given the safety arrangements in force this is considered to be extremely low) and, most important of all, are completely automated, and do not require the presence of a river or the constant supply of large amounts of fuel from outside. Among the disadvantages, on the other hand, we find the likelihood of disastrous

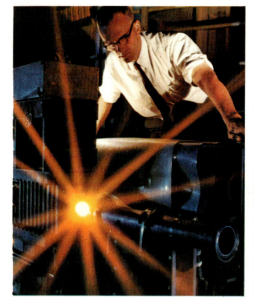

accidents, a likelihood which is considered by some to be nowhere near low enough, and the probability of extremely expensive damage caused as a result. In addition, there is the more obvious fact that thermo-nuclear power stations cause an increase in 'thermal pollution', due to the fact that the heat produced in a thermo-nuclear reactor converts less well into electricity than that produced by the combustion of coal or petroleum, and as a result needs to be cooled. The cooling process uses various types of material depending on the type of reactor (water, sodium and vaious gases) and restores these materials at a higher temperature than the ambient temperature. Lastly there is the problem of radioactive waste, which has to be buried in underground deposits or sunk to the bottom of the sea. Apart from the use described above, nuclear energy will in all probability, and before too long, play a fairly important role in at least two other fields: the propulsion of ships and inter-planetary rockets, and the use of nuclear explosions to carry out mammoth excavation projects. With regard to the first role, we can mention atomic submarines, the American atomic ship the *Savannah* and the Russian atomic ice-breaker the *Lenin* as well as other ships which have been mentioned in the previous chapter. Where the latter role is concerned, an experimental programme has been embarked on,

Top left: a device designed for the direct conversion of the energy, released in a controlled nuclear fusion, into electric energy with an intermediate transformation into thermal energy. Top right: one of the first electro-nuclear power stations, at Calder Hall in Great Britain. Below left: the Dresden (USA) electro-nuclear power station. Below right: the Latina electro-nuclear power station.

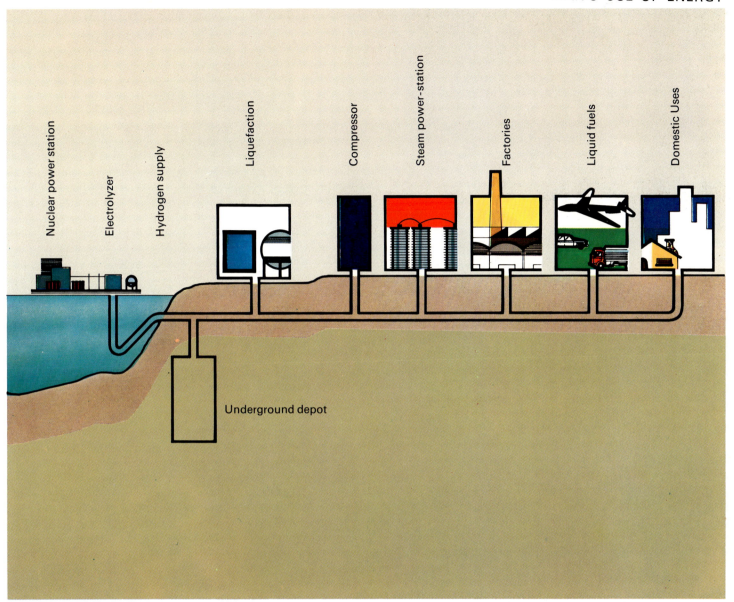

Nuclear power station

Electrolyzer

Hydrogen supply

Liquefaction

Compressor

Steam power-station

Factories

Liquid fuels

Domestic Uses

Underground depot

A diagram showing a possible way of extracting hydrogen from salt water; in this way hydrogen could become an excellent substitute for petroleum. The essential section consists of a platform on which a nuclear power station supplies the electric power necessary for electrolysis – the process whereby hydrogen is separated. Once liquified, the hydrogen obtained is distributed by canalization.

nicknamed 'Plowshare', under the auspices of the USAEC, for the creation of immense craters or pits with the use of underground atomic explosions. Of course, this latter use, which is important for the mining industry for example, entails the spread of radioactivity and seismic effect, both of which represent dangerous consequences which must be taken into due consideration.

The prospects offered by nuclear energy nevertheless go far beyond what has been achieved to date on the practical level. In-depth research is in fact under way at this moment, aimed at achieving controlled 'thermo-nuclear fusion' on a huge scale. We should mention that thermo-nuclear fusion is obtained when two atoms of light elements merge and form an atom of a heavier element. This process is accompanied by the emission of considerable amounts of energy, and the raw material which can be used to obtain this is not only extremely cheap but also virtually inexhaustible. In fact, a reaction of this type can be obtained by using the deuterium contained in water, which is present in appreciable amounts in natural hydrogen. And the energy which can in theory be produced by the fusion of the deuterium nuclei present in just one litre of water is equivalent to the energy developed by about 300 litres of petrol! But in practice fantastically high temperatures are needed to carry out a fusion

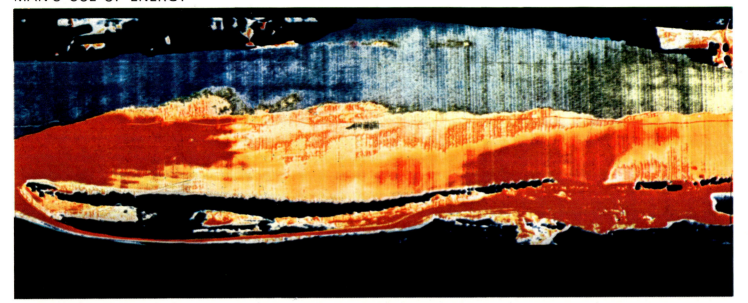

process, somewhere in the region of 100 million °C, (180 million °F) and this is the major problem into which the various research teams are looking. If and when they come up with a solution which can be applied in a practical form, mankind's demand for energy will perhaps be able to be permanently satisfied.

For the time being, however, this thirst for energy is more urgent than ever before. For this reason other ways are being tried and sought for exploiting more directly accessible sources of energy. Unfortunately the sources so far selected are less easily usable or less consistently reliable than those now in use on a large scale. And this is natural enough, and if it were not so, man would have already been making use of them for some time.

A prime example, which to some extent consists in the extension of the use of hydroelectric energy, is the proposal to exploit tidal energy. This is non-pollutive, and thus an interesting source of energy. To date only a few not very major plants have been put into operation and as a result the proportion of overall energy which can be supplied by this means is small, somewhere in the region of one per cent of all the energy that can be produced hydroelectrically.

There still remain three non-pollutive and in fact inexhaustible sources which have already been exploited in the past on a small scale and have now

Top: an aerial infra-red photograph of the Connecticut river, where the water from the condenser in the Haddam Neck nuclear plant is discharged. There is a rise in temperature of some 8°C (46°F) between the water in the tributary coming from the nuclear power station (in red) and the water in the actual river (in blue). Below: the Purex plant at Hanford, for the retreatment of radio-active nuclear fuels.

An experimental straight crater of enormous dimensions obtained by the simultaneous explosion of five 1-kton nuclear charges. The peaceful use of nuclear explosions might well be useful in, for example, the mining industry.

become the object of particular interest and investigation: wind, heat in the subsoil and the Sun's rays. As we already know, the windmill was used in the Middle Ages, in both agriculture and industry. Its use spread further and further until the last century, when it started to decline because of the competition from new devices which, at the time, embodied definite advantages. The twentieth century has seen the use of technically highly advanced windmills to produce electricity, but in about 1950 interest in them greatly declined, because at that time they were uneconomic. So much so, in fact, that the number of firms manufacturing windmills to generate electricity had dropped in 1973 to just six (this figure refers to all the non-socialist countries). Today, however, in several countries (notably the United States, Israel and Sweden) scientists and engineers are busy studying ways and means of making the use of *aeolian energy* (the technical term for the wind when applied to energy) once again possible in more favourable conditions. The technique for constructing windmills have already been highly developed. They are especially useful, and are used, wherever it is hard to erect power-lines – in polar and desert regions, for example, at high altitudes, and in isolated farmsteads. Small windmills in these cases are already at an advantage and not

least from the point of view of cost. The use of aeolian energy on a large scale was first tried out during the Second World War in the United States. On the mountain known as Grandpa's Knob in Vermont a large wind-powered generator was installed; it worked for several months, until technical and financial difficulties put an end to the experiment. Today, research programmes are underway, and have in some cases reached a fairly advanced operational stage. In particular, to solve the problem of the storage of wind-produced energy, which is obviously only available on an intermittent basis, either the use of electric batteries has been proposed, or the compression of air in large spaces, or the introduction of the electric energy obtained by aeolian means into a grid or system capable of making up from other sources of energy the short-fall which will occur when there is no wind. Research is also being carried out to establish, quantitatively, the overall amount of energy effectively obtainable by this means.

Another source of energy which gives rise to much hope is the heat trapped in the subsoil. This is what is known as *geothermal energy*, which occurs spontaneously in nature in the form of volcanoes, geysers (spasmodic spurts of boiling water) and hot springs. Geothermal energy has been exploited from

The tidal-powered power station at Rance in France. On the estuary of this river a weir makes it possible to use the tidal levels, which differ by 13.50 metres (44 feet). The power station is so built that it can use both the ebb and flow of the tides.

Left: the Sun. Solar energy is produced by thermo-nuclear reactions which occur in the mass of the star. Top right: the solar furnace at Mont Louis in France. By using solar furnaces it is possible to collect the thermal energy of the Sun. Below right: a Bell solar battery. Solar batteries are fitted in artificial satellites and other space-craft where they supply the energy necessary to operate the tele-transmitting apparatus on board.

earliest times, for a limited number of uses, such as hot baths. From the early twentieth century onwards boiling water and natural steam began to be seriously considered for the production of electricity. The first and most important geothermal power station was the Italian plant at Larderello in Tuscany. The heat from steam and water gushing from the subsoil has been exploited particularly in those countries, like Iceland and New Zealand, where the natural conditions were from the outset particularly favourable, because of the immediate accessibility of such sources. For some years now, however, the number of nations which have started to use geothermal energy has increased considerably.

More complex technical problems arise when the sources of subterranean heat are not immediately available and must be reached by drilling. The heat used is usually that of underground waters at fairly high temperatures, in particular in the form of steam at a certain pressure. It seems probable, however, that the largest reserves of thermal energy for use at some not too immediate future date are those stored in the actual hot rocks. To tap these reserves it will be necessary to cause the subterranean rocks to crack and then fill the cracks with cold water which will then return to the surface boiling, and

thus make use of the heat transported. The principal and as yet unsolved problem is that of managing to make these cracks at an economically viable depth (the temperature of the rocks increases the deeper one goes).

Finally we come to the source of energy that has, to date, been the most important of all both for mankind and for all other life on this planet: the Sun. Solar energy has been exploited from time immemorial: our actual biological energy derives from it, by way of our food; hydraulic energy, the energy contained in coal, petroleum, natural gases and wood, all originate from the Sun. But it is not our intention, here, to discuss the solar energy which is stored spontaneously by nature, chemically in organic fuels, or mechanically in river water. Rather we are concerned with the artificial use of this type of energy. In certain places the heat from the Sun's rays is used, and has been for some years, for heating homes. Wide-scale use was nevertheless generally considered to be uneconomical, until the crisis affecting the traditional sources of energy succeeded in reviving interest in it. This interest is wholly justified when one thinks that this energy source is absolutely non-pollutive and inexhaustible, and that the solar energy which falls in one year on to the Earth's surface is approximately 30,000 times the amount of energy that man uses in a year to

A solar energy laboratory. The use of solar energy on a large scale might well solve mankind's energy problems. But when this form of energy reaches Earth it has a low concentration, and it is precisely this aspect of it which poses major technical difficulties.

Three stages in the assembly of a 'solar house'. Solar energy can be most naturally applied to household heating, where the temperatures required are not too high. A specifically designed system of heat-insulating panels, tanks and other apparatus can, for example, guarantee continuous heating and save large amounts of fuel.

meet his various requirements. The drawback to this form of energy lies in its low concentration. As a consequence, it does not seem suitable to produce electric energy on a large scale. There are experimental centres designed to this end, like the one at Odeillo in France, which consists of nothing less than a solar furnace where the energy from the Sun is transformed into electric energy. But the amount of electricity so produced is in actual practice insufficient to justify the cost of producing it. The only cases where a similar use has a *raison d'être* are those where small amounts of electricity are needed in conditions where it would otherwise be hard to supply it. One such case, for example, is a space-flight. Here the device which enables the astronauts to obtain electricity from the Sun's rays is the photoelectric cell (also known as the photovoltaic or radiation cell).

The type of use, however, which is feasible, and has already been introduced in certain circumstances (for example in Japan, Israel, Australia and Florida) is for domestic heating. Houses heated with solar energy have a special appropriate design, incorporating water-tanks for storing heat, heat-insulating panels, and various other devices. In such cases the low concentration of the Sun's energy is no problem, and in some respects is even an advantage.

One idea which may sound fantastic, but which is nevertheless being looked into, is that of using solar energy to create a sort of 'thermal pump' in the sea, making it possible to produce electric energy, by appropriately varying the temperature of the water.

The best way of storing solar energy still seems to be, however, the most natural way: the production of plants which spontaneously carry on the process of 'photosynthesis', i.e. the transformation of solar energy into energy which we might call biological, and subsequently turn into fuels.

And so after a succession of technological complications and colossal undertakings, our short summary has ended up by making a peculiarly simple suggestion, and this is certainly no coincidence. The problem which we outlined at the start of this chapter, namely whether the path chosen by man to establish his relationship with the natural world has been the best possible one, now looks a little more easy to answer. In fact our admiration for man's major technological achievements cannot prevent us from making the point that the overall situation in which mankind currently finds itself does not seem any better than the situation which existed at the dawn of civilization, given that the damage caused by technological development counterbalances, and perhaps

Left: the control-room of a oil refinery. Top right: control-boards at the Gare de l'Est in Paris. Below: the control-room of a factory which produces steel alloys. These are just some examples of automation: this term applies to a series of methods for analysing and organizing production processes with the aim of achieving a better utilization of the available resources.

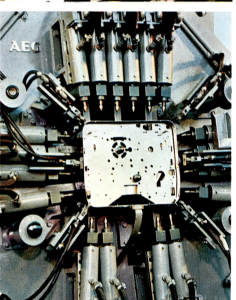

Left: an automatic spot-welding machine. Top right: automatic unloaders on Lake Erie capable of moving weights of 2000 tons. Below: remote-control press-forging.

even cancels out, the advantages brought about by it. It is our view that this has come about because man has steadily drawn too far away from nature. It should be emphasized that this is in no way written in a naïve poetic sense, but rather in a very precise and objective sense. We are, in other words, keen to emphasize the fact that the whole of human, technological creation now operates according to its own specific laws, which no longer link up directly with the conditions in which man embarked on its creation, and which are in fact liable to clash with these conditions. Take the example of pollution. Again, in spite of the progress made, for example, in the area of rationalizing industry and transportation, partly by resorting to automation, the control which man can exercise over this whole complex of activities often seems inadequate. Those massive traffic-jams in the world's great cities are dramatic phenomena, seen in this light, even if we are so used to them that we often fail to realize it. That part of mankind which considers itself privileged – the part, that is, which lives in the developed and industrialized countries – thus becomes more and more conditioned to a way of life which has little or nothing to do with the way of life for which nature made it, if for no other reason than that the pace of technological evolution has shown itself to be much quicker than the pace of

biological evolution. It is thus of paramount importance that man (and not 'man' in the abstract sense, but every single member of the human species) should learn, or re-learn, to focus his attention on himself, not only as a person with certain given psychological features, but also as a living organism, with specific anatomical and physiological characteristics, if he is in some degree to correct and modify the path or paths he has so far taken.

Top left: a windmill with canvas sails. Aeolian energy which has been used for centuries, as we have already shown, has re-emerged today as a result of the present-day energy problems and their possible solutions. Below left: a geyser. Geysers, which are natural jets of boiling water, are used in some places to produce electric energy. Right: the Larderello plant which uses soffiones (or boric acid fumaroles) to feed a power station with a capacity of 300 MW. Opposite: an example of collective human labour. The body of living beings is still more efficient than any man-made machine, and perhaps the real solution to our energy problems lies precisely in a wider and more rational use of biological energy.

THE CONQUESTS OF MEDICINE

Primitive peoples considered sickness to be a divine form of punishment and as a result of this belief medicine was practised, no matter how irrationally, by priests and witch-doctors, in the form of appeasement (of the gods), exorcism and spells. Before very long this primitive form of medicine on a magic-cum-religious basis developed a certain empirical foundation, which emerged from the observation of nature, and had as its mainstays the use of the therapeutic properties of water, heat and large numbers of plants. As century followed century and the great Mediterranean civilizations came and went, the fight against disease and sickness became something of a social commitment and the various types of therapy and cure were codified and handed down through the ages by such things as, for example, the Sumerian documents and the Babylonian codices and manuscripts dating back to 1717–1665 B.C., which were in turn followed by the Egyptian papyrus scrolls of 1500–1200 B.C. Such documents in fact represent the first known medical manuscripts. The earliest Chinese documents refer to surgical techniques, pharmacological expertise and therapeutic practices, all of which were extremely advanced. Indeed, some of them, such as acupuncture, are still used in this modern day and age in exactly the same way as they were in those bygone days. In India the fight against disease was based for the most part on surgery. In the fifth century B.C. the first schools of medicine were established in Greece, but the demoniacal concept of disease still held its ground throughout the ancient world.

Two thousand years of research and discoveries, the most important of which have been made in these last two centuries, have succeeded in pushing the nightmare of diseases considered to be incurable into the back of men's minds, and led to the disappearance of those devastating epidemics of diphtheria, yellow fever and smallpox that used literally to plague the world. There was a time when diphtheria killed off 70 per cent of its victims, and cerebro-spinal meningitis claimed the terrifying figure of 90 per cent of those so afflicted. Tuberculosis and poliomyelitis were no less fearsome diseases, but nowadays both have been almost completely eradicated by antibiotics and vaccinations. For all this, there are still many, many diseases which continue to outwit science. Modern medicine has at its disposal increasingly advanced methods of investigation, and therapeutic methods become more and more effective and long-lasting day by day, with the result that it is possible to predict that by the year 2000 certain forms of tumours will be under control, and it will be possible to cure almost all mental illnesses. All the major triumphs of medicine have been made possible by man's perfect physiological and morphological knowledge of the human body, the functions of which are linked together to form a network of interdependent relationships. This interesting aspect of our organism is best illustrated by the blood, which is without doubt the body's most important means of communication. In the blood's fluid, known as plasma, there are three different types of cells: the red corpuscles, the white corpuscles and the blood platelets. The adult human body contains something like 5–6 litres ($8\frac{1}{2}$–10 pints) of blood which is kept at a temperature of about 37 °C (98.4 °F) by the nervous system. The circulatory system is 93,000 kilometres (58,000 miles) in length, and the blood travels round it at a speed of 450 metres ($\frac{1}{4}$ mile) an hour. The red corpuscles, known as

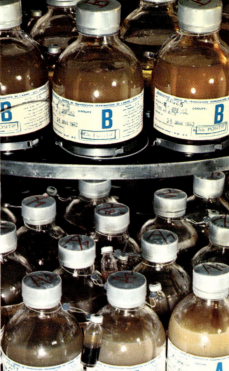

The importance of having perfect chemical and functional knowledge of the blood in our bodies is basic to the health of man. In fact of the countless laboratory analyses carried out each year, 97 diagnoses out of 100 are made on our blood. This illustration shows a blood transfusion centre where blood is taken from donors (top right) and then stored (top left and centre) in the various blood-groups (below left) and used to prepare plasma (below right).

erythrocytes, are round and biconcave. With a diameter of 7–8 1000ths of a millimetre, they are produced in the bone marrow (medulla ossium) at the rate of 150,000 million a minute; they have a life-span of 120 days and eventually die off in the liver and spleen. Each red corpuscle contains 280 million molecules of haemoglobin, a protein-bearing substance with a very complex structure which is capable of collecting the oxygen which is carried to the lungs by inhalation, boosting the cells which in this way derive the energy necessary for all the vital activities of the body, and returning to the lungs almost all the carbon dioxide which is removed by exhalation. On average, haemoglobin lasts 90 days, travels round the body about 134 times an hour, colours the blood red with a pigment known as porphyrin and, when its life-cycle is over, conveys 85 per cent of its iron to the spinal cord (medulla spinalis) which re-uses it to make more haemoglobin.

The white corpuscles, or leukocytes, are quite different from their red counterparts. Whereas the latter number five million per cubic millimetre, the white corpuscles number only 7–8,000. They are produced in the spleen, in the thymus, in the lymph nodes and in the bone marrow at a rate of seven million a minute; they die off after 48 hours and have a diameter of 12–20 1000ths of a

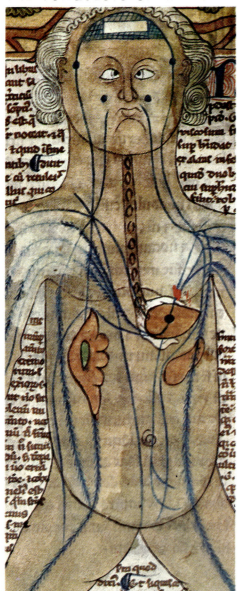

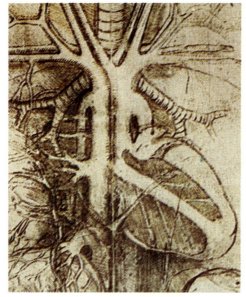

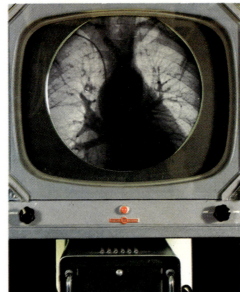

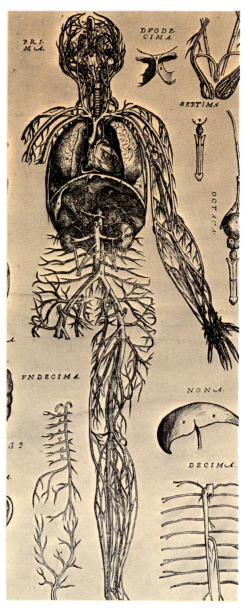

millimetre. These cells, defined by the American haematologist William Holmes Crosby as 'small, irksome cells', aroused little interest in medical circles until a few decades ago. Today they play the leading role not only in haematology but also in a whole new field of biology.

The most startling feature, in a functional sense, of the white corpuscles is in fact their defence of the organism. This defence is managed by something known as diapedesis, which is a mass movement of leukocytes towards the source of bacteria which produce toxins, and by something known as phagocytosis, which is the phase when the bacteria or any other foreign body, alive or otherwise, are encircled and destroyed. The leukocytes are responsible for the initial process of immunization and multiply each time that the body is subjected to specific pathological conditions or conditions of stress. In the case of cigarette smokers, the leukocytes even increase by as much as 30 per cent.

The smallest blood cells are the platelets. These measure 2–3 1000ths of a millimetre in diameter and number 300,000 for every cubic millimetre of blood. They are produced in the bone marrow at a rate of 300 million a minute and their task is to prevent haemorrhages when the blood comes into contact with air, should a blood vessel break, by solidifying it from the liquid state. This

It is a well-known fact that the cardiovascular system has always stimulated a huge amount of interest, from the dawn of mankind up to the present day, as is shown by (left) this oriental miniature from the Bodleian Library in Oxford; Leonardo da Vinci's drawing from his Anatomy Notebooks *(top centre) from the Royal Library at Windsor. Right: a page from* De humani corporis fabrica *by Andreas Vesalius (1514–1564) and lastly, below centre: a cardiovascular X-ray examination, using a television monitor.*

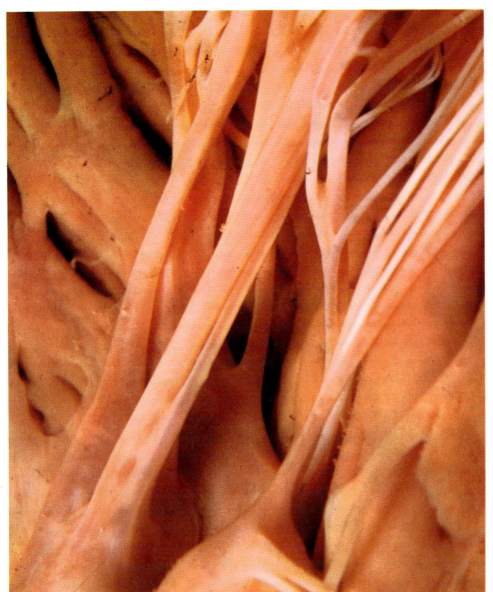

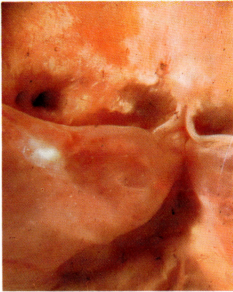

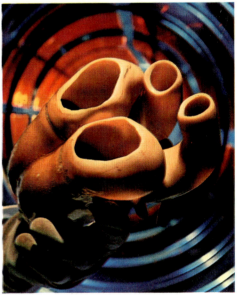

The photographs on this page show, left: the tendon-cords which connect the valves to the inner walls of the heart; and top right: the bicuspid or mitral valve which allows the blood to flow from the left atrium (or auriole) to the left ventricle and which is situated in the atrioventricular foramen. Heart surgery has advanced in leaps and bounds: the photograph below right, in fact shows the silicone atria and ventricles of an artificial heart made by Dr. Willem Kolff of Cleveland.

extraordinary and complex river which flows through our bodies, acting as a perfect and tireless biochemical laboratory, which has inspired mythologies and mysticisms, religions and witchcraft, has also inspired a contemporary poet, Salvatore Quasimodo, to write that: 'Life is a game of blood' (*La vita è un gioco del sangue*).

The bloodstream is fitted with its unflagging motor: the heart. This is a hollow muscle which weighs slightly less than 270 grammes (9.5 ounces), beats, on average, 72 times a minute, and with each beat pumps 160 cubic centimetres (9.76 cubic inches) of blood. The membrane which stretches over it is called the pericardium and the four cavities into which the heart is subdivided have the following names, based on their position: the right atrium and right ventricle, and left atrium and left ventricle. The pressure produced by the contraction of these cavities thrusts the blood around the body, along two quite different routes, one fairly short and the other much longer. In the small circulatory system the blood, which is saturated with carbon dioxide, passes through the right ventricle and the pulmonary artery, releases the carbon dioxide in the lungs and absorbs oxygen, then re-enters the left auricle (or atrium) of the heart via the pulmonary veins. In the large circulatory system, on the other hand, the

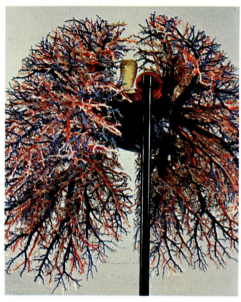

blood leaves the left ventricle, flows into the aorta, round its infinite number of branches, and is pumped from here into every part of the body. Once its nutritive function has been carried out, it passes into the veins and returns to the heart. The inferior vena cava and superior vena cava both flow into the right auricle of the heart, and all the veins in the upper part of the body lead to the upper vena cava, just as all those in the lower part of the body lead to the lower vena cava. An ingenious system of valves forces the blood to flow in the right direction and controls its inward and outward movements. The relaxation of the heart which indicates the admission of blood into the ventricles is known as diastole; the compression movement which empties them is known as systole. In addition to the arteries, which all lead off the heart, and the veins, which all flow into the arteries, there is the network of tiny capillaries, each one of which is 800 times thinner than the aorta; these are so dense and numerous that if laid together they would cover an area of 6300 square metres (68,000 square feet). The cardiac muscle is liable to contract a fairly large number of diseases, but according to Professor Ignacio Chavez, director and founder of the National Institute of Cardiology in Mexico (which is the most important such institute in the world), the most serious cardiac

Inside the lungs the bronchial division (left) creates a ramified network of canals and passages which become denser and denser and more and more numerous and run into the pulmonary alveoli (top right). The two branches of the pulmonary artery (below right) from which the capillaries originate – whose job is to oxygenate the blood – also run into the more deep-set part of the bronchial tubes. This delicately ramified network, which merges to make larger and larger vessels, enables the blood to reach the left atrium of the heart along the same route in the opposite direction.

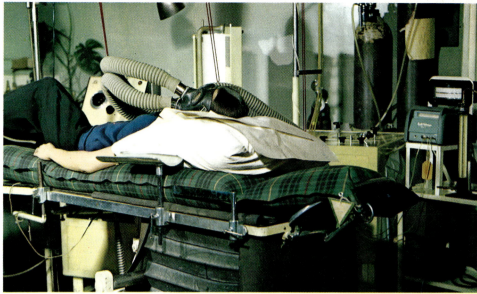

The task of breathing falls to the lungs (top left), and when the alveolar oxygen pressure is insufficient, it is necessary to resort to artificial respiration which can be carried out either manually or by machines (below left). The spirometer (top right) measures the respiratory functioning necessary to cope with man's expenditure of energy, as, for example, when any muscular effort is made (below right). Adults purify their own blood by breathing 16 times a minute; children breathe 22 times a minute.

disorders will be done away with within the next 25 years.

Two organs are of prime importance if the heart is to function properly: these are the lungs. These two sacs, situated beside the heart, weigh little more than one kilogram (2.2 lbs), and are formed by a spongy tissue subdivided into 750 million cells known as alveoli; around these we find the dense network of pulmonary capillaries, 2500 kilometres (1500 miles) in length. The junction between the lungs and the trachea or windpipe is made by the bronchial tubes, which develop a large number of ramifications; and the elasticity of the thoracic or rib cage allows the lungs to dilate. When inhaled, air enters the body through the nose or mouth, passes through the larynx and windpipe, passes through the bronchial tubes into the lungs, and is then appropriated in the alveoli and by means of the capillaries filters the oxygen in the blood. At the same time these same capillaries gather the carbon dioxide which is expelled by exhalation, going in the opposite direction along the same route, back to the nose or mouth.

Another important function carried out by the human body is the responsibility of the digestive system, which handles the digestion and assimilation of our various foods. It measures between 10 and 12 metres (33

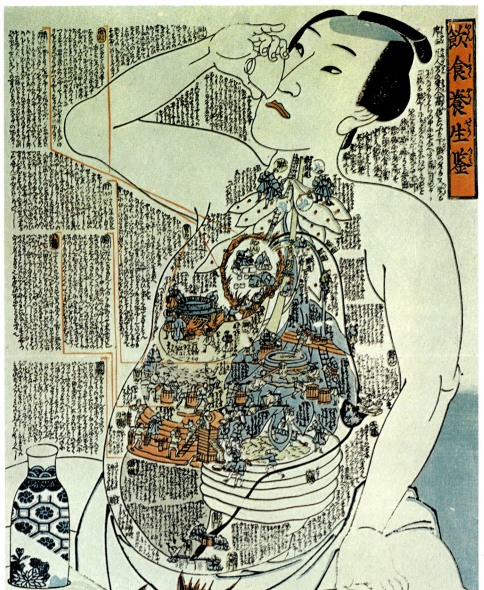

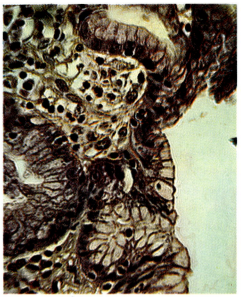

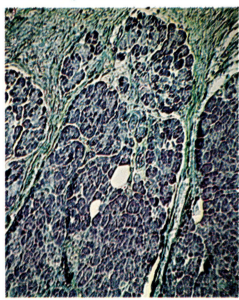

and 39 feet) in length and consists of the mouth, the pharynx, the oesophagus, the stomach, the intestine and the anus. An inestimable number of secretory glands covers the whole inner part of the alimentary canal, while certain large glandular formations, such as the salivary glands, the pancreas and the liver, are situated outside. The initial digestion takes place in the mouth itself, by means of mastication (chewing) and insalivation. After this first mechanical and chemical function has been carried out, the food passes through the pharynx, then through the 27-centimetre long oesophagus, and lastly reaches the stomach through a circular aperture known as the cardia. The stomach is a tough sac with a capacity of 1400 cubic centimetres (85 cubic inches); its walls are lined with glands which secrete the gastric juice, a liquid which can reduce foodstuffs to a mush or pulp which can be easily assimilated; it is connected to the intestine by means of a valve called the pylorus. The intestine is subdivided into three basic parts, the duodenum, the small intestine and the large intestine; here most of the alimentary substances coming from the stomach are absorbed by five million intestinal villi, which are small conical projecting elements with extraordinary absorption capacities covering the inside walls of the intestine. The pancreas is a gland with the dual function of external secretion of the

The Japanese print illustrated on the left, which is attributed to Utagawa Kunisada (1786–1865) describes in rhyme all the functions of the digestive system, in which the pylorus (top right) is one of the obligatory channels taken by food when swallowed, and governs its movement from the stomach to the intestine. Here the food is processed a last time by the liver and the pancreas (below right). Among the pancreatic juices we find trypsin, which digests meat, steapsin, which digests fats and amylopsin which digests starches and sugars.

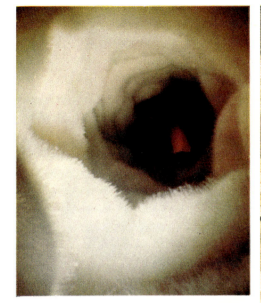

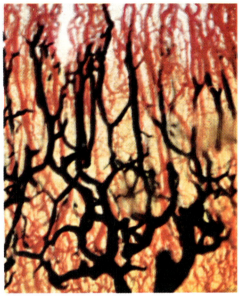

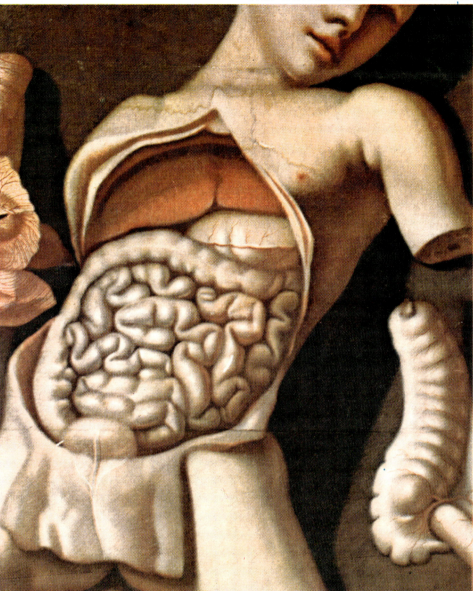

Top left shows a cross section of the small intestine which has an average gauge of three centimetres; inside it is lined with tiny swellings known as villi (below left). Their function is to absorb the various nutritive substances and as a result they are closely connected with the circulatory system. Right: a reproduction of the digestive system situated in the abdominal cavity, as it appeared in the first coloured anatomical print which was published in France in 1741.

pancreatic juice, which helps the stomach and the intestine to process food substances, and internal secretion of a hormone known as insulin, which controls the sugar exchange. A lack of this hormone gives rise to diabetes and in such cases the insulin must be introduced externally to the body. The liver weighs 1500 grammes (3 lbs), is the largest gland in the human body, and continually secretes a fluid necessary for the emulsion of fats. This fluid is called bile and uses the gall-bladder as its reservoir. In addition, the liver regulates the amount of glucose in the blood, manufactures protein, accumulates vitamins and iron, governs the exchange of water, and neutralizes a large number of toxic substances. The liver is liable to contract numerous diseases and disorders, the most serious of which is undoubtedly cirrhosis (of the liver), which was virtually incurable until a few years ago.

Man's perfect physical and psychic inclusion in the environment around him, his capacity of movement, thought, communication and reception, survival and relationships, all these gifts are due to that amazing and inimitable organ, the brain, helped in turn by the nerves, which rely closely on it.

The nervous system is subdivided into two groups of functional apparatus: the autonomic nervous system and the central nervous system. The former

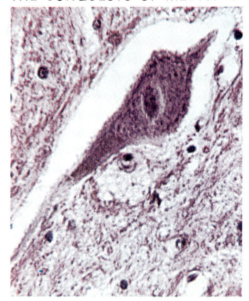

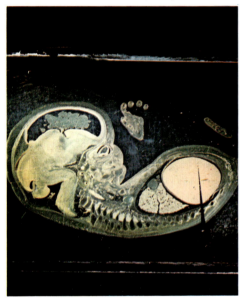

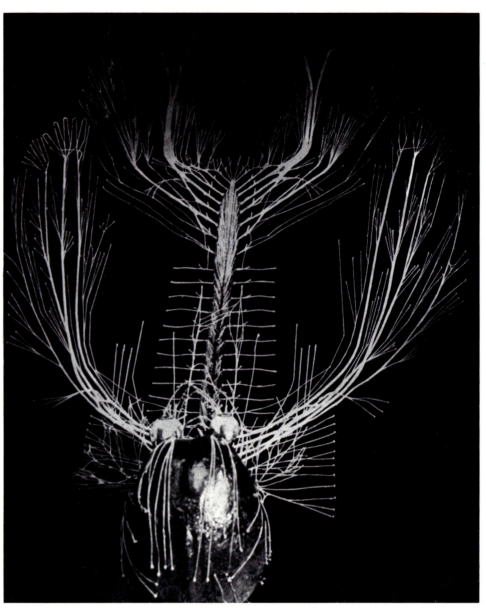

includes all the nerves which reach the organs involved in our automatic, vegetative life (breathing apparatus, circulatory system, etc), while the latter carries out its function in the locomotory apparatus, the senses, and in all those faculties which create a relationship between man and his surroundings. A basic role is played by the spinal cord which is contained in the spinal column and controls 31 pairs of nerves.

This dense communications network is stimulated, inhibited and co-ordinated by the brain. The brain is a small organ weighing less than one kilogram and a half (3 lbs) which is on average the 48th part of the organism. This control centre in charge of all man's physical and mental activities is so complex that there is nothing equivalent to it, even among the most advanced electronic computers. The most recent studies of the brain talk in terms of an organic correlation between the various areas of the brain and it is, for example, surprising to learn that a sound and healthy area has the ability to make up for the possible deficiencies of another area. Spread throughout the different parts of the body there are 500,000 touch receptors, 250,000 receptors for heat and cold, three and a half million pain receptors, all leading to the brain which can handle ten million items of information in the space of ten

The nerve-cell and its cytoplasm (top left) are the essential elements of the entire nervous system and even the human embryo is equipped with them, as shown by the section shown below left. The ramifications of the central and peripheral nervous systems are represented at right. In just a few years brain surgery (see opposite page) has made huge advances, especially since the use of improved methods of investigation such as electro-encephalography and electrocorticography, which can localize disorders in the brain.

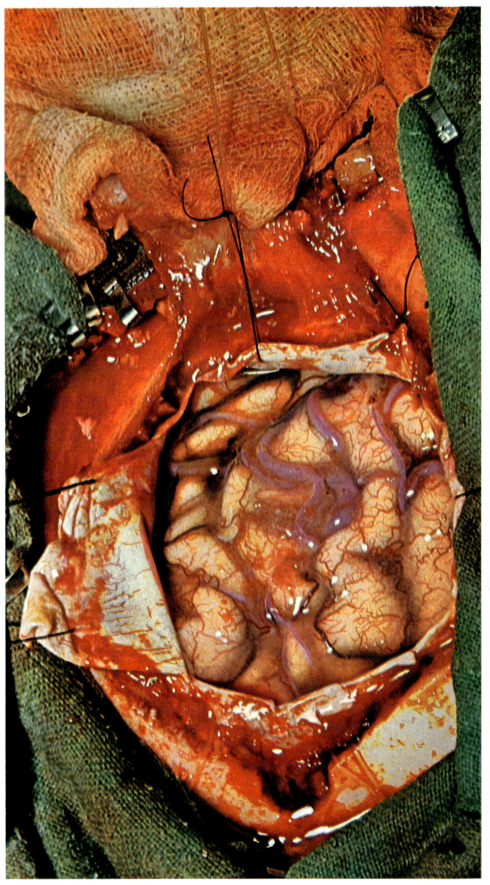
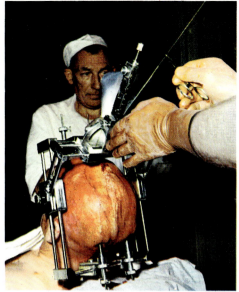

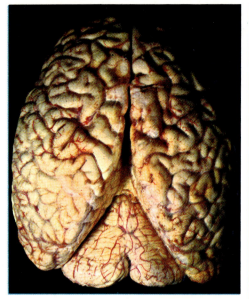

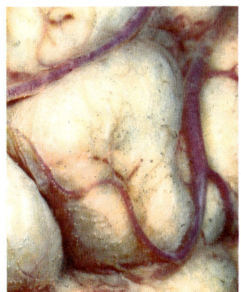

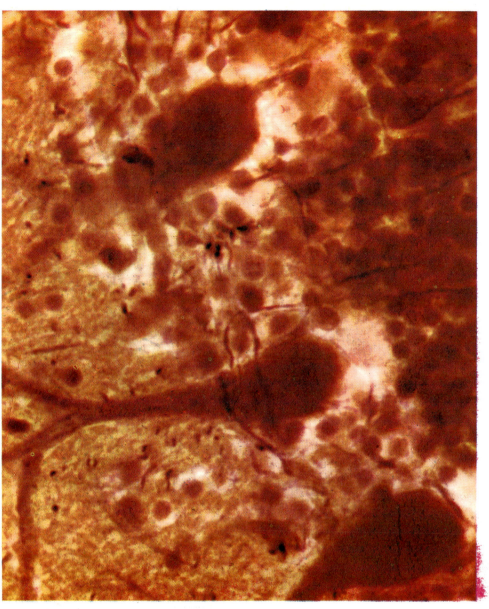

seconds. The brain is responsible for man's memory, reason, consciousness and intelligence, and as such has involved experts and scientists throughout the world in the quest to unravel its horde of still-hidden secrets. Of the latter, mention should be made – on a chronological basis – of the neurologist Robert Woodworth of the University of Columbia, who has carried out important studies dealing with learning times; Professor Pietro Mascherpa of the University of Pavia, who has established times and types for memories, while as the result of another recent scientific discovery it has been established that even the foetus in the amniotic fluid has memory; physiologist Giovanni Moruzzi, Professor at the University of Pisa, who has discovered the centre of waking; the neurosurgeon Professor Wilder Penfield of Montreal University, who has localized the centre of memory in the whole surface of the cerebral cortex; and lastly the neurologist Professor Vladimir Alexandr Negovskij of Moscow, who was the first man to restore to life a man who had been recorded as clinically dead for two hours.

The defences of our organism are countless and take on ever-changing forms, depending on the circumstances; but one of the most interesting of all our defence mechanisms is undoubtedly that triggered off in an allergic

The brain, that perfect control-desk which governs all man's activities, is formed by two cerebral hemispheres (left top), each one of which contains the cerebral cortex traversed by deep cracks which divide the hemisphere into lobes (left below, we see the occipital lobe). The cortex is formed by grey matter (right); it has a thickness of 3–4 mm and contains about 14,000 million nerve cells. The maximum weight of the brain occurs at the age of 20 and then reduces from the age of 35 onwards.

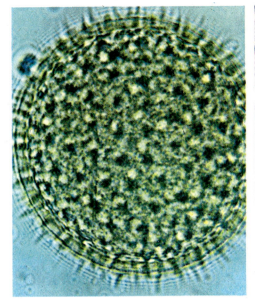

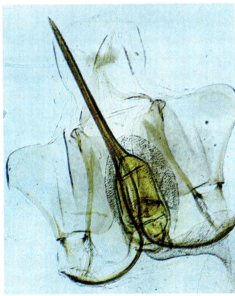

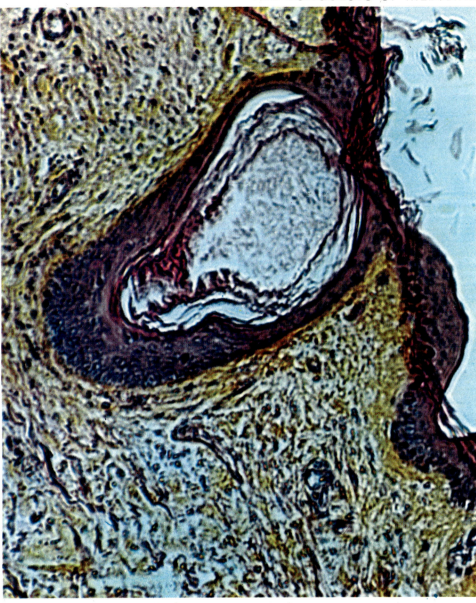

Top left: a strongly magnified pollen granule. Below: the sting of a bee, the poison from which, in sensitized people, acts as an allergen and triggers off the allergy. Right: the histological cross-section of a papule, four hours after the injection of the allergen. The allergic reaction occurs every time that the allergen is introduced into the body via the respiratory system, gastric system or parenteral system, producing defensive substances which react with the antigen.

individual when a foreign and harmful substance – or allergen – enters some part of his or her body. The substance enters the bloodstream and immediately becomes an antigen – in other words something that stimulates certain proteins known as antibodies which have the capacity to react against the antigen which has given rise to their production. The antibodies neutralize the antigen, and in some cases destroy it, but the allergic individual is still sensitized, i.e. liable to further attacks from the allergen which will recur, not in the bloodstream but via the breathing system, the blood vessels, the stomach and the intestine. The best-known allergens are pollen, shellfish, strawberries and a large number of drugs, and the difficulty resides precisely in being able to single out the substance to which the patient is allergic. Once isolated, the allergen is injected in larger and larger amounts until the patient has developed a tolerance to it. Allergies affect ten per cent of the world population, but until a few years ago the causes that trigger them off and their reactive mechanisms were unknown; today they are under control thanks to effective antidotes.

Our body is also threatened by numerous other agents whose aggressivity is far more developed and dangerous than that of the various allergens; these agents often carry diseases which were considered to be incurable until a short

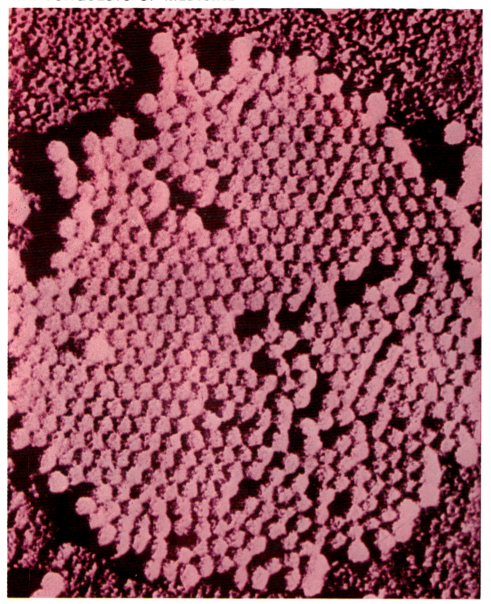

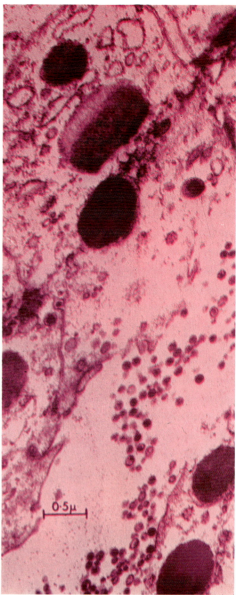

time ago; they fall into the two major categories of viruses and germs. The viruses are too small to be examined under ordinary microscopes, with dimensions ranging between 250 and ten millimicrons; it is only with the electronic microscope that they can be singled out, and they are so small that they even manage to make their way through the porcelain filters and the ultrafilters made of colloidal membranes used in bacteriological experiments. The main feature of the viruses is the fact that they live and reproduce inside the cell and as a result not only elude attacks from the antibodies circulating in the blood but even manage to outwit antibiotics and sulpha drugs (or sulphonamides). They only become vulnerable when they leave a cell which has been destroyed and move off in search of a new healthy cell; the viruses are responsible for many serious diseases and illnesses such as small-pox, yellow-fever, poliomyelitis and influenza, which usually occurs in benign forms, but does in some instances give rise to very high mortality rates, as happened in 1918 when there were 21 million victims. The germs, also known as bacteria or microbes, are single-celled organisms, and depending on the morphology in each case are subdivided into cocci, bacilli, vibrions and spirilli. They enter the body through wounds, the breathing system and in food. Most of them release

Left: leukaemia virus. Right: poliomyelitis virus. Virology, which is a branch of biology, is a modern science which was not developed until the invention of the electron microscope. At the end of the last century, when the appropriate means of research were not available, viruses were confused with all the other micro-organisms; it was only in about 1930 that the first experimental methodologies for research into and study of viruses were set in motion. In 1944, after the discovery of certain properties contained in viruses, the door to viral genetics was opened.

Top left and centre: the French chemist and biologist Louis Pasteur at work in his laboratory. Below: a phial containing rabbit medulla (or marrow) infected with rabies, to be used for the preparation of the appropriate vaccine. Top right: instruments used by Pasteur for measuring the microbic density in the air in Paris; and below: a modern pasteurization plant. Pasteur's discoveries laid the foundations for microbiology and preventative hygiene.

toxic substances which are harmful to man's health, and they behave like antigens, in other words causing antibodies to appear in the bloodstream. Germs are responsible for some extremely serious diseases, such as tuberculosis, typhus fever, tetanus and cholera. In modern medicine bacteriological examination is one of the most important diagnostic methods; it is used to establish the nature of the virus or germ responsible for a given disease, and of all the laboratory tests made in connection with infectious pathology it is undoubtedly the most reliable, thanks to the progress made in the field of microbiology, a discipline founded by Louis Pasteur.

On 5 November 1860, at the Académie des Sciences in Paris, this French scientist read his unforgettable paper on the vital capacities of yeasts in the fermentation of wine, and by isolating the micro-organism responsible for this alcoholic fermentation (defined by him as 'life without air' because oxygen plays no part in this particular fermentation process), the door to micro-biology had been opened. Having demonstrated the existence of microbes, the discoveries which then followed about certain diseases affecting animals and men (from the anthrax bacteria to chicken cholera and swine erysipelas, from osteomyelitis to puerperal infection and the rabies virus, with their respective

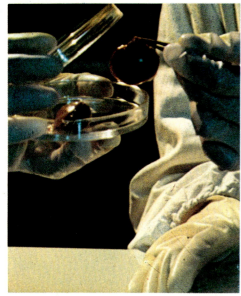

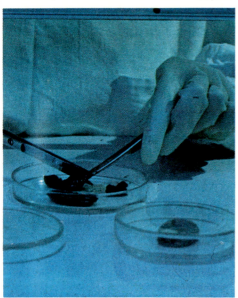

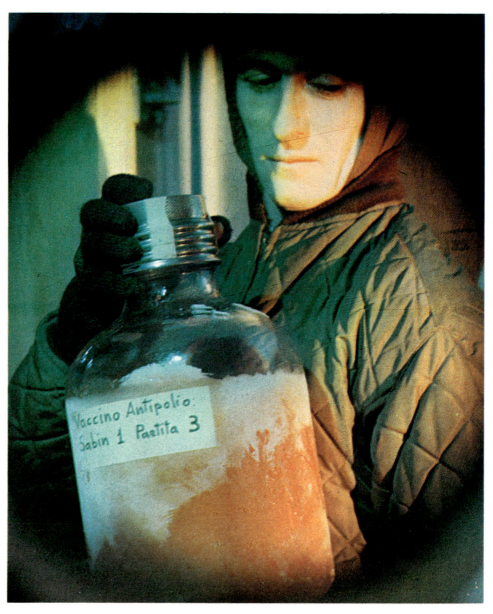

vaccines) simply confirmed the theory of contagion, and gave birth to the modern experimental studies dealing with the microbic genesis of infectious diseases. The identification of the agents responsible for these diseases, together with increasing knowledge about the reactive capacities of the organism in the face of attacks from microbes, inevitably directed research to the prophylactic and curative field of serums and vaccines, those vital elements which stimulate the body's natural defensive mechanisms. Serum, which is taken from the plasma, from which it keeps certain globulins filled with antibodies which are vital for the defence mechanism of the organism, is taken from people convalescing from infectious diseases; it is then preserved and used on people suffering from that disease, or on people who appear likely to fall victims of it. Serums can also be obtained from immunized animals, as was shown at the turn of the century by the Polish scientist Paul Ehrlich who was the first person to draw up an assessment criterion for the anti-toxic strength of animal serum.

Vaccine, on the other hand, is a preparation, the base of which is made up of dead or weakened germs; it is called autogenous when it is prepared with the actual germs of the patient, and heterogenous when the germs come from

Left, top and below: virus cultures. Unlike bacteria which can be developed in ordinary culture media, viruses need living cells or organisms, such as chicken embryos or, as in this case, the kidney cells of a monkey. Right: preparing vaccine doses, which will then be used for inoculating people (prophylactic vaccination) to prevent the outbreak of infectious diseases, and for vaccine therapy, i.e. the cure of such diseases. The anti-poliomyelitis vaccine was discovered by Albert Sabin in 1953.

Left, top and below: removing the venom from an immobilized viper which has been forced to open its mouth wide (top centre). The venom which runs through the dental canaliculi (below centre) is collected in a crystallization vessel. Top and below right: crystallizations of white and yellow venom. Antivenom serum is an antidote, in other words a substance which can neutralize the harmful effects of another substance which has been previously introduced into the organism. In this case the venom acts as an antigen.

somewhere else. The technique of vaccination, which has been known about since 1789, when the famous English physician Edward Jenner immunized a child from smallpox by inoculating him with a drop of pus taken from a cow suffering from cowpox, was taken up again a century later by Louis Pasteur, and is now very widespread as a practice throughout the world. It has warded off, and in some cases even done away with, those great epidemics which used to claim millions of human lives. In theory there is no longer a single infectious disease in existence for which it is now impossible to find the specific vaccine. The last such disease, in terms of time, to have its vaccine is leprosy. The vaccine was discovered in 1973 in Norway, homeland of the scientist Armauer Hansen who, precisely a century earlier, had isolated the bacillus for this terrible disease. Immunology has thus come about as a derivation of microbiology and for a long time was restricted to studying the defence mechanisms of the body against attacks from bacteria; today, this interdisciplinary science, which has been developed with the help of genetics, biology, medicine and mathematics, is deeply engaged in trying to resolve the problem of the biological process of life, and in the words of the Australian immunologist Gus Nossal, 'the second golden age of immunology' started just a few years back.

According to the Milanese pathologist Professor Carlo Zanussi 'the development of immunology is so rapid that clinical medicine can't keep pace with it', and this means that the discovery of certain immunological phenomena, such as the auto-immunizing diseases, immunological tolerance, the phenomenon of rejection in transplants and the role of immunizing mechanisms in the development and regression of malignant tumours is not as yet accompanied by any definitive clinical practice.

In 1939 the therapeutic research into infectious diseases underwent a radical change: the English bacteriologist Alexander Fleming managed to isolate, from a mould that was capable of inhibiting the development of staphylococci, a substance which he called penicillin. This, produced on an industrial scale, has shown itself to be the most effective medicament as yet known to man. As a result of Fleming's discovery the age of the antibiotic was born, and for the first time in the history of medicine extremely serious diseases such as pneumonia, meningitis and empyeme (or purulent pleurisy) were cured so swiftly that seeing was almost believing. In 1944 the American microbiologist Selman Abraham Waksman discovered streptomycin, which is especially effective against tuberculosis and urinary infections; four years later, in 1948,

Top left: a specimen of mould on penicillin culture and (below) the development of the same. Top right: Penicillum notatum, reproduced in pure culture. Below: streptomyces. Certain specific chemical substances produced by living micro-organisms; among them bacteria halt growth, or even destroy many micro-organisms, among them bacteria which cause infectious diseases. The discovery was made in 1929 by Alexander Fleming, who finally developed penicillin in 1942.

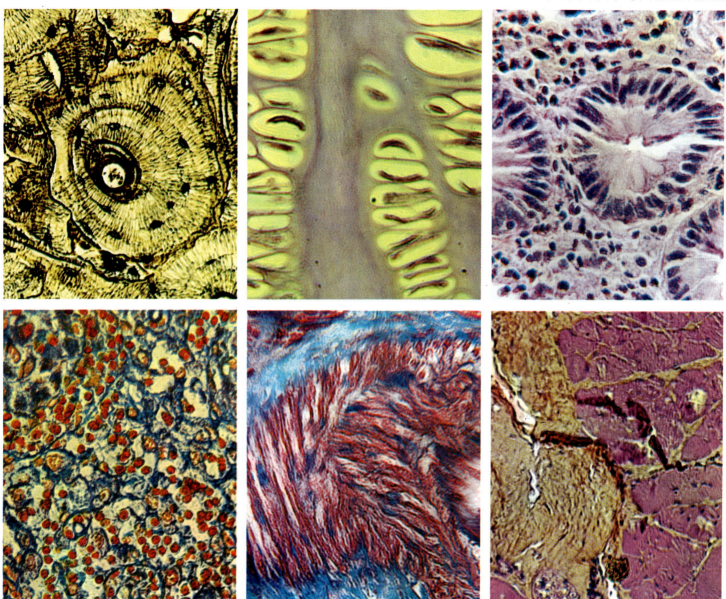

Histology is the study of the internal structure of tissues, and is a branch of anatomy which came into being with the discovery of the microscope, although its origins can be traced back to research carried out on tissues by the famous anatomist from Emilia, Marcello Malpighi. The photographs on this page show, top left: the structure of bone tissue in a microphotograph; below: the tissue of a lymph node; top centre: cross-section of cartilage; below: a prepared specimen of nerve tissue; top right: a cutaneous specimen; below: a section of striated muscle.

chlorampheny was discovered, an astounding weapon against typhus fever; and this was followed by the discovery of tetracyclin. Antibiotics stop bacterial reproduction and propagation because they act directly on the cell wall which isolates and protects the bacterium inside the cell and defends it from the outside world. It can be deduced from all this that the greater our knowledge of a disease, from the causes which give rise to it to its natural conclusion, the greater the benefit to the therapeutic treatment of it. This is precisely why that somewhat obscure disease, cancer, which oncologists define as '100 different diseases with a thousand different causes', and which continues to elude clinical control and outwit experimental pathology, is still without an effective cure.

This extremely serious disease develops from the epithelial tissue when one or more cells degenerate and multiply rapidly to the detriment of the normal cells, increasing in size and losing their cohesive capacity; but the causes of this abnormal behaviour are still unknown. The only thing that is known for certain is the viral origin of certain tumours, and this encourages many scientists to believe that the degenerative process of the neoplastic cells is in some way attributable to a slowed-down immunological control by the organism. Another theory, known as the 'oncogenic' theory, attributes the

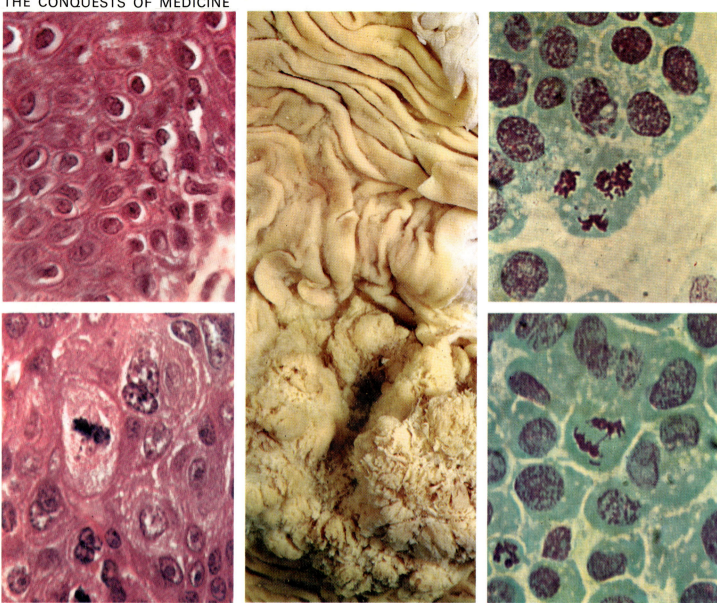

causes to a stage way back on the evolutionary ladder, presupposing that an oncogenic virus entered the cell of a remote ancestor and that via this cell handed down to successive generations its specific genetic data; but no matter how plausible they may sound, they are still no more than theories. It has been ascertained, on the other hand, that many substances encourage the formation of tumours, the best-known of which (the so-called 'environmental carcinogens') are absorbed by man along with the 2000 litres (440 gallons) of air necessary for his daily breathing, and it is these carcinogens which cause the atmospheric pollution: industrial waste and the waste from heating plants, cigarette smoke and the exhaust fumes from vehicles. Professor Bruce Ames of the University of California has discovered that benzopyrene, one of the hydrocarbons which every industrialized country releases into the air every year in hundreds of tons, causes a radical change to occur to the cells of the body when metabolically activated in the organism; and this mutation identifies itself by the appearance of those characteristic neoplasiae. Likewise the nature of leukaemia still remains somewhat shrouded in mystery and as a result the therapeutic help available is thoroughly inadequate. This terrible disease – which is also called 'cancer of the blood' – shows a dizzy increase in

Top left: sections of normal epidermic cells. Below: sections of cancerous epidermic cells. Centre: carcinoma of the stomach. In the lower part of the stomach one can see the tissue which has been invaded by the tumour, while in the uppermost part the increased number of folds indicates the way in which the disease is spreading. Top and below right: Ehrlich's adeno-carcinoma cells. The term adeno-carcinoma applies to tumours which affect the glandular tissue and spread as far as the bottom of the lymph vessels.

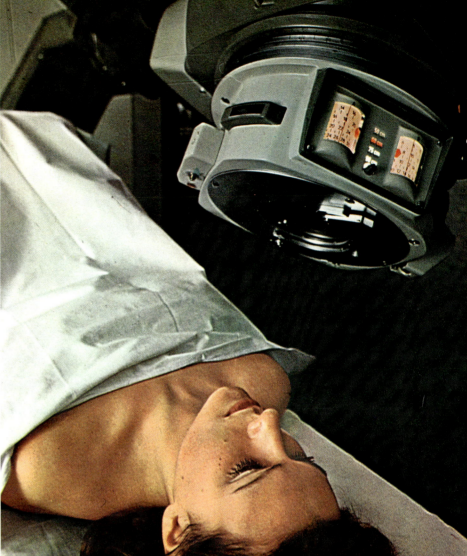

More than 60 per cent of the cobalt used in the world comes from the mines in Katanga. Top left: open-cast mining of cobalt-bearing minerals at Kamoto in the Congo. Below: substances exposed to the action of gamma rays in a chemical laboratory. Right: a patient undergoing gamma ray treatment. There are numbers of ways in which the artificial isotope cobalt-60 can be applied as a source of gamma rays. In medicine cobalt-60 is used to cure tumours by means of cobalt bombardment.

the number of conspicuously immature leukocytes which cause an irreversible upheaval of the blood and its functions.

Modern medicine has not just been limited to perfecting methods of investigation and changing therapeutic practices. On the contrary, it has made it a common practice to use instruments capable of standing in for organs which are no longer capable of carrying out their functions. This is the case with the artificial kidney, an ingenious piece of equipment which carries out the kidney's purifying function. The kidneys, which are perfectly designed chemical regulators of the body, contain slightly less than two million functional units, the nephrons, which, by means of the urine, are capable, in 24 hours, of ridding more than 1400 litres (330 gallons) of circulating blood of their impurities, by filtering them 300 times. The overall washing of the blood by the kidneys is called haemodialysis, and the process comes to a halt when these precious filters in our bodies contract some disorder or other. In the case of kidney blockage, for example, the kidneys stop secreting urine and as a result the body is very swiftly poisoned by the very harmful waste produced by it. The first rudimentary artificial kidney was experimented with – with positive results – during the Second World War, but it is only recently that this valuable

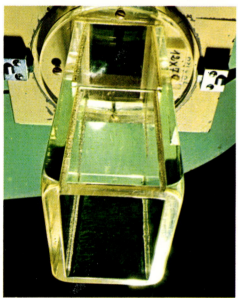

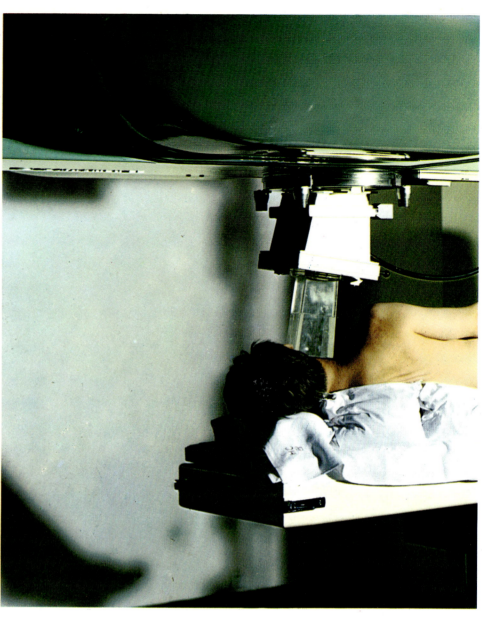

machine has made the necessary advances for it to function with extreme reliability. At the ends it has a double catheter, the first to be inserted in an artery to suck in waste-filled blood, and the second to be inserted in a vein to restore the blood, once purified, to the bloodstream; there is then the central section consisting of a tubular spiral acting as filter, which is immersed in a dialyzing solution of electrolytes, water and glucose.

Another area in which technical development combines in a coherent way with scientific application is heart surgery. The invention of the heart-to-lung machine has made it possible to suspend and substitute the activity of the heart for a period of four to five hours. In this way, and once the circulation outside the body has been safely set in motion, the business of operating on the motionless and bloodless heart is made considerably simpler. The pacemaker represents a further contribution by technology to science and consists in an electronic device, no larger than a matchbox, which contains a battery-driven energy source and a generator which generates electric impulses. An electrode joins the pacemaker to the heart and the whole piece of equipment is situated in the thoracic cavity.

Its function is to encourage the heart to contract when, for pathological

Betatron is an electron accelerator used in radiotherapy (top and below left), because of the effectiveness of its gamma rays. Right: the complex apparatus used for radiation photographed while working. The frequency with which betatron is used in medicine for therapeutic purposes is due to the success-rate shown by it in the cure of malignant tumours. The destructive action of its rays works much more vigorously on tumourous cells than on other cells.

The acute and most dangerous forms of leukaemia are divided into myeloid and lymphatic forms. Top left: bone marrow smear of a person suffering from acute leukaemia. Below: blood smear from a person with myeloid leukaemia. Top right: removing leukaemic cells with a laboratory instrument known as a pipette. Centre: concentration of the virus isolated in leukaemic blood in a test-tube. Below right: the final stage in isolating the virus by counting the number of viral particles.

reasons, the heart is no longer capable of maintaining a normal rhythm. In France in 1970 a device was perfected which can replace the short-lived batteries. This involves a generator which is powered by plutonium 238, which has the dual advantage of lasting an extremely long period of time and being extremely small (100 milligrammes of plutonium are about the size of a chick-pea). One of the greatest victories scored by heart surgery is the first heart transplant carried out on 2 December 1967 in Cape Town by the surgeon Christian Barnard; this operation has been followed by a further 200 heart transplants in every corner of the world – up to 1975. In 1975, in Houston, the American surgeon Denton Cooley carried out the first installation of an artificial heart, and in 1973 the Swedish Professor Viking Bjork, who is considered to be one of the founding fathers of heart surgery, invented a brilliant system for replacing the valves of the heart with special prostheses.

Among the host of specialized clinical fields in modern medicine, one of the most interesting is undoubtedly perinatology. This new discipline, which owes much of its existence to the contributions of obstetrics and gynaecology, studies diseases and illnesses which affect the foetus, the mother, and the newborn infant, as well as the appropriate preventive and corrective measures

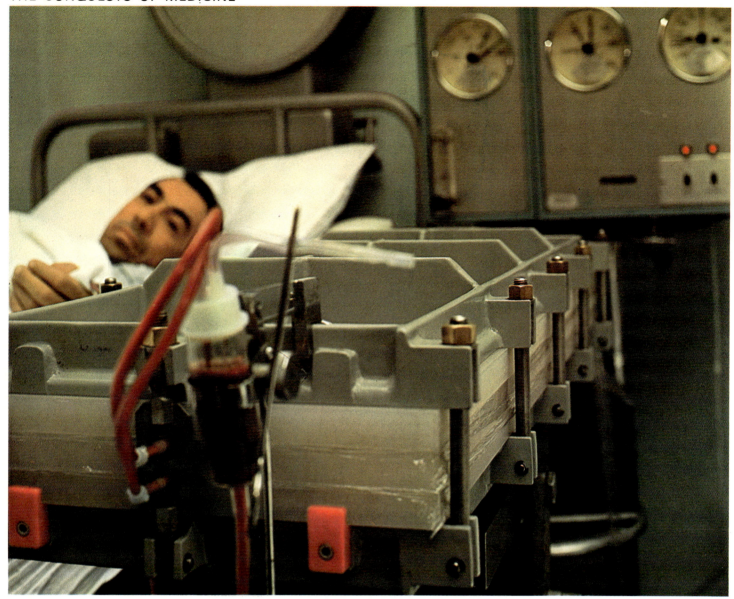

to be taken. A recent survey taken in England has shown that in cases of perinatal death, chromosomic anomalies (defects in the molecules, in other words, inside the cell which contains the various genetic data) have accounted for an incidence of 7.2 per cent as opposed to the figure of 0.71 per cent found for infants alive at birth. And if we add to this cause the various alimentary, toxic, infectious and radiation-connected causes, not to mention the hereditary diseases – which number 2185 – it is not hard to see that many diseases originate in the very first weeks of the life of the foetus, or even in the first moments after the foetus is conceived. Amniocentesis is the examination of a certain amount of amniotic fluid which is rich in foetal cells, and is the most reliable kind of examination for establishing the appropriate treatment if there is some sort of perinatal risk.

Cheek by jowl with orthodox scientific medicine, based as it is on general theories which have been proven by experiment, we find an alternative, heterodox kind of medicine which has managed to find official recognition in certain instances. Acupuncture, homeopathy, chirotherapy and phytotherapy (the use of vegetable drugs) are the major therapeutic techniques which have suffered the critical eye of traditional medicine for some years now.

The artificial kidney can stand in for the kidney's functions during surgery or in the event of renal insufficiency. This photograph shows haemodialysis treatment with Kul's kidney. In the foreground we can see the machine, and in the background the monitor which ensures that the machine will work perfectly; on the left we see the patient, lying on the 'weighing' bed. This particular type of bed is vital for checking any variations in the weight of the patient, which indicate the amount of urine being discharged.

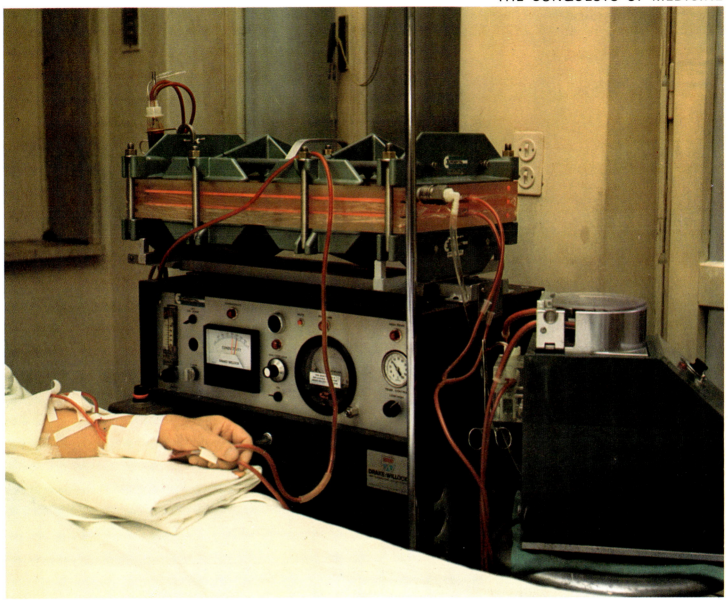

In recent years the various types of artificial kidney have increased, but the basic system of purifying the renal waste has remained the same. In this photograph we can see Drake Willock's artificial kidney, a portable device which has been especially designed for dialysis treatment at home. The dialytic bath is produced as required and guaranteed by an ingenious system of proportional pumps. Before the artificial kidney was invented, patients suffering from renal insufficiency were doomed to die.

Acupuncture has been practised in China for some 2000 years; it was swiftly included in official Chinese medicine and as early as the year 265 B.C. the Chinese physician Huang Fu Mi described this therapeutic system in a treatise entitled: *An Introduction to Acupuncture*. The technique consists of piercing certain critical points of the skin with very thin needles. These points, it is claimed, correspond to various organs; they are all interconnected by longitudinal lines known as meridians. According to the traditional Chinese theories, the rotation and manual vibration of the needles is meant to stimulate the circulation of an advantageous energy from one organ to another and, if required, to act as an effective analgesic. Although acupuncture has found its way into several European universities, opinions among scholars and researchers continue to be divided on the issue. Those who reject it maintain that the psychological element plays a decisive role in the practice of acupuncture and consider it significant that only adults (and of these 90 per cent of the Chinese patients who undergo acupuncture treatment) manage to obtain positive results, whereas children do not. Other voices, on the contrary, continue to support an eclectic form of medicine which is open to examination and experimentation by every kind of empirical means.

Homeopathy came into being in Leipzig at the end of the eighteenth century as a result of observations made by a doctor called Samuel Friedrich Hahnemann, who had remarked that if cinchona was used as an antipyretic it could actually cause fever. After many an experiment he therefore became convinced that diseases and illnesses could only be cured by remedies which were capable of producing similar diseases and illnesses, and he summed up this concept with a now famous Latin expression: *Simila similibus curantur*. Homeopathy still enjoys a certain following and although its basic principle has not been scientifically proven, it is used in certain immunization techniques, such as serum therapy and vaccine therapy.

Chirotherapy is more recent. In 1895 at Davenport, when the druggist D. D. Palmer managed to cure a customer who had been suffering from pains in the back for seventeen years, he did so by manipulating his spinal column. Like all other heterodox practices, chirotherapy – which etymologically means 'manual treatment' – has also been regarded with scepticism by official medical circles, but it has been established that excellent results have been obtained by it in cases where orthopaedics have been unsuccessful.

The most ancient form of heterodox medicine is without doubt phyto-

This photograph shows a detail of Kul's kidney. The principle underlying the way the artificial kidney works is extremely straightforward. A tube is inserted into the bloodstream between an artery in the leg and a vein in the same leg and for a period of time ranging from six to twelve hours, it replaces the kidney's task of drainage. In the meantime the blood undergoes purification treatment by being passed through a special bath in which all the waste is discharged. Many patients have to undergo haemodialysis at least once a fortnight.

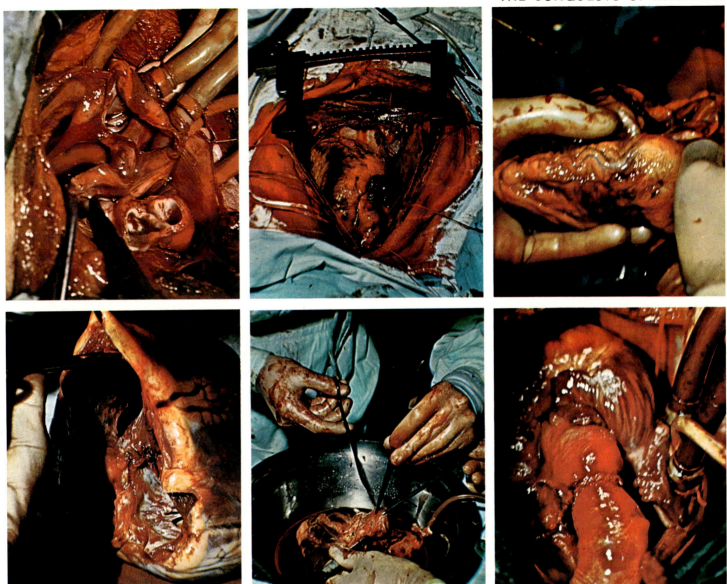

The photographs show a heart transplant in its crucial stage. Top left: detaching the ventricles from the auricles. Below: the heart that has been removed from the donor. When it has been completely removed from the patient's body (top and below centre) the circulation is restored in the new cavity in which the heart has been inserted (top and below right) and the cardiac muscle starts producing its spontaneous pulsations once again. On the following page: a specialized team engaged in a delicate open-heart operation.

therapy: curing illnesses and disease by means of plants. This medical practice probably came into being when man started to observe the behaviour and habits of animals. In fact some animals abide by specific rules governing their alimentary hygiene, such as dogs and cats, for example, which manage to avoid suffering from intestinal constipation by nibbling certain laxative plants, to which they are led by some extraordinary intuition. Some phytotherapeutic practices in use among primitive peoples still exist in this modern day and age; they have been handed down from generation to generation, and the so-called 'medicine-bag', filled with health-restoring herbs, which every member of the various North American Indian tribes keeps as a talisman is the very same bag that the nineteenth-century warrior would place in his knapsack before setting off. For the past few years modern medicine has been re-assessing phytotherapy, and at the moment it is enjoying one of its popular revivals.

For more than 50 years certain results obtained by clinical analyses and laboratory tests have only been able to be confirmed by radiography, a diagnostic technique which uses X-rays to obtain the negative of the internal organs being examined. The application of X-rays, which were discovered on 23 January 1896 by the German physicist Wilhelm Röntgen, revolutionized the

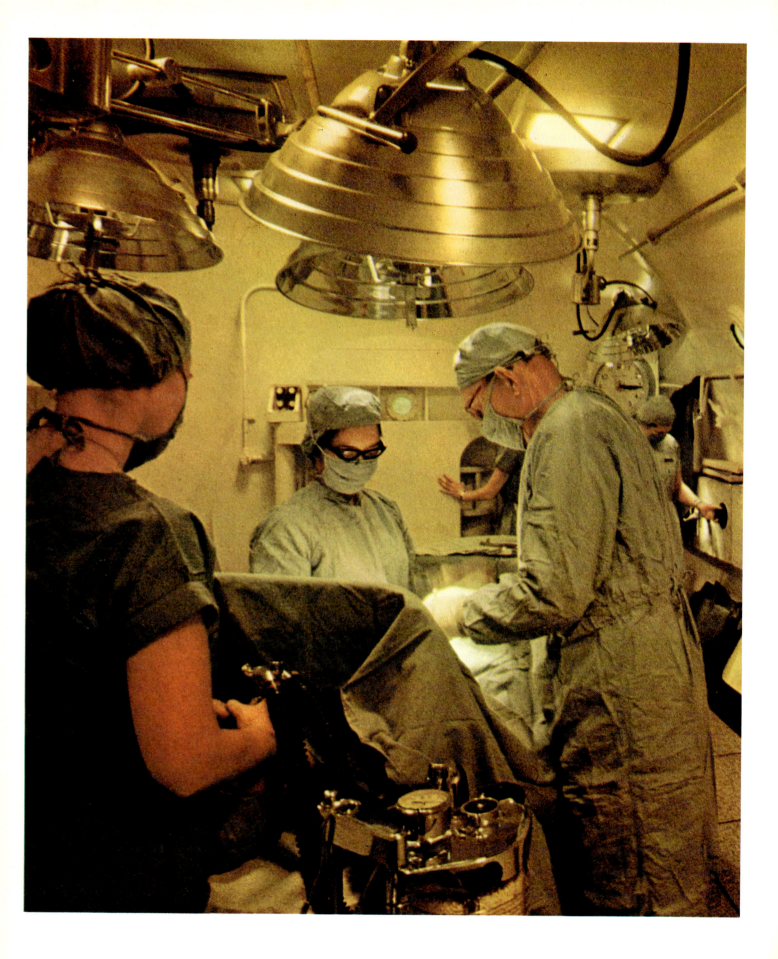

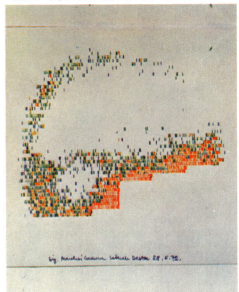

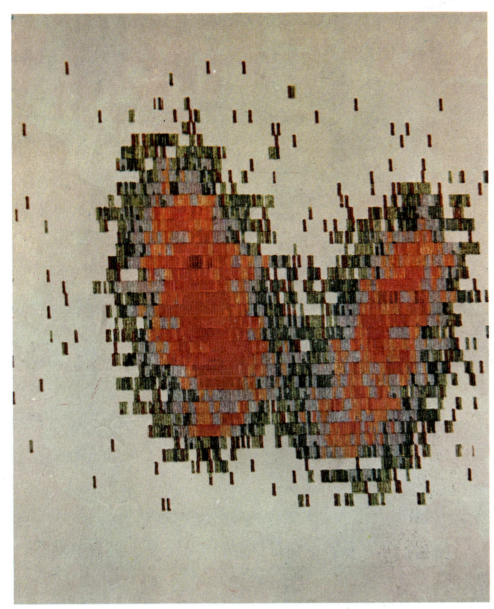

Scintigraphs of the thyroid (left) and of the brain (right). The scintigram is one of the most far-reaching developments for medical diagnostic techniques and, in particular, for the surgeon about to carry out an operation, the accurate map of the diseased organ is an indispensable guide. This modern form of exploration by means of radioisotopes is carried out on various organs in the human body, such as the liver, the brain or the thyroid. The diagnostic technique for each one of these is identical. Twenty-four hours after the administration of the radioactive iodine it is possible to study the cartogram with a special counter.

entire methodological system hitherto used on an empirical basis in the investigation of diseases, and has made a crucial contribution to surgery in particular. This band of electromagnetic rays, capable of exposing photographic films through the organs encountered on the way, soon became one of the most important techniques in modern medicine, and one which we might even call irreplaceable because of the dual possibility inherent in it of both confirming clinical and laboratory examinations and tests, and making them exhaustive and definite. X-rays have enjoyed wide popularity in the past and they are still widely used, but for the last 25 years medical interest has switched somewhat to radioactive isotopes, i.e. those elements produced artificially by means of a nuclear reaction. Atomic piles are the major source of neutrons (particles with no electric charge which are part of the nucleus of every chemical element and have formidable capacities for penetrating matter) which, by bombarding the nucleus of specific stable elements, make it possible to produce radio-isotopes which can be used for biological and medical purposes.

In medicine the use of radio-isotopes for diagnostic purposes is called scintigraphy and is based on the principle that these molecules, when

introduced into the organism, are able to remain visible during their entire journey. The emission of gamma rays is picked up by a device called a scintigraph, the technical features of which are, essentially, such that it can record the various gradations of the radioactivity emitted by the organic radioactive source; each of these emissions is accompanied by a small stroke traced on a sheet of paper. In this way the surface of the organ under analysis is outlined stroke by stroke and depending on the intensity of the signal it is possible to identify quite clearly the healthy parts and the diseased parts of the organ. For some years scintigram readings have been made considerably easier by the development of the coloured scintigraph; here, when a strongly radioactive area is encountered it is shown not only by a denser mass of strokes, but also by the colouration which tends to become bright red.

As well as carrying out the normal topographic and diagnostic functions (presence of tumours, obstruction of blood vessels, aneurism of the cerebral arteries etc.) the cerebral scintigram can establish the amount of blood-flow watering the brain, in time-units. This is an extremely crucial element which makes it possible to take the appropriate steps in good time in the event of any insufficiency, and it is such insufficiencies which, in the long term, can cause

The preparation of homeopathic medicines in the Nelson Institute in London. Homeopathy does not explain the origin and nature of diseases, faithful to its founder, Hahnemann, who defined all forms of sickness as 'unfathomable accidents of nature', but concerns itself exclusively with their cure. This unorthodox form of medicine, which does not claim to have any scientific bases, crops up in many areas of official medicine, such as serum therapy, vaccine therapy, autohaemotherapy and in all forms of specific desensitization.

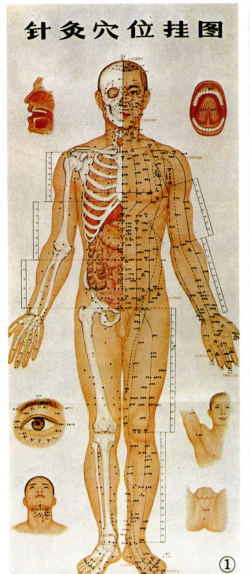

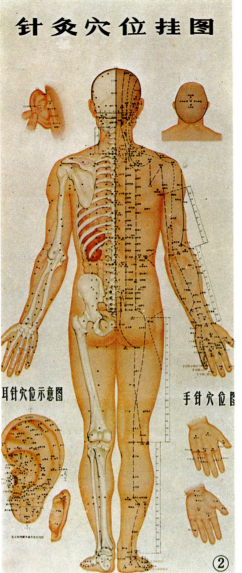

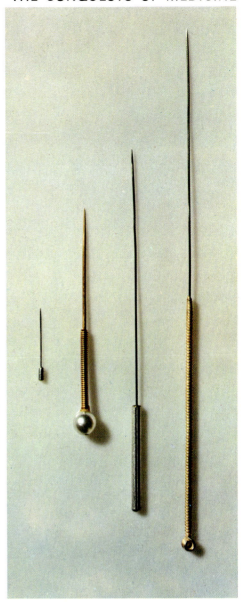

Left and centre: two Chinese posters advertising the practice of acupuncture. Right: a sample of needles for the various therapeutic operations. In China, the method of acupuncture has been practised for thousands of years and is applied together with the more evolved therapeutic techniques used in modern medicine. Based on the dogma of positive (yang) and negative (yin) principles, which govern man's biological equilibrium, acupuncture has also made its way in modern medical circles in the West and is being constantly examined and tested.

usually irreversible damage to the entire nervous system.

In 1713, after a long period during which he lived in close contact with factory workers and farm labourers, Dr Bernardino Ramazzini, a native of Emilia, laid the foundations of professional hygiene by publishing a treatise entitled *De morbis artificum*; in it he described the causes and effects of diseases contracted at work.

Occupational diseases affect workers involved in the handling and treatment of certain substances which are widely used in industry, such as lead, which causes a serious form of poisoning known as saturnismus, or lead poisoning, or benzene, one of the best-known solvents, which has even proved fatal in some instances, and lastly aniline, a dye much used in industrial dye-works which has even been known to cause tumours of the bladder, a very serious disease indeed which occurs in a high percentage of people whose job it is to be constantly handling aromatic amines. Another fearsome disease, which was widespread back in the Middle Ages, is silicosis, which affects miners, quarrymen, drillers and anyone else forced continually to breathe in silica dust, and usually develops tuberculosis as a complication. All that is needed is a two- or three-year period of exposure to this harmful dust for the disease to rear its ugly head

in its most serious forms, and it is significant that the highest incidence of silicosis is still recorded in the large mining regions of Belgium and England, whereas the percentage has dropped considerably in Sardinia where the mines at Sulcis and Argentiera have been going through a period of crisis for several years. And lastly one should not forget the occupational hazard run by radiologists and their technicians; this disease can seriously affect the locomotor, vascular and nervous systems, often degenerating into cancerogenic and leukaemic forms. Accidents suffered at work have also considerably increased with the advent of mechanization, and a recent statistic indicates that in 50 per cent of cases the accident has occurred because of psychic troubles or incapacities existing in the worker in the face of his job; 25 per cent are accounted for because of functional defects, such as giddiness, poor hearing, poor vision etc . . . and another 25 per cent because of the simple inexperience of workers with absolutely no technical qualifications for the job. It is evident that from its initial and simpler organizational structure, industrial medicine reflects the human requirement of protection and social security, and of reassurance that they are given the best protection in terms of hygiene. Medicine, in the future, reckons on reaching this goal via the two basic stages

Three photographs of a herbalist's shop with its numerous and varied wares ready for use. Therapy based on the use of medicinal plants is undoubtedly the world's oldest form of medicine, which has been practised by all primitive peoples and in many cases handed down to the present day in its original formulae. The discovery of the beneficial action of certain plants, and the dangers of others, guided man in this field of research and made it possible for him to develop extremely important medical practises.

Left: an assembly-line in the FIAT factory in Turin, and, right, two stills from Charlie Chaplin's film Modern Times, *a hard-hitting satire about manufacturing systems and the exploitation suffered by the working man at the hands of the mass-society. Work has always given rise to major social problems involving the protection and safeguarding of both the mass and the individual. Nowadays many such problems have been solved by industrial medicine which uses every technical and scientific means at its disposal to reduce and prevent risks at work.*

of community medicine and preventive medicine.

The former aims at the constant improvement of the average level of man's health by means of specific measures of a hygienic and prophylactic nature; the second can only function as a result of periodic and systematic examination and control aimed at an early diagnosis of diseases and illnesses. The WHO – World Health Organization – was founded in 1946 and has a membership of 61 countries with one common aspiration: to improve the health of everyone on this earth.

LOOKING TOWARDS THE FUTURE

Lemmings, which are small rodents found in northern Europe and the Arctic region of Asia, are smitten by a headlong suicidal urge whenever their numbers increase disproportionately. In their thousands they flee the mountains where they live and head for death in lakes, rivers and the sea. A tiny number of lemmings survives this ceremony, returns to its habitat and a balanced biological process gets under way again.

Bees thin out their hives with periodic massacres of males, and in summer, when the colony becomes overpopulated, they swarm – in other words, several tens of thousands of bees leave the hive in an organized manner.

The activity of the suprarenal glands in female mice increases considerably, thus atrophying the ovaries, when the total community exceeds a certain limit; and for the same reasons the reproductive faculties in the Japanese gazelle are suspended.

Ethology has therefore taught us that in nature there are various organic mechanisms which self-regulate the number of births. Man, however, is not equipped with such mechanisms and must endeavour to come by them by means of his extraordinary capacity of observation and reflection. As things stand now, the problem of over-population has become the problem of the survival of the human species and man has no option but to take notice of it as such, and in a responsible way. Up until now the job of demographic control has been tragically taken on by wars, epidemics and natural disasters, with man indifferently playing the role of protagonist and onlooker to his own destiny. In the view of the French demographer Gaston Bouthoul, if the two world wars had not in fact occurred, Europe would have numbered an additional 200 million people in 1945.

Today the world numbers 4000 million people, and if the current incredible rate of growth recorded in recent years (1000 million in sixteen years) is maintained, an extremely serious demographic crisis will be triggered off in not many decades to come, the consequences of which it is hard to predict. The prospect of there being 7000 million people on earth in the year 2000 has caused Henri Houerou, the FAO expert (Food and Agriculture Organization of the United Nations) to remark that within one century every square metre (nine square feet) of land will, in theory, be inhabited by 100 people. This chilling hypothesis may seem to touch the limits of science fiction, but it should also make us stop and think about the huge economic and industrial dangers, and the threat to actual physical survival, inherent in this sort of uncontrolled demographic growth, with no clear-sighted and informed planning. A balanced state of affairs can only be achieved if the world population becomes stable, which means keeping the birth rate and the mortality rate on an equal footing; but the 11,500 children born each hour in to the world clearly show how far mankind still is from reaching such a transition.

Demographers recommend that the present situation be analyzed with the functional back-up of statistics. By consulting such data, they uphold, there clearly emerges the macroscopic unfairness in the distribution of the world's population. Denmark, for example, where an impressive decrease in the number of inhabitants is recorded, is in stark contrast with India, which is strikingly overcrowded, and where the one city of Calcutta will number 60

Left: a sculpture of a woman from the Kostionki settlement near Voroner on the Don, dating back to the Upper Palaeolithic era – 30,000–25,000 B.C. (Museum of Anthropology and Ethnology in the Academy of Sciences, Leningrad). Right: the so-called Venus from Savignano sul Panaro (in Emilia), one of the oldest and most famous sculptures in human history. Both these sculptures represent the fertility of woman, this cult being very widespread throughout the ancient world. On the following pages: a vast crowd at a Temple in India.

million inhabitants within one generation – i.e. five million more than the total population of, say, Italy, or West Germany or the British Isles – if it continues to expand at the present rate. And West Germany, where the number of deaths exceeds the number of births, contrasts sharply with China with its population of 760 million, among them the 11.5 million people who live in Shanghai, the world's largest city.

India presents a similar picture, with a population increase of some fifteen million per annum (the equivalent of the entire population of Australia); India, in fact, has been unable to carry out badly needed agricultural reforms solely because of the rapid increase in the number of people which has far and away outstripped the increase in agricultural production.

The Punjab, which is one of the States in the Indian Union most hit by the population explosion, has introduced strict legislative measures in its attempt to stem this worrying phenomenon: a year's imprisonment and a heavy fine for any heads of families having more than two children, with the exception of parents of disabled children or of children of the same sex, and all parents who accept sterilization after the third child. This example will almost certainly be followed by many other States, thereby putting into practice the ancient theory

of the English economist Thomas Malthus, for whom every measure aimed at discouraging people from having children was worth taking into consideration. But the demographic problem will only be truly defused when the entire world accepts a fair system of planning, incorporating late marriages, the updating of doctors about the prescription of contraceptive devices, and the legalization of abortion.

In 1973 the head of the FAO, Addeke Boerma, pointed out a worrying situation. For the first time since the end of the last war, the world production of foodstuffs showed a drop of one per cent as compared with the 75 million extra mouths to be fed each year. Shortly before this announcement the leaders of the Third World, meeting in Algiers, found themselves in agreement over the urgent need to exploit the agricultural economy to the full, with the rider that 75 per cent of the world's population control just one quarter of the Earth's resources. In effect these food resources were probably adequate until a few decades ago, if equally distributed, and assisted by the advances of science and technology; but today, lamentably, the terms of the equation have been turned topsy-turvy by the soaring growth in population. In this respect the example of India can serve as a textbook model for the whole world. After the last great

The overpopulated countries of the world are faced with the task of solving colossal problems of social organization and housing. Hitherto public works have in no way met the needs of population expansion – this goes for waterworks, roads, hospitals and schools. The problem of housing is certainly no less pressing than these and the discomfort suffered by people who are crowded into shacks in Indonesia (above), shanty-towns in many large cities, or troglodytic hovels in Mexico, or in the wretched favelas in Argentina, is beyond description.

Publicity for the Family Plan *in the rural area of Maharashtra, one of the most heavily populated regions of India. For the past decade the campaign mounted by the government to control the birth-rate has been extremely intense. The government in fact handed out cash awards to fathers who agreed to undergo sterilization after having had at least two children. Now drastic measures have been introduced, entailing imprisonment and fines, throughout the Indian subcontinent, but the population explosion can only be checked by making the entire population aware of its responsibilities.*

famine of 1965 which claimed millions of human lives, India drew up an agricultural or 'green' plan which could offer this stricken country a chance to be agriculturally self-sufficient for the first time in its tormented history. Carefully selected sowings of Mexican wheat and crops of rice from the Philippines carpeted the great alluvial plains of India with lush green, causing Indira Gandhi to announce euphorically that before long India would become an exporter of wheat, rather than an importer. But the illusion was short-lived: it took just one ill-omened year, and the recurrent nightmare of those fifteen million extra mouths to be fed each year had returned the Indian situation back to square one. A major resource resides in the 65 million sacred cows which roam countryside and city street alike. These cows give barely half a litre of milk per head each day, as opposed to the ten litres produced on average in Western Europe; but they are protected by religious taboos which are so deep-rooted that there is no hope of changing them, either by psychological persuasion or by the blatantly obvious contingent needs.

The world situation, which was bad enough in 1975, has gone from bad to worse and predictions for the near future are nothing less than catastrophic. This year, 500 million people are likely to suffer from malnutrition, and the

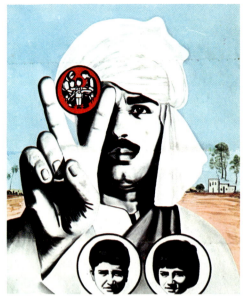

countries hardest hit will be India, Pakistan, Bangladesh, the countries of South America and the central African states. However, there is a dangerous dose of optimism inherent in the idea that only India and the handful of other Third World countries will be caught up in this dramatic spiral of hunger looming on the horizon. Suffice it to mention that in 1977 the world had only 31 days of grain-reserves, as opposed to the 105 days in 1961, to give a rough idea of the danger with which the whole of mankind is faced. A general survey of the present-day situation of the world's food can be swiftly sketched: in 1976 North America had a huge surplus of cereals – 94 million tons almost all exported – followed by Australia and New Zealand with eight million tons. China will also escape the effects of the drop in production, thanks to three factors of prime importance: the decrease in the birth rate, the prudent subdivision of its agricultural production, and the rational use of labour in the fields. But as far as the rest of the world is concerned, the situation is quite different.

Two decades ago Japan was considered to be one of the largest consumers of rice and fish. Today, having widely adopted Western ways, it has become the world's number-one importer of cereals, along with the USSR, which is

Left: schoolchildren in Indonesia. Top right: Family Planning hoarding in Pakistan. Below right: children in a Korean village. In Pakistan the government has been joined by women's organizations and numerous other social groups in the campaign to limit the number of births. It has been calculated that as a result of these information and family-aid programmes they have managed to prevent at least three million births in the past five years, a figure which is totally inadequate in the face of the rapid demographic growth in this country.

A small boy suffering from kwarshiokor, a terrible disease caused by malnutrition which, if neglected, can lead to death in 90 per cent of cases. It affects children in particular, because their body is in the process of development and suffers more acutely from harm caused by protein and vitamin deficiency. In November 1974 the FAO (the Food and Agricultural Organization of the United Nations) held a world conference in Rome to deal with the problem of food and hunger. During this meeting it was confirmed that millions of people throughout the world were suffering from hunger.

hampered by its collectivist agricultural system and which, because of disastrous harvests, has had to import vast amounts of grain from North America to meet the needs of its people. Things in Latin America are verging on the catastrophic, as is shown, for example, by the economic policies of Brazil which are rendered ineffectual first and foremost by the galloping birth-rate. Mexico is undergoing a similar situation: here the introduction of selected and more productive varieties of wheat and maize are in practice gobbled up by the tide of human mouths. The petro-dollar, that invaluable currency of the oil-exporting countries, enables these countries to import colossal amounts of foodstuffs, thus making up for the almost total absence of any agricultural economy in their arid and mainly desert territories. Lastly we come to Western Europe: here, some years ago, a drop in production had been recorded with the declining yield per hectare; today there is a drop of ten per cent as compared with the production of grain and cereals in 1974. There is thus only one conclusion to be drawn from all this: the world production of wheat, rice and maize is currently running at about 1200 million tons; this divided by the 4000 million people living in the world today gives a total of 300 kilos (675 lbs) of cereals per annum per person, an amount which in itself is barely enough to

meet the basic individual requirement, According to the American agricultural expert M. Lester R. Brown, because there is nothing less than a gulf between the levels of consumption in the rich countries and the poor countries (180 kilos [405 lbs] in Asia per annum, and 750 kilos [1700 lbs] in Canada) there can only be one solution, over and above the principle of a strict limitation of the number of births: the rich countries must introduce considerable restrictions, and dietetics must make great leaps and bounds forward if due help is to be given to those sections of mankind in need.

In the field of food research the goals reached are still in the experimental phase, as far as man is concerned, with the exception of seaweeds which are already being widely cultivated and consumed in Japan. It is, in fact, with seaweeds that economists and biologists alike, the world over, hope that the great problem of hunger will be solved. The bioproteins represent another source of hope; these are artificial flours used up until now in animal fodder, and the aim is that they should replace the natural proteins which are becoming increasingly scarce in every part of the world. These latter are vital substances for the sound functioning of the human body, and are part of its actual composition, but they are nonetheless insufficient to meet its requirements. For

Left: preparing millet-based dough in Mali. Right: a variety of edible seaweed – noctiluca scintillans, magnified 100 times. Much hope has been invested in the cultivation of seaweeds as a replacement for the numerous types of food which are becoming scarcer; it has been calculated that some 500 million people living in Asia, Latin America and Africa do not have enough to eat, or are even starving. According to FAO forecasts, in 1985, 700 million people in 34 low-income countries will be either suffering from malnutrition or inadequately fed, and this will in itself create extremely serious social problems.

Left: blue seaweed being cultivated along the Japanese beach of Hokkaido. Top right: samples of vegetables used as food for humans, with high protein content. According to food experts, the cereal requirement in 1985 will rise from the present figure of 1,200 million tons to 1,700 million, with the greatest demand coming from the under-developed countries which will increase their present consumption of 600 million tons to some 900 million. Below: sacks of bioprotein, or artificial food, ready to be put on the market.

this reason they must be introduced daily into the organism in amounts equal to two grammes for each kilogramme (0.07 oz for every 2.2 lbs) of body weight, and it is a well-known fact that the foodstuffs which contain the highest quantities of proteins are meat, eggs and fish – foodstuffs which are in some cases completely unknown in underdeveloped regions of the earth. The bioproteins, which are obtained from special yeasts which grow on certain petroleum derivatives, could easily replace the natural proteins once completely purified. Bioproteins are being produced on an intensive scale in France, Scotland, America and, before long, Sardinia, and scientists have stepped up their research in the hope of succeeding in refining this precious powder – by a three-phase process of washing, centrifugation and the extraction of hydrocarbon traces – which could save the lives of the fifteen to twenty million people who die each year of starvation in the world, and the 400,000 children in the Third World countries who are mentally retarded as a result of having been given too little protein in the first years of their lives. Autopsies made on children who have died as a result of protein deficiency have in fact shown that their brain-cells are twenty per cent fewer in number than the average.

Newborn children and children soon to be born should be given not only enough to eat, but also enough to drink, and here we come to the other problem that is causing concern: water. The world's water reserves consist of 98 per cent sea- and ocean-water, and two per cent fresh water, making up a total of 513,000 cubic kilometres. Of this colossal mass of water, of which 350,000 cubic kilometres are surface water (lakes, rivers, swamps and glaciers), 150,000 are underground and the remaining 13,000 are still in the gaseous state, only two per cent is drinkable. And this two per cent would still be an enormous amount, if it were not for the constant demands being made on it by 4000 million human beings. And with the evolution of our civilizations and the changing ways of life being witnessed, these human beings no longer use water just to quench their thirst, but rather use it for a vast number of new needs – hygienic, industrial, domestic and alimentary. In fact it is significant to note the increase in the consumption of water over a period of five or six decades. In 1915 the figure was 60 litres (13.2 gallons) per head per day; today, in some parts of the world, the figure soars as high as 900 litres (189 gallons). In addition, industrial consumption involves truly astronomical figures, if one bears in mind that ten tons of water are needed to produce one ton of steel, and

Animals raised on bioprotein feed. As far as the food problem is concerned, experts and specialists have been making huge efforts to handle the phenomenon of reducing population growth, ranging from vast schemes, aimed at utilizing the resources of the sea, to plans for changing virgin lands into areas of cultivation and plans for producing cereal crops with a yield which is at least ten times higher than that of ordinary cereals. Bioproteins are very much part of this programme and it is hoped that they will also be able to be used for man's needs.

Top left: dried up watering-places in the Algerian Sahara. The problem of the water supply for the oases has started to cause considerable concern because water is becoming constantly rarer and rarer in artesian wells and draw-wells because of the impoverishment of the water-table. Below: a view of the Algerian sahel, typified by steppe and sparse, thorny vegetation, caused by the low annual rainfall (250–300 mm) which occurs in the course of just a few days. Right: the road known as the 'cattle-track' used for transporting meat from Argentina to Chile.

1000 tons of water are needed to produce one ton of sugar; that some three tons are used to process one single barrel of crude oil into petrol, diesel oil, naphtha, paraffin and lubricating oils; and that it takes 25 litres (5.5 gallons) of water to make just one litre of beer.

It has frequently been said that the degree to which a people is civilized can be measured by the amount of water it consumes, but it does not appear that man pays much heed to this irreplaceable element which nature lavished upon him without counting the cost.

At the end of 1953 the people living on the Bay of Minamata in Japan were suddenly affected by a mysterious form of intoxication which caused paralysis, convulsions, mental disorders and in many cases death. It was many years before scientists realized that this sinister phenomenon was caused by methylmercury being dumped in the sea by one or two industrial plants situated at one end of the bay. The fish in the bay absorbed it without suffering any immediate harm, and the cycle came to its tragic full-circle when the fish were eaten by man. The Minamata disease thus became part of the history of medicine, and every so often it crops up here and there in the world, sometimes in Sweden, sometimes in Finland, and sometimes in Holland, giving added

confirmation of man's complete disregard for his water. The biological death throes of the sea can invariably be put down to human negligence and carelessness, be it oil-tankers discharging their water ballast and the water used for washing out their tanks without filtering off any residual petroleum. The French ecologist Jean Dorst holds that this practice pollutes the sea with a tide of petroleum equivalent to the cargo of 50 Torrey Canyons (the shipwrecked oil-tanker which went aground in 1967 and released 117,000 tons of crude oil) – or be it the millions of tons of often non-biodegradable detergents which find their way into the sea each year, in other words, detergents which can withstand attacks from bacteria living in water. And yet the sea should represent for mankind a more or less inexhaustible source of drinking water, and the numerous desalination plants already operating in many parts of the world are the most obvious proof of this. Suffice it to mention the gigantic plant at Key West in Florida, which can produce about twelve million litres (2.6 million gallons) of drinking water per day, and a glimpse forward to the year 2000 might cause us less apprehension, because by the year 2000 we shall need 3,785,000 million cubic metres of water to quench mankind's thirst, and this represents one tenth of all the fresh water currently available on earth.

Left: a water-purifying plant. Top right: water being desalinated. Below: water entering the settling tanks which can rid each litre of water of the 35 grammes of salt contained in it. The desalination plant shown here is at Eilat in Israel. The Israelis waste not a single drop of water and with a brilliant system of canals, wells and tanks they have managed, in a very few years, to increase their agricultural yield by 60 per cent. To do this they have transformed an arid desert into one of the world's most fertile and luxuriant gardens.

PICTURE SOURCES

Museums, Libraries, Research Institutes and Learned Societies

American Museum of National History (Bruce Hunter), New York; Arizona University (Richard Orville); Armed Forces Institute of Pathology, Washington; Astrofysisk Institutt (C. Störmer), Oslo; Biblioteca Estense, Modena; Biblioteca Nazionale, Turin; Imperial College of Science & Technology, London; Bodleian Library, Oxford; British Museum, London; Brookhaven National Laboratory; California Institute of Technology; Cavendish Laboratory, Cambridge; Carnegie Institution, Washington; C.N.R.S., Paris; Compactors Engineering; Defense Electronic Division Radio Corporation of America; Fondation Curie, Paris; Fondation Saint-Thomas; Graphische Sammlung Albertina, Vienna; Helsinki University of Technology, Otaniermi; Institut de France, Paris; Istituto di Anatomia Umana, Milan University; Istituto di Chimica Industriale, Milan Polytechnic; Istituto di Patologia Generale, Milan University; Lawrence Radiation Laboratory, University of California; Lockhead Missiles and Space Company; Louvre, Paris; Météorologie Nationale, Paris; Musée des Arts Décoratifs, Paris; Musée Pasteur, Paris; Museo Civico di Storia Naturale, Milan; Museo Nacional de Antropologia, Mexico City; National Institute of Oceanography (Peter David); National Portrait Gallery, London; Nelson Gallery, Kansas City; Nicolini-Istituto di Onde Elettromagnetiche del C.N.R., Florence; Oxford University (D. C. Phillips); Royal College of Surgeons, London; Seattle Art Museum Eugene Fuller Memorial Collection; Smithsonian Institution, Washington; St. Bartholemew's Hospital (A. J. Salsbury), London; Stanford University, California (Forbman); Sweizerisches Institut für Nuklearforschung, Villigen; University of Michigan; University of Milan; University of York (A. Chambers); Verkehrshaus der Schweiz, Luzern; Yale University Art Gallery (F. Richards)

Commercial, Industrial and Other Sources

A.E.G.; A.G.I.P.; A.N.I.C.; Australian Embassy, Rome; C.E.A. (Commissariat à l'Energie Atomique), Paris; C.E.R.N.; C.G.E. Laboratories – Ansaldo San Giorgio; COMSAT; C.W.F.; Dunstall, Ontario Department of Highways; E.N.E.L.; E.N.I.; Carlo Erba; ESSO Standard Italian; FIAT; Ford Motor Company; General Electric; Hughes Aircraft Company, Los Angeles; Hughes Tool Company – Oil Tool Division; Humble Oil Refining Company; IBM World Trade Corporation; Imperial Chemical Industries; International Telephone and Telegraph Corporation; I.R.I.; Katanga Mining Union; KLM Royal Dutch Airlines; KODAK; Lavigliano Publicity; Lenning Chemicals; Montecatini; Montedison; NASA; National Accelerator Laboratory; Nickel Information Centre; Piave Laboratory, Milan; Shell; Sikorski Aircraft, Stratford, Connecticut; Societé Général Métallurgique de Hoboken; Società Generale Semiconduttori, Agrate; Swiss National Tourist Office; Texas Eastern Transmission Corporation, U.S.A.; United Aircraft; Union Pacific Railroad; United Steel Companies; Ward's National Science Establishment

Artists

Giorgio Arvati; G. B. Bertelli; Luciano Corbella; Alessandro Fedini; Raffaello Segattini

Photographers, Photographic Agencies, Publishers, Literary Agents, Photographic Archives, Press Agencies, Magazines

ATES; Aldus Books; Alpenland; G. Christofer Angeloglon; O. V. Antisari; Archivio E.D.E.R.A.; Archivio Mondadori; Ardea Photographics; Associated Press; Badische Anilin und Soda Fabrik; A. Barone; Basf; Batell Northwest; Bell Telephone Laboratories; Bethleham Steel Company; Bevilacqua/Ricciarini; Boeing, U.S.A.; Lee Boltin; W. Bonatti; Borella/Ricciarini; Herbert Bridge, Massachusetts Institute of Technology; British Crown Copyright; Camera Press; Carrese; Cerasoli; Cirani/Ricciarini; Gene Cox; Corsi, Milan; A. Cozzi; R. Crespi; Crown; E. J. Cyr/Shostal; De Biasi; W. Disney; Ray Delvert; Geoffrey Drury; H. Edgerton – Bell System; EG & G Inc.; Harry Engels; R. Everts/Rapho; Farabola; Feature Pix; H. Fernandez-Moran; Fiore; Fiorentini; A. Foroni; Port of New York Authority; Port of New York; Fototeca Est; Fox Photo; John Freeman; Franco Frezzato; Genovesi; Giraudon; Guido Gregorietti, Private Collection, Milan; Harry Gruyaert; Halik; Hamlyn; David Harris; Hewlett – Packard Ltd; Michael Holford; Jacana S.A.R.L.; R. C. Jennings/Frank Lane; J. A. Kitchener; Ursula Kohler Kurze; John Launois/Black; Ledoux-Lebard Private College; Leoni; Erich Lessing; Lessing/Magnum; LIFE; Tony Linck – University of California, Stanford; Lod; Lod/Kister; G. Lotti; Lotti-Kodak; Louvegnies; Rex Lowden; Foto Jean-Luc Lubrano; Magnum Photos; Mairani; Malak/Annan Photo; Malayan Rubber Fund Board, London; Mansell Collection; Marelli S.p.a.; Marka; Paola Martini/Ricciarini; Michael Mellish; Merril; Metro-Goldwyn-Mayer; Edizioni Scolastiche Mondadori; W. Mori; H. Morozumi; G. Motto; Joseph Muench; Newsweek; L. Nilsson; Sam Nocella; Novosti Press Agency; Oak Ridge National Laboratory; Orion Press; Osservatorio Astronomico di Monte Mario, Rome; Raffaele Ostuni; P2/Ricciarini; P.V.R.; Palnic; Mount Palomar & Mount Wilson Observatory, California; Panicucci; Di Paolo; Elena and Lino Pellegrini; Philips; Phillips and Thomas; Pictor; C. F. Powell; Mirella; Prato/Ricciarini; Pris-Match; Publifoto/Carrese; Rastellini; RCA; Rheinisch – Westfälisches Elektrizitätswerk; A. Rich, Massachusetts Institute of Technology; Luisa Ricciarini; E. Robba/Ricciarini; Rossetti; Science Journal, 1965; Sateri Osakeyhtio; André Sauret; Scala; Scarnati – Institut du Radium; Roberto Schezen/Ricciarini; E. Schroder; Schub; Schulthess; Schulthess/Carrese; R. S. Scorer; S.E.F.; Selenia; Siemens; Simion/Ricciarini; Sipa-Lardarello; Smith; Sochurek/Carrese; Howard Sochurek/John Hillelson Agency; Sorci; Sulzer Bros Ltd, London; P. Summ; TASS; J. Taylor; Termier, Paris; Three Lions Inc.; The Times; Tomsich/Ricciarini; Julian Trevelyan; U.S.I.S.; U.K.A.E.A., London; Van Phillips; Vantier-Decool; Vantier – De Nanxe; Robert Viollet; René Vital; J. A. Viollet; J. B. Watson Ltd.; John Webb; S. Weber; Weidenfeld & Nicolson; Westinghouse; Henry Wiggin & Company; Mab Wilson; Windsor Royal Library; Winston, New York; Sidney Books; World Books

A special thanks to Walt Disney Productions, Burbank (California), who have kindly allowed us to use photographic material published in their series

We apologize if for reasons beyond our control we have omitted or incorrectly credited any of the above bodies